CODE DIVISION MULTIPLE ACCESS COMMUNICATIONS

Edited by

SAVO G. GLISIC
and
PENTTI A. LEPPÄNEN
University of Oulo,
Oulo, Finland

KLUWER ACADEMIC PUBLISHERS
BOSTON / DORDRECHT / LONDON

A C.I.P. Catalogue record for this book is available from the Library of Congress.

ISBN 0-7923-9553-0

Published by Kluwer Academic Publishers,
P.O. Box 17, 3300 AA Dordrecht, The Netherlands.

Kluwer Academic Publishers incorporates
the publishing programmes of
D. Reidel, Martinus Nijhoff, Dr W. Junk and MTP Press.

Sold and distributed in the U.S.A. and Canada
by Kluwer Academic Publishers,
101 Philip Drive, Norwell, MA 02061, U.S.A.

In all other countries, sold and distributed
by Kluwer Academic Publishers Group,
P.O. Box 322, 3300 AH Dordrecht, The Netherlands.

Printed on acid-free paper

Printed in the Netherlands

TABLE OF CONTENTS

PREFACE

Code division multiple access has become a main candidate for the next generation of mobile land and satellite communication systems. Although spread spectrum technique has been used for military applications for a half of the century only recently it has been recognised that this technique combined with some additional steps can provide higher capacity and better flexibility for the mobile cellular radio communications. In these systems N users use the same frequency band and assuming that all signals are of the same power S, the bit energy per equivalent noise density become $y_b = E_b/N_e = ST_b/[S(N-1)/R_c] \cong G/N$ where $G = R_c/R_b$ is the system processing gain (ratio of the sequence chip rate and bit rate).In other words the system capacity (the maximum number of users that can operate in the same frequency band for a given signal to noise ratio y_b needed for a given quality of communications) is proportional to the system processing gain and inversely proportional to y_b i.e. $N \cong G/y_b$.The first step to increase the system capacity was to use powerful FEC that will decrease y_b needed for a given probability of error. Shannon's limit suggests that theoretically in limiting case if coding is powerful enough y_b might be even less than one which makes $N > G$ and automatically capacity of the CDMA becomes larger than capacity of TDMA even at this stage.

In practice utilisation of convolution coding (e.g.constraint length K=7 and rate R=1/3) would bring this parameter to $N \cong 0.5G$. If voice activity monitoring is used the average level of interference will be reduced for a factor $\alpha=3/8$ so that we have $y_b \cong G/\alpha N$ and the system capacity will become now $N = G/\alpha y_b > G$. Further improvements are due to cell sectorization (factor 3) and frequency reuse (factor 7) so that the capacity of CDMA in cellular network becomes considerably higher than the capacity of any other multiple access technique.

Based on these results several systems have been already developed and some standardisation documents (IS-95 and a number of proposals for the common air interface in 2GHz region) are in the final stage of preparation.

Having in mind the importance of this techniques and being in charge of organising The Third IEEE International Symposium on Spread Spectrum Techniques and Applications - IEEE ISSSTA'94 the editors of this book have invited the most distinguished world scientist in the field to present review type papers on some selected topics in spread spectrum. These papers are collected in this book.

As an introduction, evolution of code division multiple access technique is summarised by R. Scholtz.

In chapter 2 information theory aspects of spread spectrum are discussed by J.Massey from the fundamental viewpoint of Shannon's information theory. Performance limits of "error preventing" or "error protecting" methods in multi cellular CDMA systems with and without interference cancellation is discussed by A. Viterbi. A considearable work done by many researchers in the past in the field of PN sequences for CDMA systems is summarised by D.Sarwate. The approach is based on the fact that most known methods for the design of codes in frequency hopping and direct sequence spread spectrum systems generate sequences that can be viewed as codewords selected from low rate Reed Solomon codes. An overview of block demodulation techniques is presented by E. Biglieri et al.

Chapter 3 deals with interference suppression. Work in the field of adaptive multiuser detection is summarised by S. Verdu. Since 1986 a considerable number of the papers has been published in this field demonstrating a huge improvement of the optimum performance compared with the performance of the conventional single user detector which neglects the presence of multiaccess interference. Spatial and temporal filtering of co-channel interference in CDMA network is presented by R.Kohno. Interference suppression for CDMA overlays of narrowband waveforms is presented by L.Milstein et. al. Using this concept a standard communication network with narrowband users and CDMA network can coexist in the same frequency band. Such a system has been already built and field tested. Successive interference cancellation that is assumed to reduce hardware complexity is summarised by J.Holtzman.

Chapter 4 deals with performance analysis. Rake reception for CDMA mobile communication system with multipath fading is discussed by M.Pursley and D. Honeaker. Frequency hopped systems for PCS are discussed by D. Borth et al and code synchronisation by A. Polydoros and S. Glisic.

The most of the work is concentrated in chapter 5 dealing with different applications of CDMA. Design aspects of a CDMA cellular radio network are presented by W.Lee. Consumer communications based on spread spectrum techniques are summarised by M. Nakagawa. These applications include power line communications, data carrier, radio remote control ISM wireless LAN, vehicle to vehicle communications, digital TV broadcasting and different devices and systems. Application of CDMA in satellite communications is discussed in two papers. E.Geraniotis et al discuss optimal polices for multimedia integration in CDMA networks and R.Pickholtz et al discuss CDMA for mobile LEO satellite communications. Nowadays a number of projects are carried out in this field example being Globalstar, Odyssey, Constellation, Ellypso, Archimedes, etc. Finally progress in standardisation for wireless communications is presented by D. Schilling. The main message from this work is that multiple standards including TDMA and CDMA should be expected and already are a reality.

We believe that material presented in this book summarises the main problems in modern CDMA theory and practice and represents a solid starting point for studying this complex and still challenging field.

Editors

Like any other technical field, Spread Spectrum (or code Division Multiple Access) has its own pioneers. One of them is Solomon Golomb. We are glad to have his contribution within this book too.

Editors

SHIFT-REGISTER SEQUENCES AND SPREAD-SPECTRUM COMMUNICATIONS

Solomon W. Golomb

Forty years ago, when I began to study shift register sequences, digital technology was in its infancy. The most advanced electronic computers still used vacuum tubes. The integrated circuit was not even on the horizon. In that environment, a two-tap linear shift register of length n, producing a randomlooking binary bit stream of period $2^n - 1$ was an incredible device. With only twenty active delayline positions, and only two of these positions accessible, using no logical circuitry except a single "half adder", a binary sequence with a period of more than one million bits could be generated!

The first problem I addressed was how to predict the periodicity of a linear shift register from the feedback tap connections. I quickly discovered the equivalence of this question with the primitivity of the roots of the corresponding polynomials over the field of two elements. Gradually I learned of the long mathematical history of this problem, in which connection the names of L. Euler (ca. 1760), E. Lucas (ca. 1875), and Ø. Ore (ca. 1933) deserve special mention .

I also noticed that these "maximum-length linear shift register sequences", named *m-sequences* by Neal Zierler, had several properties suggestive of randomness. Three of these, which I designated "R-1", "R-2", and "R-3", were the following:

S.G. Glisic and P.A. Leppänen (eds.), Code Division Multiple Access Communications, ix-xii.
© 1995 *Kluwer Academic Publishers. Printed in the Netherlands.*

R-1. In a binary sequence of period 2^{n-1}, there are 2^{n-1} *ones* and 2^{n-1} -1 *zeroes*. [The "balance property.]

R-2. In each period (of length 2^n-1), there are 2^{n-2} *runs* of ones alternating with 2^{n-2} *runs* of zeroes. Half the runs of each kind have length 1, one-fourth of the runs of each type have length 2, and in general $\frac{1}{2^k}$ of the runs of each type (i.e. 2^{n-k-2} runs of each type) have length k, for $1 \leq k \leq n-2$. In addition, there is a single run of $n - 1$ zeroes, and a single run of n ones. [The "run property."]

R-3. Compared with every non-identical cyclic shift of itself, the sequence has 2^{n-1}-1 "agreements" and 2^{n-1} "disagreements." If we regard the sequence as consisting of +1's and -1's (instead of 0's and 1's), then its normalized autocorrelation function $C(\tau)$ satisfies $C(\tau) = 1$ when τ is a multiple of the period $p = 2^n$-1, and $C(\tau) = -1/p$ for all other values of τ. [The "two-level correlation property."]

These "randomness properties" made the *m*-sequences particularly useful in many applications which have subsequently been referred to as "spread spectrum", and more specifically "direct sequence spread spectrum." In the last few years, in the context of digital cellular communications, these sequences now form the basis of *code diuision multiple access* (CDMA) technology.

There are several other properties of *m*-sequences which are worth noting. One of these is:
The cycle-and-add property: "If an *m*-sequence is added, term-by-term modulo 2, to any non-identical cyclic shift of itself, the result is another cyclic shift." This property actually *characterizes* the *m*-sequences. It can be restated as follows: "The 2^{n-1} cyclic shifts of an *m*-sequence of period $p = 2^n$-1, together with the sequence of 2^n-1 *zeroes,* regarded as a set of 2^n vectors of leneth 2^n-1 over the field GF(2) of two elements, form a *subspace* of the space of all 2^p binary vectors of length $p = 2^n$-1." [The *subspaoe property.*]

The "two-level correlation property", R-3, follows immediately from the "cycle-and-add property" of *m*-sequences. However, the binary seqllences of period p (not necessarily of the form $p = 2^n$-1) with two-level autocorrelation ($\frac{p-1}{2}$ agreements and $\frac{p+1}{2}$ disagreements with all non-identical cyclic shifts) are a larger class, and correspond to the combinatorial objects called "cyclic Hadamard difference sets." All *known* examples of cyclic Hadamard difference sets have $p \equiv 3 \pmod 4$ where either i) $p = 2^n$-1, $n > 1$, ii) $p=4t-1$ is a *prime,* $t \geq 1$; or iii) $p = r(r + 2)$ where r and $r + 2$ are both primes (the *twin-prime* examples). Over thirty years ago, with little direct evidence, I conjectured that all cyclic Hadamard difference sets must have periods of one of these three types. The experimental evidence for this is now quite impressive, though there is still little theoretical basis for this conjecture. Even in the case of cyclic Hadamard difference sets with period $p = 2^n$-1, which includes all the m-sequences, we do not yet know all of the inequivalent constructions which yield examples. Several member of my group (H.-Y. Song; D. Rutan; etc.) at USC, as well as my long-time colleague Lloyd Welch, are actively investigating these unresolved questions concerning the existence of two-level-correlation sequences.

The "run property", R.-2, follows easily from the fact that in an m-sequence of period $p = 2^n$-1, all possible subsequences of lellgth n, except for n consecutive zeroes, occur within each period, each exactly once [the "span-n" property]. There are only $\phi(2^n$-1$)/n \approx 2^n/n$ different m-sequences of period $p = 2^n$-1, but there are $2^{2^{n-1}-n}$ different span-n sequences with this period, all obtainable from *nonlinear* shift registers of length n. (These differ from the "de Bruijn sequences" of span n simply by omitting a single *zero* from the unique run of n zeroes in the de Bruijn sequence.)

In their book *Cipher Systems,* H. Beker and F. Piper introduce the term *G-randomness* for sequences with all three properties R-1, R-2, and R-3. It was shown by U. Cheng that G-randomness is insufficient to characterize m-sequences. (In particular, there is a sequence of period $p = 127$ which has G-randomness but is not an m-sequence.) However, the "span-n" property is more restrictive than the "run property" R-2, and I have long conjectured that the span-n property (modified de Bruijn sequences) together with R-3 (the two-level correlation property) can be satisfied only by m-sequences. This conjecture has now been verified for $n \leq 9$ (period $p \leq 2^9 -1 = 511$), but no proof is yet in sight.

Shift register sequences have becn used in both pulse and CW radar systems for several decades. The first attempt at radar contact with another planet, Venus, conducted by Lincoln Laboratories in the late 1950's, used pulse radar modulated by an m-sequence of period 2^{13}-1 = 8191. The JPL interplanetary ranging system, developed in 1959 - 60, used a CW signal with binary phase modulation specified by a long sequence obtained as a Boolean combination of several short-period shift register sequences. Incidentally, it was at JPL that we had the first *successful* radar contact with Venus, on March, 10, 1961.

Much of the early impetus for the use of "direct sequence spread spectrum" was to make military communications relatively resistant to jamming. Using only m-sequences for this purpose assumes a very unsophisticated jammer. The "cycle-and-add" property enables the jammer, without even "deciphering" the sequence, to generate a forward time-shift of the intended modulating sequence, which might be used successfully to fool the receiver. A trivial exercise in linear algebra over GF(2), "rediscovered" in numerous algebraic coding/decoding contexts, enables one to determine the span and the recursion of any linear sequence from a small number of its terms. To achieve more jam resistance, or any degree of resistance to deciphering, it is necessary either to subject linear sequences to nonlinear operations, or to generate nonlinear sequences to begin with.

To the extent that the sllccessive bits of a shift register sequence (linear or nonlinear) are sufficiently random for the application, consecutive blocks of k bits may be interpreted as k-bit binary numbers which are then used to specify 2^k different frequencies in a pseudo-random frequency-hop spread spectrum system. I am not aware of any nonmilitary motivation for employing frequency hopping to achieve spread-spectrum communications, but there may be some *naturally* hostile communication environments for which this type of system would be appropriate.

Short m-sequences have been employed as Synchronization patterns in a variety of applications, including such use for initial lock-up in spread spectrum systems. Many other uses of shift register sequences unrelated to spread spectrum applications could be enumerated, but that is beyond the scope of the present paper.

The commercial use of shift register sequences in CDMA cellular communications closely resembles the "direct sequence spread spectrum" military systems, but the justification is different. A hostile jammer is not assumed to be present in the cellular communication application. Instead, CDMA packs more calls into the same bandwidth, with a lower power level per call, than the principal alternatives which have been proposed. Other speakers at this symposium, however, are both better qualified and more strongly motivated financially than I to elaborate on the virtues of CDMA for cellular communications applications.

Chapter 1

Introduction

The Evolution of Spread-Spectrum Multiple-Access Communications

Robert A. Scholtz

Abstract

This paper contains a brief history of electrical communications, emphasizing the emergence of multiple-access and spread-spectrum techniques.

I. Introduction

The history of spread-spectrum and multiple-access systems is intertwined with the history of electrical science in many intriguing and curious ways. At some times, developments have been shrouded in veils of secrecy, and at other times the latest ideas have been open to public scrutiny. Here is one engineer's sketch of this history, including the enabling technologies for communication, the emergence of multiple-access concepts and spread-spectrum techniques, and events that have led to current developments in the commercialization of spread-spectrum signalling.

With out a doubt, even in the eighteenth century, electrical devices were envisioned as the enabling technology for communication systems that could operate at night and in bad weather, replacing the semaphore (visual telegraph). Communication may very well be the first major use of electrical technology. As concepts were refined and electrical telegraphy became a commercial success, inevitably the pressure to use resources efficiently led to many improvements, including the concepts of frequency-division and time-division multiplexing (see Sections II and III).

The most pressing need at the beginning of the twentieth century for the infant wireless technology was for mobile communication with ships at sea. The development of tuned circuits allowed different radios to asynchronously frequency-access the air waves with little or no interference (see Sections IV and V). As more applications of radio technology were considered, the advantages of using more than the minimum bandwidth necessary to communicate an information-bearing signal were uncovered (see Section VI).

By the end of the World War II, a remarkably complete mathematical theory of communication was put forth by Claude Shannon, and the seed concepts of code-division multiple-access (CDMA) communications were sown (see Sections VII and VIII). The last forty years have seen this initially military communications and navigation technology gradually transfer to the commercial world (see Section IX).

S.G. Glisic and P.A. Leppänen (eds.), Code Division Multiple Access Communications, 3-26.
© 1995 Kluwer Academic Publishers. Printed in the Netherlands.

4

What follows is a reasonable historical time line (summarized in Tables 1-3) for the introduction of basic concepts, and gives representative references. Certainly the list is nowhere near complete, and the events of recent years related here are biased toward activities within the U.S.A. of which the author is aware. This broad setting is a background for the detailed papers [1],[2],[3],[4] on the origins of spread-spectrum communications.

II.   Early Electrical Communication [5],[6]

The first authenticated suggestion of communication by electricity is described in [5] as follows:[1]

> A letter written at Renfrew, Scotland, on February 1, 1753, signed merely "C.M." and published in the *Scots Magazine* of Edinburgh, under the caption, "An Expeditious Method of Conveying Intelligence," proposed the stringing between two distant points of as many insulated wires as there are letters in the alphabet, through which "electrical discharges should separately exhibit themselves by the diverging balls of an electroscope, or the striking of a bell by the attraction of a charged ball."

The first telegraph instrument, indeed similar to the above description, was constructed by George Le Sage at Geneva in 1774.

The first electrical telegraph using a single wire was demonstrated in England in 1816 by Francis Ronalds, whose dial telegraph required *synchronized wheels* at transmitter and receiver, carrying alphabetical symbols past a slit. The wire was charged via Leyden jar as the desired letter passed the slit.

Numerous ingenious telegraphic designs were proposed in the 1800's. Certainly Count Alessandro Volta's discovery in 1799 of the "voltaic pile," the first source of constant current electricity, was a major contribution to telegraphic experimentation. Among the many who produced telegraphic equipment designs were several of the founders of the field of electromagnetic science, whose contributions influenced the development of a means of reliable communication at a distance. In 1820, based on Gian Domenico Romagnosi's discovery that a magnetic needle can be deflected by a nearby electrical current, Hans Örsted invented the galvanometer. André Ampére proposed using Örsted's galvanometer as the sensor in a telegraphic system similar to Le Sage's, and attributes the idea to a suggestion by Pierre Laplace. In 1833, Karl Friedrich Gauss and Wilhelm Weber constructed a galvanometer telegraph at Gottingen. The further development of this equipment was given to Karl Steinheil, who increased power and added two bells of different tones and a pen recording system. In

---

[1]References cited at the beginnings of sections
indicate more extensive treatments and sources of
background references.

1838, Steinheil proposed that the Earth could be used as a return conductor in telegraphy if wire terminals were buried in the ground. In Britain in 1837, Sir Charles Wheatstone and Sir William Cooke patented a chronometric telegraph, using synchronously rotating wheels at both ends of the line, and arrested simultaneously by a magnetic armature.

In 1828 the American Joseph Henry first exhibited his electro-magnet at the Albany Academy in New York, and in 1831 he demonstrated a means of signalling using the *attractive* power of an intermittent magnet. Henry, who later became the first director of the Smithsonian Institution, never capitalized on his inventions. Samuel Morse claimed to have conceptualized the telegraph in 1832, and by 1837 had filed a caveat with the U.S. Patent Office, the patent being granted in 1840. The succeeding decades saw the commercialization and litigation of the
electromagnetic telegraph.

In 1854, Charles Bourseul wrote in *L'Illustration* about a technique that suggests the replacement of a telegraph key with a voice ([5], p. 342):
"Imagine that one speaks near a mobile plate flexible enough not to lose any of the vibrations produced by the voice; that the plate establishes and interrupts successively the communication with a battery. You would be able to have at a distance another plate which would execute at the same time the same vibrations. It is true that the intensity of the sounds produced would be variable at the point of departure, where the plate is vibrated by the voice, and constant at the point of arrival where it is vibrated by the electricity. But it is demonstrable that this would not alter the sounds. ... At any case, it is impossible to demonstrate that the electric transmission of sounds is impossible. ... An electric battery, two vibrating plates, and a metallic wire will suffice."
In the early 1860s, the German Philip Reis built several telephone models along these lines with limited success. The next 10 years were the formative years for many claims to invention of an electrically operated telephone. But yet in 1874, the year that Reis died, the public still was skeptical over the possibility of voices being carried by wires, as seen from this quote from a Boston newspaper ([7] , p. 214):
"A man about forty-six years of age, giving the name of Joshua Coppersmith, has been arrested in New York for attempting to extort funds from ignorant and superstitious people by exhibiting a device which he says will convey the human voice any distance over metallic wires so that it will be heard by the listener at the other end. He calls the instrument a "telephone" which is obviously intended to imitate the word "telegraph" and win the confidence of those who know of the success of the latter ... were it possible to do so, the thing would be of no practical value. The authorities who apprehended this criminal are to be congratulated, and it is to be hoped that his punishment will be prompt and fitting."

From this author's viewpoint, some nearly simultaneous telegraph-related inventions stand out. As the telegraph business expanded, the need to get more than one message at a time onto the wire became apparent. As early as

1853, Wilhelm Gintel of Vienna had developed a duplex system, and by 1874 the one-time telegraph operator Thomas Edison had achieved quadruplex signalling. At the time of Edison's invention, both Elisha Gray and Alexander Graham Bell were separately working on the design of a harmonic telegraph. Gray was first to achieve a patent on a scheme that assigned different tones to different messages being signalled simultaneously on the line. Here is an early version of frequency-division multiplexing, a forerunner of frequency-division multiple access (FDMA), albeit on a wire. Coincidentally, at nearly the same time (1874), Baudot introduced his five-unit code for use with his manual time-multiplexing system for telegraph traffic. Certainly this system was a primitive form of time-division multiple access (TDMA) [8].

It evidently occurred to both Gray and Bell that the human voice is similar to a frequency-multiplexed system, and both worked intensely to get the human voice directly onto the telegraph lines. Bell applied for his landmark telephone patent 174,465, "Improvement in Telegraphy," in 1876 on the same day that Gray filed a caveat on his invention, and three weeks later, the twenty-nine year old Bell received one of the most valuable patents ever awarded. A more modern patent in response to the need to multiplex voice messages on a single telephone line is given in [9].

Table 1
Electrical Communication Prior to the Electric Light

1729	Stephen Gray charges an electroscope through 293 feet of wire.
1745	Leyden jar for the storage of charge discovered.
1753	Letter in *Scots Magazine* suggests electric communication.
1774	Le Sage constructs first electric telegraph instruments.
1785	Coulomb publishes law of force between charged particles.
1799	Volta announces discovery of the electrical cell.
1816	First single-wire dial telegraph by Ronalds.
1820	Örsted invents the galvanometer.
.1826	Ohm publishes voltage, current, resistance relation.
1828	Henry exhibits the electro-magnet.
1831	Faraday discovers law of induction and develops a homopolar generator.
1832	Hyppolyte Pixii constructs an ac generator with permanent magnets.
1833	Gauss and Weber construct galvanometer telegraph.
1837	Wheatstone and Cooke granted patent for a chronometric telegraph. Morse files U.S. Patent Office caveat on telegraph.
1838	Steinheil proposes closing telegraphic circuit through the earth.
1847	Von Helmholtz deduces oscillatory nature of Leyden jar discharge.
1850	Telegraph cable laid across English Channel.
1853	Gintel develops duplex telegraph system.
1854	Bourseul suggests on-off telephony. Faraday suggests capacitive model for submarine cable.
1857	Kirchoff develops velocity-of-propagation along telegraph wire.

1858   First Atlantic cable message received.
1862   Reis constructs primitive telephone models.
1865   Maxwell publishes *A Dynamical Theory of the Electromagnetic Field*.
1869   Hale considers "The Brick Moon" as a navigational aid!
1874   Edison develops quadraplex telegraphic signalling. Gray and Bell separately design harmonic telegraph. Baudot introduces time-multiplexing telegraphic system.
1876   Bell files and receives patent for telephone. Heaviside produces equation of telegraphy.

III. Mathematical Electricians

Oscillatory electrical phenomena were observed and predicted in the early-19th century. In 1827 F. Savary conjectured about the discharge of a Leyden jar that "the electric motion during the discharge consists of a series of oscillations." Henry stated a similar conjecture in 1842 ([6], pp. 226-227). Hermann Von Helmholtz theoretically deduced the existence of oscillatory behavior of the discharge of a Leyden jar in 1847; the mathematical demonstration of this was given by William Thomson (later to be Lord Kelvin) in 1853 and experimental verification was provided by W. Fedderson in 1859 ([10], pp. 178-179). This might be viewed as an early demonstration of the resistance, inductance, capacitance equations of electrical circuit theory.

In the mid-1840s, the insulating qualities of gutta percha in low temperature environments were discovered, and it finally was possible to effectively insulate submarine cable. By 1850 the first gutta-percha-coated cable was laid from Dover to Calais ([5], p. 220). In 1854, Michael Faraday demonstrated that such a cable "may be assimilated exactly to an immense Leyden battery; the glass of the jars represents the gutta-percha; the internal coating is the surface of the copper wire" ([6], p. 228). This demonstrated that the electrostatic capacity must be considered in the theory of electrical signals on the wire. In the same year, correspondence between G. G. Stokes and Thomson analyzed the effect of capacity on submarine cable signals. In 1857 Gustav Kirchoff also took into account the self-inductance of the telegraph wire, and showed that an electrical disturbance is propagated along the wire with a specified velocity.

In the mid-1860s, James Clerk Maxwell published his mathematical theory of electromagnetism, in which he predicted that the speed of propagation of electromagnetic waves would agree with the then estimated speed of light, and hypothesized that light is an electromagnetic phenomenon ([6], chap. VIII).

In 1876, Oliver Heaviside produced the following form of the mathematical relations for propagation of electrical signals along a wire:

8

$$\frac{1}{C}\frac{\partial^2 V}{\partial x^2} = L\frac{\partial^2 V}{\partial t^2} + R\frac{\partial V}{\partial t}$$

where L and C are inductance and capacitance per unit length of cable, R is resistance, and V is the electrical potential at distance x from one terminal. This relation became known as the *equation of telegraphy* ([6], p. 229).

IV. The Development of Wireless Communication

The physical demonstration of propagation of electromagnetic waves in the aether,[2] without the benefit of a conductor, is generally credited to Heinrich Hertz in 1888. While demonstrating an experiment with two short flat coils of insulated wires, Hertz noticed that the discharge of a small Leyden jar through one was able to induce currents in the other, provided that a small spark-gap was made in the first coil. In one experiment, Hertz had employed both an exciter of electromagnetic waves and a reasonably matched detector of such waves ([11], p. 18).

Remarkably, seven years earlier in 1879, David Hughes had successfully demonstrated virtually the same spark transmitter with a coherer (a receiving device employing metal filings, usually credited to Edouard Branly, ca. 1885-1891) and telephone as a receiving system separated by as much as a few hundred yards, but could not convince learned observers that it was wave propagation, not mere electromagnetic induction effects. Hughes never published his work ([10], Appendix D).

In early 1896, twenty-two year-old Italian Guglielmo Marconi left his home for London carrying with him apparatus with which he had successfully communicated at distances over a mile. He had achieved this remarkable distance by combining the spark-gap transmitter of Hertz with the Branly coherer as a detector, and most importantly, elevated antennas similar to those used in electrical storm detection experiments ([11], p. 26). With the assistance of his cousin Henry Jamieson Davis, Marconi received British Patent 12,039 in June 1896 on his invention, and one year later the Wireless Telegraph and Signal Company limited was formed to develop the Marconi apparatus commercially. There was skepticism, as well as technical problems, to overcome. Even Lord Kelvin once said ([12], p. 46),[3] "Wireless is all very well, but I'd rather send a message by a boy on a pony."

---

[2]This Latin form of the word *ether* was used in early publications to denote the medium permeating all space, whose existence was supposed necessary to support electromagnetic wave propagation.

[3]The date of this quote is not apparent, but in the context in which it was presented in [12], it was before Marconi's trip to Britain in 1896.

Successful demonstrations of *mobile* communication with ships at sea supplied the niche opening to a market dominated by wire-guided telegraph signals.

Marconi's spark-gap transmitter emitted energy over a broad spectrum of frequencies, and successful reception when more than one station was transmitting was nearly impossible. Oliver Lodge had demonstrated *syntony* in a wired circuit (a tuned circuit) in 1889, and later Lodge had seen the utility in wireless telegraphy and received Patent 11,575 for such an improvement. Marconi and others both received further patents on such systems, culminating in Marconi's Patent 7777 on a system using *adjustable* tuned circuits at both the transmitter and receiver, thereby allowing simultaneous (i.e., multiple-) access to the air waves on different frequencies ([11], Chap. 5).

Remarkably, Edison's experiments with electric lamps in the early 1880's led to an inadvertent discovery of the vacuum tube diode, on which he applied for a patent in 1883. However, with the electron not yet identified, let alone electron emission discovered, the underlying phenomena were not well understood. Further work on the *Edison effect* led to J. Ambrose Fleming's patent application in 1904 of the "oscillation valve" for conversion of wireless signals to direct current. Lee de Forest added a third element, the *grid*, to the Fleming valve in 1906, and was awarded U.S. Patent 879,532 in 1908 on the triode, which he called the "Audion" (*aud*ible *ion*s).

The year 1906 also saw the first voice broadcast, on a mechanically generated high-power carrier at 100 kc., from R. A. Fessenden's station at Brant Rock, Massachusetts. Charles Proteus Steinmetz had recruited Ernst Alexanderson for this high-speed (20,000 rpm) alternator design at General Electric, after previous attempts suffered mechanical failures at speeds above 4,000 rpm ([13], pp. 72-73).

In 1912, AT&T engineers realized that no residual ionized gas was needed for the Audion to function properly, and the tube was improved. In the same year, De Forest and the youthful Edwin Howard Armstrong each discovered the regenerative circuit with feedback from the output to the grid, that became the basis for high gain amplifiers ([14], Chap. 4; [13], Chap. 3). Litigation over this invention lasted for two decades.

V.   Controlling the Radio Spectrum in the U.S.A. [15]

In the U.S.A., the Wireless Ship Act of 1910 put regulation of the infant radio industry under the Secretary of Commerce and Labor, through the Radio Service of the Bureau of Navigation. The first US commercial broadcast was made on November 2, 1920, by Westinghouse Electric's KDKA in Pittsburgh. Interference was already a problem in 1923, when

10

Herbert Hoover, then Secretary of Commerce, stated at the Second Radio Conference that ([15], p. 6)

> "When this committee met a year ago there were 60 broadcasting stations in the United States; today there are 588. It was estimated then that there were between 600,000 and 1,000,000 receiving sets, today it is believed there are between 1,500,000 and 2,500,000 persons listening.
>
> "Public broadcasting has been limited to two wave-lengths, and I need not dilate to you on the amount of interference there is and the jeopardy in which the whole development of the act stands."

The result of this conference was to spread the spectrum available for commercial broadcasting to the range from 550.9 kc. to 1351.4 kc. In addition, the first step toward orderly *frequency reuse* was made when the United States was divided into five zones, with carrier frequencies of different stations in the same zone at least 50 kilocycles apart, stations in adjacent zones at least 20 kilocycles apart, and more distant stations at least 10 kilocycles apart.

In 1926, the regulatory power of the Secretary of Commerce was challenged by the Zenith Corporation which ignored the wavelength and time limitations set forth in its license. Eventually the question reached the Acting Attorney General, whose opinion removed the last vestiges of control by the Secretary of Commerce by declaring that the Secretary could not limit the transmitter power or wavelength of a broadcaster. In 1927 an act of the U.S. Congress established licensing powers and created the Federal Radio Commission to regulate the broadcast industry. In 1934, this entity was replaced by the Federal Communications Commission (FCC), which still oversees access to the air waves.

VI. Emerging Applications of Spectrum Spreading [1],[2],[3],[4]

The quantity that we call transmission *bandwidth*, generally under control of the FCC in the U.S.A., is undoubtedly the single most important parameter in most communication system designs. The many aspects of this fact were discovered during the years after the First World War.

Bandwidth and Transmission Rate

In 1924, Harry Nyquist [16] considered the transmission of one of m^n symbols, each symbol represented by a length n sequence of basic waveforms, and each waveform amplitude-modulated by one of m distinguishable signal levels. He then noted that the symbol transmission rate W is directly proportional to the line speed s (waveforms per second) divided by the number n of waveforms per symbol. In the process of eliminating n to show W proportional to $s \log m$, Nyquist had to use logarithms, thereby hinting at logarithmic measures of information rate.

In 1928, R. V. L. Hartley, published a paper [17] whose avowed purpose was to "set up a quantitative measure whereby the capacities of various systems to transmit information may be compared." Taking as a "practical measure of information the logarithm of the number of possible symbol sequences," Hartley went on to show that

> "the total amount of information which may be transmitted over such a system is proportional to the product of the frequency-range which it transmits by the time during which it is available for the transmission."

Wideband FM --- Willful Spectrum Spreading

In 1933, Edwin Armstrong received five patents on frequency modulation (FM) radio (the first being 1,941,066). John Renshaw Carson, the inventor of single-sideband modulation, had studied FM with the idea of narrowing the bandwidth required for a single telephone signal, and declared in 1922 that narrowbanding FM "distorts without any compensating advantages whatsoever." Armstrong's breakthrough was to *spread the transmission bandwidth* beyond that occupied by the original information signal and to add a limiter prior to the FM detector to eliminate the effects of amplitude modulation on reception ([13], Chap. 10).

Table 2
Signals in the Ether

1879	Hughes propagates signals from spark-gap transmitter to coherer.
1880	Edison and Swan invent the incadescent lamp. Edison effect discovered at about this time.
1888	Hertz announces the production of \ae ther waves.
1889	Lodge demonstrates syntony in a circuit.
1897	Marconi files Patent No. 12,039 on wireless telegraphy. Thomson identifies the electron as a fundamental charged particle.
1900	Marconi granted Patent No. 7777 on an adjustable syntonic wireless.
1904	Fleming applies for oscillation valve patent. Hülsmeyer receives British patent for telemobiloscope.
1906	De Forest produces the triode. Voice broadcast using mechanically generated 100 kc. carrier.
1910	Passage of the Wireless Ship Act.
1912	Armstrong and De Forest separately discover regenerative circuits.
1920	First licensed radio broadcast by Westinghouse station KDKA.
1921	Detroit Police install private mobile phone system.
1923	Second Radio Conference (USA) defines AM band.
1924	Nyquist paper hints at logarithmic measure of information flow. Goldsmith files for patent on anti-multipath FM spectrum-spreading.
1927	Federal Radio Commission created to regulate radio communication.
1928	Hartley finds information transmitted proportional to time-bandwidth product.
1933	Armstrong receives five patents on FM circuitry.
1936	First station in British coastal early-warning radar line completed.
1938	Guanella files for Swiss patent on noise radar.

12

1941   Markey and Antheil file for anti-jam frequency-hopping patent.
1943   Hansen files for patent on secret communication using both
        transmitted and stored reference techniques with FM modulation.
1946   Mobile radio public correspondence system created in St. Louis
        area.
1948   Shannon publishes *A Mathematical Theory of Communication*. Bell
        Telephone Labs announces development of the transistor.

RADAR and Navigation - Spreading for Finer Time Resolution

The possibility of radio wave echos being used for the location of targets
was an idea that had been around even before De Forest's Audion tube
detector. In 1904 Christian Hülsmeyer of Düsseldorf received a British
patent for an anti-collision device which he called a Telemobiloscope. It
was successfully demonstrated in Cologne but never developed ([18], p.2).

Experimental research in Britain, Germany, and the United States, on radio
echo location techniques began in the early 1920s and continued into the
1930s. At the beginning of 1935, under pressure from Winston Churchill,
his friend F. A. Lindemann, and the Royal Air Force, the British
government initiated a study of air defenses. In the spring of 1936, the first
experimental coastal early warning pulse radar was in operation and a year
later the first British shipborne radar was in sea trials.

One remarkable radar patent was filed in 1938 by Gustav Guanella of
Brown, Boveri, and Company in Switzerland. The radiated signal in
Guanella's continuous-wave radar was "composed of a multiplicity of
different frequencies the energies of which are small compared to the total
energy." His examples of such signals included acoustic and electrical *noise*,
and an oscillator whose frequency is "wobbled at a high rate between a
lower and upper limit." Ranging was accomplished by adjusting an internal
signal delay mechanism to match the external propagation delay. Delay-
matching errors were detected by cross-correlation of the internally delayed
signal with a 90° phase-shifted (across the band) version of the reflected
signal. Many of the *synchronization* system concepts (e.g., phase-locked
loops, tracking loop S-curves, and delay-locked loops) are hidden in this
patent [19].

Certainly radar engineers understood at the time that by narrowing the
transmitted pulse width (and hence expanding bandwidth) finer time
resolution and hence better ranging accuracy would result. Furthermore,
near the end of World War II, the Germans were developing a linear-FM
(chirp) radar system called Kugelschale, and hence had discovered that
bandwidth expansion without pulse narrowing could also provide finer time
resolution.

Anti-Jam Communications - Defensive Bandwidth Spreading

In World War II, when radio and radar systems aided the delivery and
guidance of weapons of destruction and the defense of regions, creative

minds came up with ingenious ways to disrupt the enemy radio systems. R. V. Jones, who eventually became Director of Intelligence under Churchill near the end of World War II, commented on these times in the epilogue of his remarkable memoirs ([20], p. 529):

> "Just as the impact of radio in the 20's gave a unique chance for everyman to dabble in the 'marvels of science' by making his own receiver - a task complex enough to be fascinating without being so complex ... that it was beyond the competence of the average man - so it was with Scientific Intelligence in World War II, and for much the same reason. The very development in science and technology that led to everyday radio in the '20's also led to the radar and radio navigational systems of World War II, and these were relatively simple to understand and, if necessary, frustrate."

Upholding Jones' assertion is undoubtedly the most celebrated spread-spectrum radio invention by non-experts. A patent filed in 1941 by Hedy Keisler Markey and George Antheil describes a frequency-hopping system for the guidance of torpedoes. The sequence of frequencies used by the proposed radio-guidance link is stored on piano rolls that are read in synchronism at the transmitter and receiver. The disclosure declares that, lacking the exact synchronization pattern between sender-ship and torpedo, the radio link cannot be jammed. Inventor Antheil had used synchronized player pianos in the 1920's to perform his work *Ballet Mechanique*, and inventress Markey was none other than the MGM film star Hedy Lamarr! [2],[4],[21].

With regard to anti-jam radar systems, the many experiences of World War II were summarized by Frederick E. Terman when he said,

> "In the end, it can be stated that the best anti-jamming is simply good engineering design and the spreading of operating frequencies."

Certainly this observation applied equally well to communications as to radar, and the Markey-Antheil patent is one of several manifestations of anti-jam communications during the war years [1],[2].

VII. Technological and Conceptual Post-War Revolutions

The hasty and intense wartime design efforts were followed in the post-war period by fresh looks at communication system design with an eye to engineering as a science. At the same time, two monumental revelations were made in 1948 at AT&T's Bell Telephone Laboratories.

In June 1948, BTL announced that they had developed the transistor, much through the efforts of three men, William Shockley, Walter Brattain, and John Bardeen. This invention, which earned the 1956 Nobel Prize in physics for the three leaders, marked the advent of a sweeping solid-state technological revolution and the digital revolutions that followed in communication and computer technologies.

14

The Birth of Information Theory

Also in 1948, the U.S. mathematician Claude Shannon published a Mathematical Theory of Communication as a monograph in the *Bell System Technical Journal*. This paper is remarkable because of its elegant theorems, derived from statistical characterizations of *both* the information source and the channel effects. Roughly speaking, Shannon proved the existence of a scheme for transmitting an information source over a communication channel, that achieves error-free communication whenever a source-dependent quantity called the *source information rate* is less than a channel-dependent quantity called the *channel capacity*.

By this time considerable effort had already gone into characterizing statistical properties of signals and receivers, e.g., the works of Norbert Weiner and Steve Rice in the U.S.A. and of V. A. Kotel'nikov in the U.S.S.R. The remarkable ideas put forth in Shannon's work made both the abstract concept of information and the capacity to communicate it measurable and concrete. The following two quotes, from scientists who themselves may be considered giants in the field, reflect the magnitude of this work:

> "Information theory is one of the youngest branches of applied probability theory; it is not yet ten years old. The date of its birth can, with certainty, be considered to be the appearance in 1947-1948 of the by now classic work of Claude Shannon. Rarely does it happen in mathematics that a new discipline achieves the character of a mature and developed scientific theory in the first investigation devoted to it. ... so it was with information theory after the work of Shannon."
> - *A. I. Khintchin, 1956* [22].

> "It is hard to picture the world before Shannon as it seemed to those who lived in it. In the face of publications now known and what we now read into them, it is difficult to recover innocence, ignorance, and lack of understanding. It is easy to read into earlier work a generality that came only later." - J. R. Pierce, 1973 [23].

Perhaps it was no coincidence that communication system patents, once described in terms of elementary devices, soon were being described in terms of block diagrams and mathematical models, reflecting new ways of creative engineering thinking. The development of correlation computing devices that would implement the inner product computations of abstract theory was one of the enabling technological developments of that time.

Direct-Sequence Spread-Spectrum Modulation

Certainly Shannon (and previous researchers) viewed signalling as being done in a function space whose finite dimension grew linearly with the duration of the signal. The scale factor between time duration and dimensionality is one definition of bandwidth, and follows naturally from the sampling theorem for strictly band-limited signals. In fact, other useful bandwidth definitions can also be viewed as providing such a scale factor.

Shannon's famous equation for the communication capacity of the band-limited additive Gaussian noise channel is [24]

$$C = \int_0^W \log_2\left(1 + \tfrac{P(f)}{N(f)}\right)\ df \ \text{bits / second}$$

where W is the available/allowed bandwidth of the channel, $P(f)$ is an optimally chosen signal power density, and $N(f)$ is the noise power density at frequency f. In fact, the optimal choice of $P(f)$, subject to a total signal power constraint, makes the minimum value over f of the total power density $P(f)+N(f)$ as large as possible in the allowed band. Within the assumptions which were the foundation for this formula, the capacity equation suggests that the more available bandwidth W, the larger the capacity C. Hence, the allowed bandwidth should always be used. In [24], Shannon goes on to comment that the stationary Gaussian noise process which minimizes capacity is the one that spreads its available power uniformly across the given band. These arguments indicate that, with both sides in a Gaussian noise jamming game using their optimal strategies for spending communication power P and jamming power N_J in a receiver with thermal noise density N_O, the capacity of the jamming channel is

$$C = W \log_2\left(1 + \tfrac{P}{WN_0 + N_J}\right)$$

Certainly there is motivation here in a very abstract context for expanding bandwidth W in jamming situations until the total receiver noise power WN_O dominates the jamming power N_J. A early unclassified discussion of jamming and bandwidth expansion, based on the above capacity formula, was given by John Costas of General Electric in 1959 [25].

The random signalling arguments that were used in Shannon's proof of the noisy channel coding theorem, also may have further motivated engineers to carefully consider noise-like waveforms for communication signalling. Certainly it is a measure of the then growing understanding of the nature of anti-jam communications that the first direct-sequence spread-spectrum communication systems were designed within the first year or two following Shannon's 1948 publication. The history of anti-jam spread-spectrum systems developed in the USA ca. 1948-60 (covered in detail in [1]) includes many novel approaches to the problems encountered in storing and detecting wideband noise-like waveforms, the first fielded direct-sequence and frequency-hopping spread-spectrum systems, the first written explanation of a processing gain notion, possibly the first use of error-correction in communications, etc. Virtually all of this work was performed in secret for the Department of Defense and did not appear in the open literature.

In the mid-1950's, motivated by the multipath-impaired testing of the anti-jam spread-spectrum F9C teletype communication system over a transcontinental HF link, Robert Price and Paul Green of Lincoln

Laboratory developed a signal processing technique called Rake [26]. The Rake processor uses the fine time-resolution capability of wideband signals to resolve signals arriving over different propagation paths, and inserts them into a diversity combiner to coherently construct a stronger received signal. The disclosure in [26] is probably the first unclassified discussion of the potential of direct-sequence signals to improve communication over multipath channels.

Curiously enough, as with Ronald's dial telegraph, synchronously rotating wheels were used in one of the earliest direct-sequence systems to store and play back a pseudorandom carrier signal ([1], pp. 833-836). However, the early 1950's saw the development of the shift-register technology for generating pseudorandom number sequences, and this has over time become a standard component of most spread-spectrum communication systems. The origins of this idea are explored in [1]. The mathematical study of pseudonoise (PN) generators was assigned to graduate student Sol Golomb after his supervisor at the Glenn L. Martin Co. brought back the notion from a 1953 M.I.T. summer course. After graduation, Golomb continued to work on PN generators at the Jet Propulsion Laboratory, and his books [27],[28] reflecting these efforts have made this subject accessible to many engineers.

VIII. Code-Division Multiple-Access (CDMA) Techniques

In a 1949 technical memorandum [29], John Pierce describes a multiplexing system in which a common medium carries coded signals that "need not be synchronized in any fashion." Its technological underpinnings probably evolved from pulse code modulation (PCM), although we might now classify this system as time-hopping spread-spectrum multiple-access. Pierce noted that as "... an increasing number of channels are transmitted over the medium, there is a gradual degradation of quality." Furthermore, Pierce credited Shannon as having earlier suggested that this sort of performance "could be obtained by using as 'code functions' voltages which are approximately orthogonal functions of time." It was declared that in a given frequency band, any number of noise functions could be found that were approximately orthogonal over a long enough time period. Many of these ideas later appeared in open publication in 1952 [30]. Undoubtedly, the quasi-orthogonality of two long noise signals was realized by engineers much earlier, e.g., possibly by Guanella when working on noise radar design [19].

In a paper [31] submitted in 1949, Warren White references early work on Loran in which asynchronous transmissions of different Loran station pairs were separated on the basis of "slight differences in pulse repetition frequency," and mentions the existence of interrogator-responder systems possessing an asynchronous multiplexing feature. Referencing the recent work of Shannon, White goes on to analyze an asynchronous time-hopping multiple-access system, and concludes that "Asynchronous multiplex techniques seem most applicable in cases where the nature of the service is such that the bandwidth tends to be much wider than is justified by the information to be transmitted over a single channel, ..."

Code-division multiple-access is elegantly described in a conversation between Claude Shannon and Robert Price over a decade ago [32]. Reflecting on his thoughts ca. 1949, Shannon stated that
> " [CDMA] seemed like a very democratic way to use up the coordinates that you have, and to distribute the "cost of living," the noise, evenly among everyone. The whole thing seemed to have a great deal of elegance in my mind, mathematically speaking, and even from the point of view of democratic living in the world of communications."

Shannon then made the comparison to a party:
> "... More and more people can come, and they would all pay equally, so to speak. If more people were there, gradually the noise level would increase on each channel. But everyone could still talk, even though it might be a pretty noisy 'cocktail party' by that time."

Price commented that this was "what we now call 'graceful degradation' in military jargon." (The term graceful degradation was in use as early as 1963 [33].)

The two-user multiple-access communication model was carefully investigated in several information-theoretic papers during the early 1970's [34],[35],[36],[37]. The result of this investigation was a description of the possible pairs of information rates that two transmitters might use to reliably communicate simultaneously with one receiver. Generally the resulting rate region exceeded what one might achieve with TDMA in this case. The effects of code asynchronism on the useful rate regions has been investigated in [38],[39].

Louis De Rosa and Mortimer Rogoff's 1950 proposal, in addition to containing perhaps the earliest description of the processing gain concept for a direct-sequence stored-reference spread-spectrum system, describes a means of multiplexing a number of data signals onto time-shifted versions of a single noise carrier [40]. Another noise-carrier multiplexing concept, although transmitted reference in nature, was demonstrated by graduate student Bernard Pankowski at M.I.T. [41].

The first direct sequence spread-spectrum systems were built during the 1950's. As designers developed means for storing pseudo-random sequences and solving the synchronization problems associated with detecting the pseudorandom carriers, the concept of asynchronous code-division-multiplexing (what is now called code-division multiple-access)[4] on the air

---

[4]The distinction between *multiplexing* for co-located signal sources and *multiple-access* for spatially separated signal sources was probably first made in the early 1960s. The term CDMA, as used by many, includes frequency hopping, time-hopping, direct-

waves by independently operating radios was explored with these systems. Using Shannon's formula, the classic paper [25] in 1959 by Costas evaluates the capacity of a spread-spectrum multiple-access (SSMA) system under the section title "The Question of Channels." Here he concluded that SSMA is better than FDMA over comparable bandwidths when the offered signal traffic is *intermittent*.

Other early spread-spectrum multiple-access papers include those by J. E. Taylor of General Electric [42], W. J. Judge of Magnavox Research Labs [43], and H. Magnuski of Motorola [44]. Magnuski's paper is notable because of its early reference to the near-far problem, i.e., the adverse signal-to-noise ratio condition that can occur when a nearby SSMA transmitter provides a high level of interference to a signal from a distant transmitter in the same frequency band. Schwartz [33] mentions several multiple-access spread-spectrum systems under development in the early 1960's.

Judge's paper reflects the then current views of the designers of the ARC-50 radio at Magnavox. Bob Dixon recalls that the concept of code-division addressing was taken for granted during the development of the ARC-50 spread-spectrum radio in the late 1950's and early 1960's. The operator could adjust the transmitter power of the ARC-50 over a 90 dB range as required. The system was designed to send/receive any one of several short PSK-modulated m-sequences (later replaced by Gold codes [45]) which, after code synchronization was accomplished, were switched to corresponding long m-sequences. Magnavox also used spread-spectrum techniques in their satellite communications radios developed in the 1960's. Many of these concepts were used in several operational satellite radios, including the OM-55 and USC-28 radios, which were designated as multiple-access equipment by the Army.

As early as 1960, military satellite communications experts were concerned about two aspects of the power-control problem occuring in satellite multiple-access systems. The first was controlling the total signal power processed by the satellite's multiply-accessed nonlinear transponder (to avoid intermodulation effects), and the second was the problem of apportioning the total signal power among the signals processed by the transponder in a way that would adequately support their individual data rate requirements. A comparison of multiple-access techniques, including frequency division (FDMA), time-division (TDMA), spread-spectrum (SSMA), and pulse-addressing (PAMA), is given for the hard-limiting satellite repeater in [46], and may be the first open-literature use of these terms.

---

sequence (DS), and other modulation formats, and is virtually synonymous with SSMA. Others tend to specialize the term CDMA to DS-SSMA.

The Department of Defense has built several spread-spectrum satellite systems, the most visible in the early literature being the tactical communications satellite TATS [47] and somewhat later the Global Positioning System (GPS) navigation satellite system [48]. In GPS, whose development began in 1973 [49], the navigation receiver (not the satellite) is the multiply accessed resource. To receive the GPS signal from one of several satellites, one must first synchronize to that satellite's short *C/A* code which is 1023 chips long with a period of 1 msec. Then if allowed access, navigation receiver can demodulate information that allows synchronization handoff to the satellite's long *P* code which has a chip rate of 10.23 Mcps and a period of exactly one week. The first hint of commercial application of spread-spectrum signalling is present, because the *C/A* portion of the satellite's transmitted signal is available for use by commercial navigation receivers. (Remarkable as a GPS satellite is, it's novelty as an orbiting navigational aid is challenged by E. E. Hale in his farsighted 1872 short story, *The Brick Moon* [50].)

IX. Commercialization of Spread-Spectrum Communications

The first extensive private mobile radio systems were used for public safety purposes, e.g., New York City communications to harbor patrol boats in 1916, and the Detroit Police Department's patrol car communications, tested in 1921 [51]. The first public correspondence system for land mobile radio was created in St. Louis just after the Second World War. Almost immediately, the cellular radio concept was [52] "verbalized in 1947 by D. H. Ring of Bell Laboratories in unpublished work." After thirty years of research and development contributions, the first cellular system (AMPS) in the USA went into developmental system trials. Certainly, the advantage of frequency reuse and the "roamer problem" of mobility tracking and handoff were uncovered in the design process (e.g., see [53]). Even as AMPS was going into system trials in 1978, George Cooper and Ray Nettleton were suggesting spread-spectrum signalling as potentially providing a more efficient (i.e., traffic per unit bandwidth per unit area) cellular communication system than other signalling means [54].

Several spread-spectrum studies and development efforts during the late 1970's and early 1980's are notable. Among them, Hewlett-Packard developed an indoor wireless terminal communication system [55], but did not pursue it as a commercial product. George Turin studied the application of Rake reception techniques to overcome urban multipath [56]. The first widespread commercial use of spread-spectrum techniques was in Equatorial Communications Company's C-band receive-only small satellite earth stations [57]. Founded by Edwin Parker, Equatorial introduced its first micro earth-station product in 1981 and by 1984 had sold over 10,000 such items.

By 1985, the FCC adopted a policy that encouraged experimentation with spread-spectrum communications. Interim Standard 95 was adopted by the TIA in 1993, specifying the spread-spectrum modulation formats and protocols for communication between a cell's base station and mobile telephone. The IS-95 is written along the lines of the CDMA system designed by Qualcomm, Inc., a company headed by Irwin Jacobs and Andrew Viterbi. The current Qualcomm hardware has several unique features, including a patented soft handoff procedure [58] that implements an adaptable Rake receiver [26] employing diversity reception of signals both from multipath and from more than one base station, and a patented power control algorithm [59] to ameliorate the near-far problem. The first commercial cellular base-station employing Qualcomm's CDMA design went into operational trials in the Seattle area in January 1994.

Qualcomm's original commercial CDMA product was OmniTRACS, a vehicle communication system operating via a satellite channel, which first went on the market in 1988 and by the fall of 1993 had shipped over 57,000 units. Satellite networks will soon break into the forthcoming wave of personal communications. Both TRW's Odyssey satellite communications system and Loral/Qualcomm's Globalstar use CDMA modulation techniques, with the latter's modulation format being consistent with IS-95 (e.g., see [60]).

X.   Some Thoughts

In this history I've used the term revolution only in regard to events in 1948. Even there, an in-depth look will indicate that somewhat more gradual evolutionary progress was occurring on several fronts, and the linking of ideas at that time was especially fruitful. It is a comforting thought to people now entering the engineering and science professions and trying to find out how they can contribute, that giant strides in technological development are usually composed of many smaller positive steps. Perhaps Isaac Newton made a case for scientific evolution when he said, "If I have seen farther than others it is because I stood on the shoulders of giants."

The concepts of frequency-division and time-division access to common transmission resources are ingrained in the history of communication, and perpetuated both by capital investment and by regulatory authority. It was indeed a remarkable event when, without external pressure, the Federal Communications Commission in the USA decided to encourage a new access algorithm to the radio spectrum. Will the commercialization of spread-spectrum multiple-access be a revolution? Only time will put the commercial uses of spread-spectrum techniques into a true historical perspective.

Table 3
Spread-Spectrum Communications --- After Shannon

1949	Pierce suggests asynchronous time-hopping multiple access in BTL memo.
1950	De Rosa - Rogoff proposal includes direct-sequence spread-spectrum system, processing gain equation, noise multiplexing idea.
1952	First secret theses on NOMAC systems completed at M.I.T.
1954	First transcontinental field trials of F9C direct-sequence spread-spectrum HF teletype system.
1956	Price and Green file for anti-multipath "Rake" patent.
1957	Sputnik 1 launched into orbit. Sturgeon publishes *The Pod in the Barrier*.
1959	Costas reveals spread-spectrum concepts in unclassified paper, suggests SSMA superiority in intermittent traffic situations.
1961	ARC-50, first all solid-state airborne DS system, goes into production.
1962	Frequency-hopping system BLADES, with interleaving and error-correction, delivered for shipboard testing.
1965	First commercial satellite, Intelsat 1 launched.
1971	Information-theoretic work suggests time-sharing not best multiple-access strategy.
1973	Phase I of GPS development begins.
1978	AMPS Developmental system trial at 850 MHz.
1979	Four GPS satellites up, testing phase begins.
1980	Direct-sequence indoor wireless system disclosed by HP.
1985	FCC Part 15 rule encourages SS experimentation.
1993	IS-95 adopted by the CTIA for spread-spectrum cellular service.
1994	First CDMA cellular base station undergoes operational trials in Seattle.

XI. Acknowledgement

I am grateful for the comments and information provided by Joe Aein, Ed Bedrosian, Charlie Cahn, Bob Dixon, Gaylord Huth, Bill Lindsey, John Pierce, Bob Price, Eb Rechtin, Herb Taylor, Andy Viterbi, and Aaron Wyner concerning this history of CDMA systems. I also want to acknowledge the assistance of Thomas Tsui and Fernando Ramirez whose literature searches have aided in this work.

XII. References

[1] R. A. Scholtz, "The origins of spread-spectrum communications," *IEEE Trans. Commun.*, vol. COM-30, pp. 822-854, May 1982.

[2]   R. Price, "Further notes and anecdotes on spread-spectrum origins," *IEEE Trans. Commun.,* vol. COM-31, pp. 85-97, January 1983.

[3]   R. A. Scholtz, "Notes on spread-spectrum history," *IEEE Trans. Commun.,* vol. COM-31, pp. 82-84, January 1983.

[4]   M. K. Simon, J. K. Omura, R. A. Scholtz, and B. K. Levitt, *Spread Spectrum Communications.* Rockville, MD: Computer Science Press, 1985; New York: McGraw Hill, 1994.

[5]   A. F. Harlow, *Old Wires and New Waves.* New York and London: D. Appleton Century Co., 1936; reprinted by Arno Press and the New York Times, 1971.

[6]   E. T. Whittaker, *A History of the Theories of Aether and Electricity: I The Classical Theories.* London, New York: Nelson, 1951-53. Reprinted with new material by American Institute of Physics, 1987.

[7]   G. P. Oslin, *The Story of Telecommunications.* Macon, GA: Mercer University Press, 1992.

[8]   Ronald Brown, "Telegraph," *Encyclopedia Americana,* Grolier Incorporated, 1989.

[9]   G. A. Campbell, "Basic types of electric wave filters," U.S. Patent 1,227,113, May 22, 1917.

[10]   J. J. Fahie, *A History of Wireless Telegraphy.* New York: Dodd, Mead and Co.; Edinburgh and London: William Blackwell and Sons, 1901; reprinted by Arno Press and the New York Times, 1971.

[11]   W. J. Baker, *A History of the Marconi Company.* London: Methuen & Co. Ltd, 1970.

[12]   D. Gunston, *Marconi: Father of Radio.* New York: Crowell-Collier Press, 1965.

[13]   Tom Lewis, *Empire of the Air.* New York: Edward Burlingame Books, 1991.

[14]   J. A. Hijiya, *Lee de Forest and the Fatherhood of Radio.* Bethlehem: Lehigh University Press; London and Toronto: Associated University Presses, 1992.

[15]   L. F. Schmeckbier, *The Federal Radio Commission: Its History, Activities, and Organization,* Washington, D.C.: The Brookings Institution, 1932.

[16]   H. Nyquist, "Certain factors affecting telegraph speed," *Bell Sys. Tech. J.,* vol. 3, no. 2, pp. 324-346, April 1924.

[17] R. V. L. Hartley, "Transmission of Information," *Bell Sys. Tech. J.,* vol. 7, no. 3, pp. 535-563, July, 1928.

[18] D. Howse, *Radar at Sea: The Royal Navy in World War 2.* Annapolis MD: Naval Institute Press, 1993.

[19] G. Guanella, "Distance Determining System," U.S. Patent 2,253,975, August 26, 1941 (filed in U.S. on May 27, 1939, in Switzerland on September 26, 1938).

[20] R. V. Jones, *Most Secret War: British Scientific Intelligence 1939-1945.* London: Hamish Hamilton, 1978.

[21] D. Kahn, "Cryptology and the origin of spread spectrum," *IEEE Spectrum,* vol. 24, no. 9, pp. 70-80, Sept. 1984.

[22] A. I. Khintchin, "On the Fundamental Theorems of Information Theory," *Uspekhi Matematicheskikh Nauk,* vol. XI, no. 1, 1956.

[23] J. R. Pierce, "The early days of information theory," *IEEE Transactions on Information Theory,* vol. 19, January 1973.

[24] C. E. Shannon, "Communication in the presence of noise," *Proc. IRE,* vol. 37, pp. 10-21, January 1949.

[25] J. P. Costas, "Poisson, Shannon, and the radio amateur," *Proc. IRE,* vol. 47, no. 12, pp. 2058-2068, December 1959.

[26] R. Price and P. E. Green, Jr., "A communication technique for multipath channels," *Proc. IRE,* vol. 46, pp. 555-570, March 1958.

[27] S. W. Golomb, *Shift Register Sequences.* San Francisco, CA: Holden-Day, 1967.

[28] S. W. Golomb, ed., *Digital Communications with Space Applications.* Englewood Cliffs, NJ: Prentice-Hall, 1964.

[29] J. R. Pierce, "Time division multiplex system with erratic sampling times," Technical Memorandum 49-150-15, Bell Telephone Laboratories, June 15, 1949.

[30] J. R. Pierce and A. L. Hopper, "Nonsynchronous time division with holding and with random sampling," *Proc. IRE,* vol. 40, pp. 1079-1088, September 1952.

[31] W. D. White, "Theoretical aspects of asynchronous multiplexing," *Proc. IRE,* vol. 38, pp. 270-275, March 1950.

24

[32]   ---, "A conversation with Claude Shannon," interview conducted by R. Price, edited by F. Ellersick, *IEEE Communications Magazine,* vol. 22, no. 5, pp. 123-126, May, 1984.

[33]   L. S. Schwartz, "Wide-bandwidth Communications," *Space/ Aeronautics,* pp. 84-89, 1963.

[34]   R. Ahlswede, "Multi-way communication channels," delivered at the 2nd Int. Symp. on Information Transmission, U.S.S.R., 1971.

[35]   H. Liao, *Multiple Access Channels,* Ph.D. dissertation, Dept. of Electrical Engineering, U. of Hawaii, Honolulu, Hawaii, 1972.

[36]   D. Slepian and J. K. Wolf, "A coding theorem for multiple access channels with correlated sources," *Bell Syst. Tech. J.,* September 1973.

[37]   A. D. Wyner, "Recent results in Shannon theory," *IEEE Trans. Inform. Theory,* vol. IT-20, pp. 2-10, January, 1974.

[38]   T. M. Cover, R. J. McEliece, and E. C. Posner, "Asynchronous multiple-access channel capacity," *IEEE Trans. on Inform. Theory,* vol. IT-27, no. 4, pp. 409-413, July 1981.

[39]   J. Y. N. Hui and P. A. Humblet, "The capacity region of the totally asynchronous multiple-access channel," *IEEE Trans. on Inform. Theory,* vol. IT-31, no. 2, pp. 207-216, March 1985.

[40]   L. A. De Rosa and M. Rogoff, Sect. 1 (Communications) of "Application of statistical methods to secrecy communication systems," Proposal 946, Fed. Telecommun. Lab., Nutley, NJ, Aug. 28,1950.

[41]   B. J. Pankowski, "Multiplexing a radio teletype system using a random carrier and correlation detection," M.I.T. Res. lab. and Lincoln Lab., Tech. Rep. 7, May 16, 1952 (ATI 168857; not available from M.I.T.).

[42]   J. E. Taylor, "Asynchronous Multiplexing," *AIEE Transactions (Communications and Electronics),* vol. 79, pp. 1054-1062 January 1960.

[43]   W. J. Judge, "Multiplexing using quasiorthogonal functions," *AIEE Winter General Meeting,* January, 1962.

[44]   H. Magnuski, "Wideband channel for emergency communication," *IRE Int. Conv. Rec.,* New York, NY, Mar. 20-23, 1961, part 8., pp. 80-84.

[45] R. Gold, "Optimal binary sequences for spread-spectrum multiplexing," *IEEE Trans. Inform. Theory,* vol. IT-13, pp. 619-621, October, 1967.

[46] J. W. Schwartz, J. M. Aein, and J. Kaiser, "Modulation techniques for multiple access to a hard-limiting staellite repeater," *Proc. IEEE,* vol. 54, pp. 763-777, May, 1966.

[47] P. R. Drouilhet, Jr., and S. L. Bernstein, "TATS -- A bandspread modulation system for multiple access tactical satellite communication," *EASCON Convention Record,* 1969.

[48] J. J. Spilker, Jr., "GPS signal structure and performance characteristics," *Journal of the Institute of Navigation,* vol. 25, no. 2, pp. 121-146, Summer 1978.

[49] C. Sherod, "GPS positioned to change our lives and boost our industry," *Microwave System News,* vol. 19, no. 7, pp. 24-33, July, 1989.

[50] E. E. Hale, "The Brick Moon," *His Level Best and Other Stories.* (Originally published in 1872 by James R. Osgood & Co.) New York: Garrett Press, 1968.

[51] E. W. Chapin, "The past and future techniques of vehicular communications," *IRE Trans. Veh. Tech.,* vol. VC-9, no. 2, pp. 6-10, August, 1960.

[52] W. R. Young, "Advanced Mobile Phone Service: Introduction, Background, and Objectives," *Bell Syst. Tech. J.,* vol. 58, no. 1, pp. 1-14, Jan. 1979.

[53] W. D. Lewis, "Coordinated broadband mobile telephone system," *IRE Trans. Veh. Tech.,* vol. VC-9, no. 1, pp. 43-48, May, 1960.

[54] G. R. Cooper and R. W. Nettleton, "A spread-spectrum technique for high-capacity moblie communications," *IEEE Trans. Veh. Tech.,* vol. 27, no. 4, pp. 264-275, Nov. 1978.

[55] J. P. Freret, "Applications of spread-spectrum radio to wireless terminal communications," *Proc. of NTC'80,* vol. 4, pp. 69.7.1-69.7.4., 1980.

[56] G. L. Turin, "Introduction to spread-spectrum antimultipath techniques and their application to urban digital radio," *Proc. IEEE,* vol. 68, no. 3, pp. 328-353, March 1980.

[57] E. B. Parker, "Micro earth stations as personal computer accessories," *Proc. IEEE,* vol. 72, no. 11, pp. 1526-1531, Nov. 1984.

[58] K. S. Gilhousen, R. Padovani, and C. E. Wheatley, "Method and system for providing a soft handoff in communications in a CDMA cellular telephone system," U.S. Patent 5,101,501 (filed Nov. 7, 1989), March 31, 1992.

[59] K. S. Gilhousen, R. Padovani, and C. E. Wheatley, "Method and apparatus for controlling transmission power in a CDMA cellular mobile telephone system," U.S. Patent 5,056,109 (filed Nov. 7, 1989), Oct. 8,1991.

[60] Session 39, Specific Personal Communications Systems, *1994 AIAA 15th International Communications Satellite Systems Conference,* San Diego, CA, February 27-March 3, 1994.

Chapter 2

Information Theory & Spread Spectrum

Towards an Information Theory of Spread-Spectrum Systems

James L. Massey

Abstract–A novel definition of a spread-spectrum signal as a signal whose Fourier bandwidth is much greater than its Shannon bandwith (one-half the number of dimensions of signal space required per second) is proposed. Six different communication systems are analyzed in terms of this definition. It is shown that there is a fundamental difference between the bandwidth expansion due to coding and that due to "spectrum spreading". It is further shown that spectrum spreading plays no role in increasing channel capacity, but can perform other useful roles such as providing low probability of interception of the signal, good electromagnetic compatibility, and a multiple-access capability. The effects of linear and nonlinear filtering on bandwidth are considered and seen to be quite different for Fourier bandwidth and for Shannon bandwidth. The concepts developed are used to resolve two paradoxes in spread-spectrum communications: the apparent increase in capacity when users become unsynchronized in a code-division multiple-access (CDMA) system and the fact that a heavily loaded CDMA system is as energy-efficient for transmitting information as a single-user system with the same (total) average power constraint. Areas of spread-spectrum communications where further information-theoretic development is needed are indicated.

1. INTRODUCTION

The main purpose of this paper is to consider, from the fundamental viewpoint of Shannon's information theory [1], systems that employ spread-spectrum signals . To do this requires that we carefully define what we mean by a spread-spectrum signal. This is done in Section 2 in which we give a rather unconventional definition of a spread-spectrum signal, but the only one that we were able to formulate that we ourselves found to be satisfactory. To illustrate the implications of this definition, we consider the transmitted signals in six different communication systems in Section 3 to see which qualify (under our definition) to be called spread-spectrum signals. In Section 4, we consider various reasons why one might wish to use a spread-spectrum signal. In Section 5, we make a more strictly information-theoretic investigation of single-sender systems where we show that spreading the spectrum of the transmitted signal can never increase capacity but also that such spreading need not decrease capacity significantly. In Section 6, we consider the quite different effects of linear and nonlinear filtering on the Shannon bandwidth and the Fourier bandwidth of a signal. In Section 7, we use the

S.G. Glisic and P.A. Leppänen (eds.), Code Division Multiple Access Communications, 29-46.
© 1995 *Kluwer Academic Publishers. Printed in the Netherlands.*

theory that has been developed in the previous sections to resolve two paradoxes that arise in spread-spectrum communications, namely the apparent increase in capacity when users become unsynchronized in a code-division multiple-access (CDMA) system and the fact that a heavily loaded CDMA system is as efficient for transmitting information as a single-user system with the same (total) average power constraint. In Section 8, we conclude with some remarks as to what more must be done to reach an information theory of spread-spectrum systems that can be used as a basis for making sound practical judgements and choices.

Throughout this paper, we have limited ourselves for simplicity to baseband signals, but the reader should have no difficulty in adapting our approach to passband signals.

2. WHAT IS A SPREAD-SPECTRUM SIGNAL?

In his brilliant treatise [1] that established the field, Shannon called information theory the "mathematical theory of communication". We have often maintained that, in a very real sense, mathematics is definitions. Once the definitions are in place, all the lemmas, theorems and corollaries are determined; one has only to find them and prove them. If we wish to say something about the information theory of spread-spectrum systems, it follows that our unavoidable first task must be to define such systems. Of course, it is "signals" rather than "systems" that have spectra so that our task, more precisely formulated, is to define spread-spectrum signals. This task may well strike the reader as either superfluous or quixotic. Like the U.S. supreme court justice who admitted the difficulty of defining pornography but claimed that he knew it when he saw it, many communication engineers might maintain that a definition is not needed; they know a spread-spectrum signal when they see it. One such friend described a spread-spectrum communication signal to us as "one that uses much more bandwidth than it needs". There seems to be a certain coarse truth in this description, but it will hardly do for mathematical purposes. After some futile attempts to make this description more precise, our friend concluded that a satisfactory general definition of a spread-spectrum signal is not possible, which whetted our appetite to take a stab at formulating one.

Every communication engineer is familiar with the ordinary notion of bandwidth, which we will call *Fourier bandwidth* both to honor the French pioneer in this field and to distinguish it from a less familiar but no less important type of bandwidth. The "sinc pulse" $m(t) = \text{sinc}(2Wt)$, where $\text{sinc}(x) = \sin(\pi \cdot x)/(\pi \cdot x)$, has a Fourier Bandwidth of W Hz, as one sees immediately from its Fourier transform $M(f)$ shown in Fig. 1. For less dichotomous spectra, there are many options for calculating the precise Fourier bandwidth (rms bandwidth, 3 dB bandwidth, 99% energy bandwidth, etc.), but they are all roughly equivalent and any is good enough for our purposes. The notion of Fourier bandwidth extends easily from deterministic signals to stochastic processes (such as modulated signals) in a way familiar to all communication engineers.

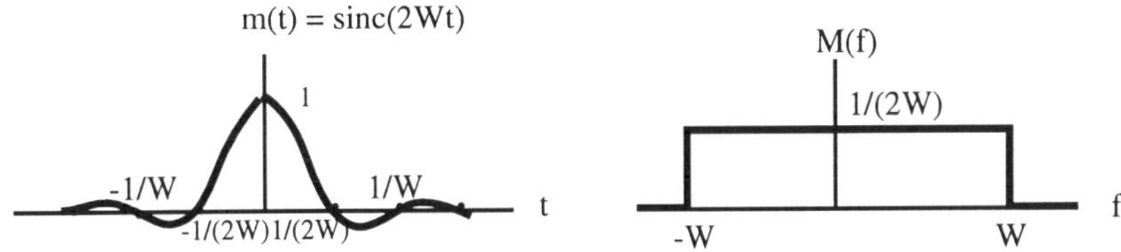

Fig. 1: The Sinc pulse m(t) = sinc(2Wt)
and its Fourier Transform M(f).

The second type of bandwidth, which we will call *Shannon bandwidth* because Shannon [2] was the first to appreciate its importance, makes no real sense for a deterministic signal since it always zero for a single time function. Non-zero Shannon bandwidth implies a "variable" signal (or a stochastic process) such as a modulated signal s(t) that can take on any of a multiplicity of time functions as its value. To determine the Shannon bandwidth of such a signal, one must in principle consider a signal-space representation of s(t) over some very long time interval, say the interval $0 \leq t < T$. By this we mean that one must find orthonormal functions, $\phi_i(t)$, i = 1, 2, .., N, so that one can represent (or very well approximate) every possible realization of s(t) by some choice of the coefficients s_1, s_2,..., and s_N in the linear combination

$$s(t) = \sum_{i=1}^{N} s_i \phi_i(t) \tag{1}$$

for $0 \leq t < T$. One says then that one has a signal-space representation of s(t) as a vector $\mathbf{s} = (s_1, s_2,..., s_N)$ in N-dimensional Euclidean space. When one does this in such a way as to minimize the dimensionality N of the signal space, i. e., to minimize the number of orthonormal functions used, then one has arrived at the *Shannon bandwidth* B, which we now define as

$$B = \frac{1}{2} \frac{N}{T} \text{ (dim/sec).} \tag{2}$$

Equivalently, the Shannon bandwidth is *one-half the minimum number of dimensions per second required to represent the modulated signal in a signal space.* [In earlier papers [3], [4] where we used the notion of Shannon bandwidth, we omitted the division by 2 in (2). Emboldened by Emerson's dictum that "a foolish consistency is the hobgoblin of small minds" [5], we have now opted for the the factor 2 in the denominator of (2) in order to avoid many such factors elsewhere.]

We now state what might be called *the fundamental theorem of bandwidth:*

> The Shannon bandwidth B of a modulated signal is at most equal to its Fourier bandwidth W; [Rough] equality holds when the orthonormal functions are $\phi_i(t) = \sqrt{2W}$ sinc(2Wt - i) [or any orthonormal functions whose spectra are nearly flat in magnitutde for -W < f < W and nearly zero elsewhere].

There are many proofs of this theorem; Shannon [2] gives a conceptually simple proof whose essence he credits to Nyquist [6] and Gabor [7]; essentially one shows that one can construct 2WT orthonormal functions of Fourier bandwidth W or less that are confined within the time interval $0 \leq t < T$ when WT >> 1, but that one can construct no more than this. See the insightful book of Wozencraft and Jacobs [8] and the penetrating paper of Slepian [9] for further discussion of this theorem.

We are now ready to offer our *definition of a spread-spectrum signal as a signal whose Fourier bandwidth is substantially greater than its Shannon bandwidth.* If one considers the Shannon bandwidth to be the amount of bandwidth that the signal *needs* (and we will offer arguments to this effect later) and the Fourier bandwidth to be the amount of bandwidth that the signal *uses*, then we are back at our friend's pithy characterization of a spread-spectrum signal as "one that uses much more bandwidth than it needs".

It is an obvious next step to define the *spreading factor*, γ, of a modulated signal as the ratio of its Fourier bandwidth to its Shannon bandwidth, i.e.,

$$\gamma = W/B. \tag{3}$$

For every modulated signal, $\gamma \leq 1$. A spread-spectrum signal is a modulated signal with "large" γ, say $\gamma \leq 5$, but of course the precise dividing line between a spread-spectrum signal and an unspread signal is rather arbitrary.

We now have our definitions. It remains to show that they make sense, i.e., that they lead to interesting and useful conclusions.

3. SOME EXAMPLES AND THEIR LESSONS

In digital communication systems (to which we will mostly confine our discussion), the modulated signal in an interval $0 \leq t < T$ can assume only finitely many values. In this case, one can in principle always apply the familiar Gram-Schmidt orthogonalization technique (see, e.g., [10, p. 277]) to these signals to obtain an orthonormal basis $\phi_1(t)$, $\phi_2(t)$, ..., $\phi_N(t)$ for the signal space of smallest dimension N containing these signals and can thus determine the Shannon bandwidth according to (2). In most cases of practical interest (as the following examples will illustrate), however, one can find such a basis (and hence find N) by inspection. For an analog system, the modulated signal must generally be treated as a stochastic process. In this case, one can use the Karhunen-Loéve expansion [9, pp. 96-99] to determine N, which will be the number of orthonormal functions in the expansion that have coefficients of non-negligible energy.

We now give several examples of modulated signals, whose analysis will give insight into our definition of a spread-spectrum signal.

Example 1: The modulated signal corresponds to that for one of K users in a time-division multiple-access system in which each user sends L data symbols during each TDMA frame of duration T seconds. Choosing sinc pulses to make the Fourier bandwidth unambiguous, we can write the selected user's modulated signal as

$$s(t) = \i\su(i=1,n, b_i) \ \text{sinc}(\F(KL,T) \ t - i), \quad 0 \le t < T \tag{4}$$

where b_1, b_2, ..., b_L are the data symbols. We see that the Fourier bandwidth W must satisfy $2W = KL/T$ and hence that

$$W = \F(KL,2T) \ . \tag{5}$$

Example 2: The modulated signal corresponds to that for one user in a code-division multiple-access (CDMA) system in which each user modulates a user-specific binary (± 1) code sequence of length L with one data symbol in each symbol period of duration T. We can write the selected user's modulated signal as

$$s(t) = b_1 \ \i\su(i=1,L, a_i) \ \text{sinc}(\F(L,T) \ t - i), \ 0 \le t < T \tag{6}$$

where b_1 is the data symbol and where $(a_1, a_2, ..., a_L)$ is the binary (± 1) code sequence. The Fourier bandwidth W satisfies $2W = L/T$ and hence

$$W = \F(L,2T) \ . \tag{7}$$

Example 3: A user sends an M-ary pulse-position modulated signal in each T second interval, i.e.,

$$s(t) = A \ \text{sinc}(\F(M,T) \ t - b_1), \ 0 \le t < T \tag{8}$$

34

where $b_1 \in \{1, 2, ..., M\}$ is the single data symbol and where A is some fixed amplitude. The Fourier bandwidth satisfies $2W = M/T$ and hence

$$W = \F(M,2T) . \tag{9}$$

Example 4: A user employs binary antipodal signalling to transmit random binary (± 1) data symbols in such a manner as to send n such symbols in each T second interval, i.e.,

$$s(t) = \i\su(i=1,n, b_i) \, \text{sinc}(\F(n,T) t - i), \ \ 0 \le t < T. \tag{10}$$

The Fourier bandwidth satisfies $2W = n/T$ and hence

$$W = \F(n,2T) . \tag{11}$$

Example 5: Same as example 4 except that now the "data symbols" are the output of a powerful rate $r = 1/n$ (bits/symbol) trellis encoder fed by random "information bits". Equations (10) and (11) apply unchanged.

Example 6: Same as example 5 except that the code is a trivial $r = 1/n$ code with two binary (± 1) codewords, $(a_1, a_2, ..., a_n)$ and its negative. Then

$$s(t) = b_1\i\su(i=1,n, a_i) \, \text{sinc}(\F(n,T) \, t - i), \ \ 0 \le t < T \tag{12}$$

where b_1 is the information bit encoded. The Fourier bandwidth is

$$W = \F(n,2T) , \tag{13}$$

except for a few pathological choices such as $a_1 = a_2 = ... = a_n$ that cause the Fourier bandwidth to collapse from its typical value as given by (13).

We now consider in which of the above six systems the transmitted signal is in fact a spread-spectrum signal (by our definition). The task reduces essentially to finding the Shannon bandwidth B of each signal.

Example 1 (concluded): By inspection of (4), we see that the signal space is minimally spanned by the orthonormal functions $\phi_i(t) = \sqrt{2W}$ $\text{sinc}(2Wt - i)$, $i = 1, 2, ..., L$, where W is given by (5). Thus, its dimension is $N = L$ so that (2) gives the Shannon bandwidth as $B = L/(2T)$. The spreading factor according to (3) is then just $\gamma = K$. The transmitted signal in this TDMA system is a spread-spectrum signal when the number K of users is large.

Example 2 (concluded): By inspection of (6) and because a_1, a_2, ..., a_L are fixed, we see that the signal space is one-dimensional, i.e., N = 1. Thus the Shannon bandwidth is only B = 1/(2T) and the spreading factor is γ = L. The transmitted signal in this CDMA system is indeed a spread-spectrum signal when L is large (and our definition of a spread-spectrum signal would be totally indefensible if it were not!).

Example 3 (concluded): From (8), we see that the signal space is minimally spanned by the orthonormal signals $\sqrt{2W}$ sinc(2Wt - i) for i = 1, 2, ..., M, where W is given by (9). Thus N = M and the Shannon bandwidth is B = M/(2T). The spreading factor is γ = 1, the minimum possible. It follows that such an M-ary PPM signal is *never a spread-spectrum signal*, even though according to (9) it requires *Fourier bandwidth very much larger than* $log_2(M)/T$ *Hz,* when M is large, *to carry at most* $log_2(M)$ *bits of information in each T second interval.* This s(t) is a very wideband signal when M is large, bit it is not a spread signal.

Example 4 (concluded): From (10), we see that the signal space is minimally spanned by the orthonormal signals $\sqrt{2W}$ sinc(2Wt - i) for i = 1, 2, ..., n, where W is given by (11). Thus, N = n, B = n/(2T) and γ = 1. Binary antipodal modulation, not surprisingly, never produces a spread-spectrum signal.

Example 5 (concluded): For virtually any nontrivial trellis coding system, the encoded symbols b_1, b_2, ..., b_n that appear in (10) will take on such a variety of different possible binary (± 1) patterns that one cannot imbed the set of possible s(t) in a signal space of smaller dimension than that required when all n binary symbols can be independently chosen [even though the code, whose binary {0, 1}codewords have been mapped to {+1, -1} in the manner $0 \rightarrow +1$ and $1 \rightarrow -1$, may be linear over the finite field GF(2)]. Thus both the Fourier and Shannon bandwidths are the same as for the uncoded system of example 4 and γ = 1. The transmitted signal resulting from binary antipodal modulation of a non-trivially coded information sequence is *never a spread-spectrum signal*, even though, when the code rate r = 1/n is very small, it follows from (11) that it utilizes *a Fourier bandwidth very much larger than 1/T Hz to carry one bit of information every T seconds.*

Example 6 (concluded): Because the trivial code consists of only two codewords, (a_1, a_2, ..., a_n) and its negative, we see from (12) that the signal space has collapsed to a one-dimensional space, i.e., N = 1. The Shannon bandwidth is thus B = 1/(2T) and hence γ = n. Trivial coding of low rate r = 1/n restricts the input to the binary-antipodal modulator in such a way that its output becomes a spread-spectrum signal. In fact, comparing (6) and (12) we see that such trivial coding gives us precisely the same modulated signal as in the CDMA system of Example 2.

If one accepts our definition of a spread-spectrum signal, then the above examples allow us to draw the following two conclusions:

• From examples 3 and 5, we see that *a large ratio of Fourier bandwidth to information rate does **not** imply that a signal is spread-spectrum.*

• Example 5 teaches us the, perhaps surprising, lesson that modulating a coded sequence produced by a coding system that expands Fourier bandwidth by a large factor n generally does *not* produce a spread-spectrum signal. *The Fourier bandwidth expansion due to nontrivial coding is fundamentally different from the kind of Fourier bandwidth increase that produces a spread-spectrum system,* although example 6 shows that trivial coding can indeed produce this latter type of expansion.

It is common when considering coded CDMA systems for communication engineers to speak of doing part of the spectrum spreading with coding and part with direct-sequence multiplication—we have not infrequently so spoken ourselves. But we see now that such statements are nonsensical. These are not two parts of some single bandwidth-expansion process, but rather two very different ways of increasing Fourier bandwidth. Much of the remainder of this paper will be devoted to investigating the reasons that one might wish to do one or both of these kinds of bandwidth expansion.

4. WHY SPREAD THE SPECTRUM?

The original motivation for transmitting spread-spectrum signals was a military one, viz., to hide from an enemy the very fact that one was transmitting a signal. Today one speaks of the *low probability of interception* (LPI) of a spread-spectrum signal. The argument that spectrum spreading should provide the possibility for achieving LPI goes as follows. If the signal is confined to a small number N of dimensions within the global signal space of dimension $2WT = N\gamma$ in which all signals of bandwidth W and time-limited to $0 \leq t < T$ must lie, and *if there are parameters of the signal that can be varied to create a very large number of possible choices for the N-dimensional signal space occupied by the transmitted signal,* then one can achieve LPI by selecting the values of these parameters at random. [We ignore here the role of the signal power and of the intensity of the noise in the enemy's receiver, which together essentially determine how long it takes to search a given number of dimensions simultaneously for the presence of signal.]

For the CDMA signal of example 2, there are 2^L possible choices of the binary (± 1) parameters a_1, a_2, ..., a_L, but changing the sign of all parameters leaves the signal in the same one-dimensional signal space. Fixing $a_1 = +1$ leaves us with 2^{L-1} choices of a_2, a_3, ... , a_L, each of which gives a different one-dimensional signal space. For large L, the enemy's task of finding the signal space actually used by the sender is thus akin to "looking for a needle in a haystack". A CDMA signal with a large number L of "chips" per symbol period does indeed afford low probability of interception.

For the TDMA signal of example 1, however, the only parameter that can be varied is the choice of the L consecutive symbol periods (out of a total of KL such periods) in which data symbols will be transmitted. There are only KL possible choices so that a low probability of interception can be achieved only if the product $KL = \gamma L$ is very large. [Spies in World War II frequently used this method to hide their transmission; they transmitted Morse code for only a few seconds and then went silent for very long periods.] The point is that *a large spreading factor γ alone is not enough to provide LPI capability, there must be many different ways to choose the signal space.* For the same spreading factor γ, a TDMA signal is much easier to intercept than a CDMA signal because there is much less freedom to choose its signal space.

The twin brother of low probability of interception is *electromagnetic compatibility* (EMC). If it is hard to determine whether a signal is present, then it is obvious that this signal cannot be interfering substantially with other commonly present signals. [If it did interfere strongly with some commonly present signals, we could use those signals in a process to detect its presence.] The excellent EMC capability of a CDMA signal is perhaps the strongest argument that we have today for preferring it to a TDMA signal. [We leave to the reader the task of showing the EMC superiority of a CDMA signal over a frequency-division multiple-access (FDMA) signal.]

The first cousin of low probability of interception is small *inter-user interference* (IUI), which is the prerequisite for a good multiple-accessing capability. If it is difficult to detect the presence of a signal known to be in some class of signals, then shouldn't two such signals hardly interfere with one another? The answer is "yes, at least in a statistical sense!". If two such signals are selected independently at random, then the probability of substantial IUI between them will be small. But care must be taken when the total number of interfering signals is large or when users persist for a long time in using the signal determined by one random choice of parameters. Of course, the K users of a TDMA system with signals as in example 1 will experience no IUI at all when they are well synchronized. But K users (say, K on the order of L) of a CDMA system with signals as in example 2 and with the signal parameters frequently varied will experience IUI with roughly the same statistics no matter whether they are well synchronized or not. This robustness of a CDMA system with regard to inter-user interference is, of course, an important practical consideration.

5. CODING, SPREADING AND NOISE

It is time now to take a more strictly information-theoretic look at the advantages, if any, provided by spectrum spreading. To keep matters simple, we consider the single user system shown in Fig. 2. Of fundamental interest are (1) the *information rate*, R, measured in information bits (i.e., random binary digits) per second at the modulator input; (2) the *capacity*, C_W, also measured in information bits per second, of the channel created by the modulator and the band-limited additive white Gaussian noise (AWGN) waveform channel, which is the upper limit of rates R for which reliable (in the sense of arbitrarily small positive probability of error) recovery of the information bits is possible at the receiver when an appropriate coding system is used; (3) the *average power*, S, of the modulated signal; (4) the *one-sided noise power spectral density*, N_O, of the AWGN; (5) the *Fourier bandwidth*, W, of the bandlimited AWGN waveform channel (which we take without loss of essential generality as equal to the Fourier bandwidth of the modulated signal, as there is no point in transmitting anything outside this band and, if one transmits in a smaller bandwidth, then one might as well reduce the channel bandwidth accordingly); and finally (6) the *Shannon bandwidth*, B, of the modulated signal. Because W is the Fourier bandwidth of s(t), it follows from the fundamental theorem of bandwidth that B ≤ W.

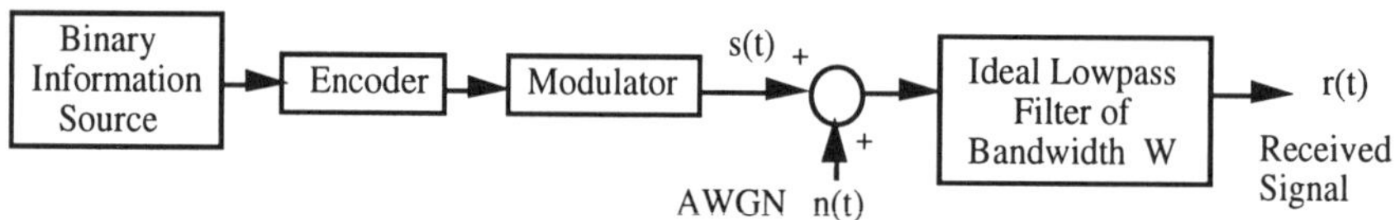

Fig. 2: The single-user communication system under study.

Shannon [1], with his penchant for getting to the heart of the matter, has given us the key relationship among these quantities, namely,

$$C_W = B \log_2(1 + \tfrac{S}{N_O B}) \quad \text{(bits/sec)}. \tag{14}$$

The reader may be surprised to see the Shannon bandwidth B rather than the Fourier bandwidth W in this expression for C. But he or she will find that (14) is precisely the equation that Shannon derives in [1]. It is easy to check that the right side of (14) increases monotonically with increasing B; because $B \leq W$, it follows that

$$C_W \leq W \log_2 (1 + \frac{S}{N_0 W}) \quad \text{(bits/sec)} \tag{15}$$

with equality if and only if $B = W$, i.e., *if and only if there is no spectrum spreading!* The reason that Shannon wrote (15) with an equality sign, rather than (14), in his final expression for the capacity of the AWGN channel is that he assumed that the choice of the modulated signal was up to the sender and that thus the sender would choose a signal with $B = W$ to obtain (maximum) capacity.

Here we must in honesty point out that we have been somewhat cavalier in writing (14) without stating the precise condition for which this expression gives the true capacity. This condition is that all the coefficients s_1, s_2, ... s_N in the expansion (1) must be independently controllable by the choice of the modulator input symbols. This is indeed the case for all of the signals in the above six examples with the exception of the PPM modulated signal in example 3. We see from (8) that in fact for this signal only one of the $N = M$ coefficients can be non-zero so they are certainly not independently controllable. For such modulation systems, the expression in (14) gives only an upper bound on capacity–which is why PPM modulation is not "energy efficient" except for high "signal-to-noise ratios."

It is important not to draw the wrong conclusion from (14) and (15). The real question is not whether spectrum spreading can increase capacity (it never can!), but whether spectrum spreading, which may be desirable for other reasons such as those considered in the previous section, necessarily entails a substantial loss of capacity for the used Fourier bandwidth W. This time the answer is more subtle and more interesting. As B increases without limit, the right side of (14) tends to 1.44 S/N_0. Thus,

$$C_W \leq C_\infty = 1.44 \frac{S}{N_0} \quad \text{(bits/sec)} \tag{16}$$

with near equality when the Shannon bandwidth B (and hence also the Fourier bandwidth W) is very large. For instance, when

$$B \geq 4 \frac{S}{N_0}, \tag{17}$$

the capacity given by (14) is at least 90% of C_∞. As long as the Shannon bandwidth is large enough to satisfy (17), then no matter how large a spreading factor is used, the capacity will be at least 90% of the maximum capacity achievable with the given Fourier bandwidth W. Spectrum spreading cannot increase capacity, but it need not reduce it significantly.

The Shannon bandwidth B is always proportional to what we will call the *symbol rate*, R_m, of the modulator, which we define as the number of symbols per second that are output from the modulator. But the information bit rate, R, *cannot exceed the capacity, C, of the channel* if the system is to operate reliably, and C, in turn must satisfy (16). How then with this fixed upper bound on R can we always achieve the necessary large Shannon bandwidth B or, equivalently, necessary large R_m? The answer is to use a code with sufficiently low *code rate* r, measured in modulation symbols per information bit, because $R = R_m \cdot r$ and hence

$$R_m = \backslash F(R,r) \quad \text{(symbols/sec)}. \tag{18}$$

Because the Shannon bandwidth is proportional to R_m, we see that *coding (except in trivial cases such as that in Example 6) increases both the Shannon bandwidth and the Fourier bandwidth by a factor proportional to the reciprocal of the code rate r.* This is the true nature of the "bandwidth expansion" due to coding. *Spectrum spreading*, on the other hand, *increases the Fourier bandwidth but not the Shannon bandwidth*, which is the true nature of its "bandwidth expansion".

6. EFFECTS OF FILTERING ON BANDWIDTH

Filtering has interestingly different effects on Shannon bandwidth and Fourier bandwidth. Consider first linear filtering. If the input signal is given as in (1) by the linear combination of the orthonormal functions $\phi_i(t)$, $1 \leq i \leq N$, then the output signal of a linear filter, whether time-invariant or time-varying, must be the same linear combination of the functions $\lambda_i(t)$, $1 \leq i \leq N$, where $\lambda_i(t)$ is the response of this filter to $\phi_i(t)$. The signals $\lambda_i(t)$, $1 \leq i \leq N$, will in general be neither orthogonal nor normalized to unit energy, but they nonetheless span the signal space needed to represent the filter output signal so that this space has dimension at most N. We conclude that *linear filtering can never increase Shannon bandwidth*.

Depending on how one defines Fourier bandwidth, e.g., if one uses the "3 dB bandwidth," linear filtering can increase Fourier bandwidth by emphasizing parts of the spectrum where the input signal is weak but substantially non-zero, but this is inessential. More essentially, *linear filtering generally reduces the Fourier bandwidth* of a signal by suppressing parts of the spectrum where the signal is strong. If such a signal had Shannon bandwidth B essentially equal to its Fourier bandwidth W, then it follows from the fundamental theorem of bandwidth that its Shannon bandwidth would also be reduced by such linear filtering. Suppose, however, that s(t) has a flat spectrum and is a well-spread signal (i.e., $\gamma \gg 1$) with good LPI capability so that the fraction $1/\gamma$ of the $2WT\gamma$ available dimensions that are actually used for the signal space appear to be quite randomly chosen. For the linear filtering of this signal performed by its transmission through a channel with multipath propagation, it is hard to imagine that the filtered versions $\lambda_i(t)$, $1 \le i \le N$, of the original basis functions $\phi_i(t)$, $1 \le i \le N$, of the signal space would not still be linearly dependent. With high probability, one expects that the signal space still has dimension N, i.e., that the Shannon bandwidth of s(t) is unchanged, even though the Fourier bandwidth may have been substantially decreased. *Linear filter generally does not decrease the Shannon bandwidth of a well-spread signal with a good LPI capability*, which is why such signals tend to be robust against the deleterious effects of multipath propagation.

Nonlinear filtering is quite a different matter. We saw in Example 5 that (nontrivial) coding of a binary information sequence increases, by a factor equal to the reciprocal of the code rate, both the Shannon and the Fourier bandwidths of the signal at the output of a binary antipodal modulator fed by that sequence compared to that for an uncoded sequence with the same information rate (i.e., as in Example 4 with n = 1). But linear (over the finite field GF(2)) coding of a binary (0 or 1) sequence is a discrete-time nonlinear (over the field of real numbers) filtering of that sequence considered as a binary (± 1) sequence, which implies that the modulator output for the coded input sequence is a nonlinearly filtered version of the modulator output for the uncoded sequence. This illustrates that *nonlinear filtering of a signal generally increases both its Shannon and its Fourier bandwidth*. Aside from the nonlinear filtering resulting from coding, it appears difficult to say much more in general about the effects of the kinds of nonlinear filtering that one might encounter in typical spread-spectrum systems.

7. TWO PARADOXES

We now consider two paradoxes that arise in CDMA systems and show how the concepts previously developed, particularly the notion of Shannon bandwidth, can be used to resolve them.

The first paradox concerns a two-user CDMA system (with no multipath propagation) where both users have as their spreading waveform a simple rectangular pulse of duration equal to the symbol period T, which they modulate with binary (± 1) data. Typical transmitted signals for users 1 and 2 are shown in Fig. 3a.

Suppose first that the two users are perfectly synchronized so that the received signal is the sum of the two signals (shown in Fig. 3b) plus Gaussian noise that we suppose to be negligibly small. This continuous-time channel is precisely equivalent to the discrete-time noiseless binary adder channel in which each user sends +1 or -1 and the output is the real sum of the inputs. The symmetric capacity of this channel (i. e., the upper limit of the total rate at which arbitrarily reliable operation is possible with all users sending information at the same rate) is well-known to be 3/2 bits/symbol [12, p. 392-393]. The users, by proper coding of their information, can simultaneously communicate as reliably as desired to the receiver, each at an information rate as close to 3/4 bits/symbol as desired, but can do this at no higher common rate.

Suppose next that the two users are out of synchronization by one-half a symbol period so that the sum of their two signals is as shown in Fig. 3c. We note that, at any time instant where the two signals reinforce one another, the receiver can correctly detect the sent (±1) symbols of both users. Moreover, at the next time instant thereafter when the receive signal changes value, the receiver detects the location of one symbol edge for one user and hence knows the locations of all symbol edges for that user; the next subsequent change of the received signal not located at a symbol edge for that user locates all the symbol edges of the other user. The receiver can now detect perfectly the entire sent (±1) data sequences of both users, since knowing the symbol-edge locations tells the receiver which user's signal has changed at each subsequent instant of change in the received waveform. Because with probability 1 the two users' symbols will reinforce at some point, it follows that the two users can now simultaneously communicate reliably to the receiver, each at the information rate of 1 bit per symbol. *The lack of synchronization has caused the symmetric capacity to increaes from 3/2 bits/symbol to 2 bits/symbol!*

Any communication engineer who has struggled to obtain synchronization in some digital communication system will have trouble accepting the fact that lack of synchronization can actually increase channel capacity--but it is true here! Even if we had taken the Gaussian noise to be small but not negligible, we would have been led to the same conclusion–the capacity with synchronization would be nearly 3/4 bits/symbol, but the capacity without synchronization would be close to 1 bit/symbol; including noise does not resolve the paradox. The resolution of the paradox comes from consideration of Fig. 3(c) where we see that, to represent the sum of the two unsynchronized signals, we need two orthonormal functions (each a rectangular pulse of width T/2 seconds) per symbol period compared to only one such function (a rectangular pulse of width T seconds) for the sum of the two synchronized signals. *The Shannon bandwidth of the sum signal in the unsynchronized case is twice that in the synchronized case!* This increase in Shannon bandwidth occurs because rectangular pulses do a rather poor job of filling out their Fourier bandwidth. To test this reasoning, we replaced the rectangular functions as illustrated in Fig. 3 with sinc functions (as in Fig. 1) of Fourier bandwidth $W = 1/(2T)$, i.e., equal to the Shannon bandwidth $B = 1/(2T)$ of each user's signal. In this case, the sum signal has this same Fourier bandwidth W regardless of the synchronization between the two users and hence, by the fundamental theorem of bandwidth, the sum signal must also have the same Shannon bandwidth $1/(2T)$. We then calculated capacity for various time shifts between the users' signals by a Monte Carlo method and found that *capacity was virtually independent of this time shift*. The communication engineer may still be somewhat surprised that the lack of synchronization has not decreased capacity in this case, but he or she should be gratified to hear that at least it does not increase capacity.

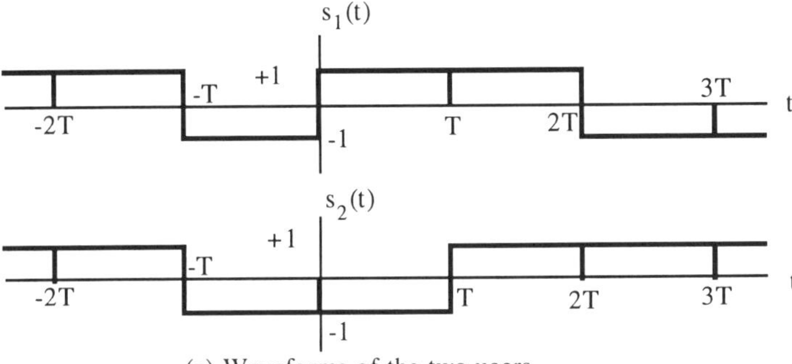

(a) Waveforms of the two users.

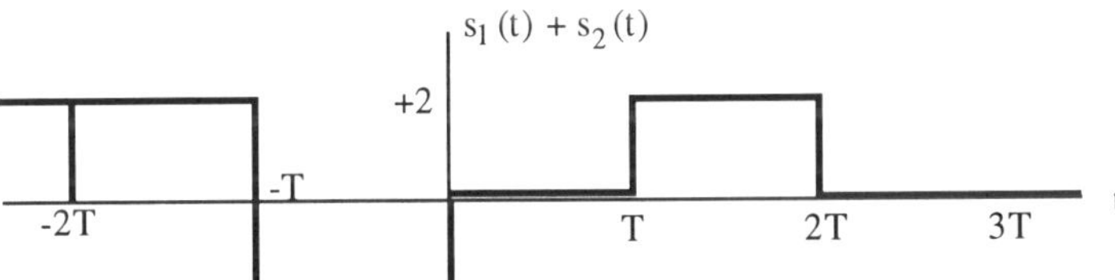

(b) Sum of the waveforms of the two synchronized users.

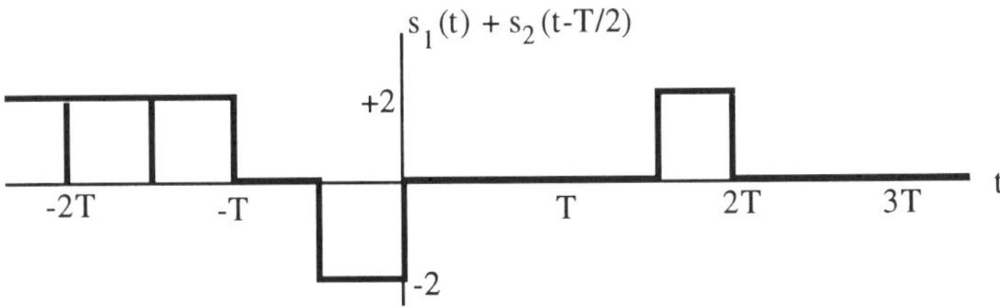

(c) Sum of the waveforms of the two users when out-of-synch by half of a symbol period.

Fig. 3: The paradox where lack of synchronization increases capacity.

The second paradox that we will consider is that of a fully synchronized CDMA system with additive white Gaussian noise (AWGN) for which each user has a (± 1) signature sequence that determines his symbol waveform in the manner described for one user in Example 2. We suppose that L is large so that each user sends a well-spread signal with γ = L. It follows then from our discussion in Section 5 that if only one user were active then he could not be using the channel at the maximum capacity consistent with his Fourier bandwidth. But, it is shown in [13] that when the signature sequences of the K active users meet the Welch bound with equality, then the symmetric capacity of this CDMA system is exactly the same as for a hypothetical single user with average power equal to the total power of the K users and who exploits the Fourier bandwidth to its fullest. *The individually bandwidth-inefficient users combine to create a system with maximum bandwidth efficiency!* Again this seems counterintuitive at first. However, again we see that the sum signal will have the same Fourier bandwidth (because of the sinc functions used a chip waveforms) as the signal of an individual user but will have a much greater Shannon bandwidth. In fact, as soon as the number K of users is on the order of L or greater, the Shannon bandwidth will generally be the same as the Fourier bandwidth–*the sum of the well-spread signals of the users is not a spread-spectrum signal at all*, which is why it can potentially make optimum use of the Fourier bandwidth. A closer study of the condition for meeting the Welch bound with equality for such a CDMA system shows in fact that this is just the condition for the sum signal to have Shannon bandwidth exactly equal to the Fourier bandwidth and for the sum signal to have the same power in each dimension of the signal space, which is what is needed in the signal of a single user for the AWGN channel to obtain the maximum capacity for a given Fourier bandwidth.

8. CONCLUDING REMARKS

It should be apparent that we have in this paper barely scratched the surface of the information theory of spread-spectrum systems. At best, we have pointed out the starting direction for a long journey. In particular, much additional thought needs to be given to spread-spectrum multiple-access systems, i.e., multiple-access systems in which each of several users sends a spread-spectrum signal in the same band and the sum of these signals is received. It is hardly a guess that, as in our resolution of the paradoxes described in Section 7, a very interesting quantity will be the Shannon bandwidth of this sum signal. We are far from being able to offer a coherent information-theoretic treatment of spread-spectrum multiple-access systems, even when we restrict the channel to be the bandlimited additive white Gaussian channel for the sum signal. And we have not even begun to take into account matters of paramount practical interest such as multipath propagation of each signal in the sum, time variation of the multipath channels, unequal user signal powers, and imperfect synchronization. Nonetheless, it is our conviction that until the information theory of spread-spectrum systems is worked out in enough generality to deal with such issues, the many arguments about which type of spread-sprectrum multiple-access system is better than another (say, offers greater "bandwidth efficiency") will continue to generate more heat than light.

ACKNOWLEDGEMENT

This paper profited materially from old fashioned personal discussions with Prof. I. Bar-David of the Technion–Israel Institute of Technology during a period when he was visiting at the ETH Zürich and from our e-mail discussions with Prof. S. Shamai, also of the Technion..

REFERENCES

[1] C. E. Shannon, "A Mathematical Theory of Communications," Bell System Tech. J., vol. 27, pp. 379-423 and 623-656, July and October, 1948.

[2] C. E. Shannon, "Communication in the Presence of Noise," *Proc. IRE,* vol. 37, pp. 10-21, Jan. 1949.

[3] J. L. Massey, "Coding and Modulation for Code-Division Multiple Accessing", in *Proc. Third Int. Workshop on Digital Signal Processing Techniques Applied to Space Communications*, ESTEC, Noordwijk, 1992, pp. 3.1-3.17.

[4] J. L. Massey, "Deep-Space Communications and Coding: A Marriage Made in Heaven," in *Advanced Methods for Satellite and Deep Space Communications* (Ed. J. Hagenauer), Lecture Notes in Control and Information Sciences No. 182. Heidelberg and New York: Springer, 1992, pp. 1-17.

[5] R. W. Emerson, "Self Reliance, in *Essays*, 1841.

[6] H. Nyquist, "Certain Topics in Telegraph Transmission Theory," AIEE Transactions, p. 617, April 1928.

[7] D. Gabor, "Theory of Communication," J. IEE, vol. 93, part 3, no. 26, p. 429, 1946.

[8] J. M. Wozencraft and I. M. Jacobs, *Principles of Communication Engineering.* New York: Wiley, 1965.

[9] D. Slepian, "On Bandwidth," *Proc. IEEE,* vol. 64, pp. 292-300, 1976.

[10] D. L. Kreider, R. G. Kuller, D. R. Ostberg and F. W. Perkins, *An Introduction to Linear Analysis*. Reading, MA: Addison-Wesley, 1966.

[11] W. B. Davenport, Jr., and W. L. Root, *Random Signals and Noise*. New York: McGraw-Hill, 1958.

[12] T. M. Cover and J. A. Thomas, Elements of Information Theory. New York: Wiley, 1991.

[13] M. Rupf and J. L. Massey, "Optimum Sequence Multisets for Symbol-Synchronous Code-Division Multiple-Access Channels," to appear in *IEEE Trans. Inform. Th.*, vol. IT-40, 1994.

Performance Limits of Error-Correcting Coding in Multi-Cellular CDMA Systems With and Without Interference Cancellation

Andrew J. Viterbi

Abstract: Capacity limits, with and without interference cancellation, are obtained for an idealized multi-cellular network model.

Introduction

The term "error-correcting code" highlights the misconception that most engineers, sometimes even communication specialists, still harbor concerning the central role of channel coding in communication system performance. More properly descriptive terms are "error-preventing," "error-protecting," or the more common "error-control." The reason is that in modern communication systems, designed according to the norms of information theory, coding is an integral part of the modulation process and the optimum modulator and demodulator combine the coding and modulation, as well as the decoding and demodulation, processes into one function at both the transmitter and the receiver. This viewpoint has become commonplace for modulation systems which operate in additive Gaussian noise, with relatively high signal-to-noise ratios, at bit rates considerably greater than their transmission bandwidth. For such systems and channels, the evolution and general acceptance of trellis coded modulation (TCM) has demonstrated how modulation and channel coding can be integrated to provide superior performance [1].

Theoretical Performance Limits of Interference Cancellation

For spread spectrum multiple access systems, with each user's data rate much less than the total spread bandwidth, which is occupied in common by all users, the situation is even more favorable. In an additive white noise background, with constant (unfaded) coherently received signals, theoretically the multiple users can be received error-free as long as the total rate, summed over all users, does not exceed Shannon capacity in terms of total received signal power summed over all users [2]. Thus, if there are M users transmitting at rates R_j ($j=1,2,\ldots M$) with received signal-to-noise density ratios S_j/N_0 over a spread bandwidth, W, the total rate,

47

S.G. Glisic and P.A. Leppänen (eds.), Code Division Multiple Access Communications, 47-52.
© 1995 *Kluwer Academic Publishers. Printed in the Netherlands.*

48

$$R_T = \sum_{j=1}^{M} R_j \tag{1}$$

is related to the total signal-to-noise density ratio,

$$S_T / N_0 = \sum_{j=1}^{M} S_j / N_0 \tag{2}$$

by Shannon's formula for the AWGN channel,

$$R_T < W \ \log_2\left(1 + \frac{S_T}{N_0 W}\right). \tag{3}$$

To approach the bound, it may be necessary to employ either unequal R_j for equal powers S_j, or unequal powers S_j for equal rates R_j. Reference [3] shows how to select R_j, or S_j, to approach the bound (3), by employing successive interference cancellation of each demodulated and decoded user at the common receiving (base) station. If we require all users to employ equal rates $R_j = R$, so that $R_T = MR$, the maximum number of users is obtained from (3) to be

$$M < \frac{W}{R} \log_2\left(1 + \frac{S_T}{N_0 W}\right) \tag{4}$$

Two particularly interesting conclusions of [3] were that:

(a) processing gain W/R, per user, is not sacrificed or reduced by employing coding down to rates as low as the inverse processing gain, R/W, so that the entire spreading factor can also be allocated to the channel coding function. This corroborates the exposure of the "first myth" of spread spectrum systems [4];

(b) the asymptotically optimum, and finitely quasi-optimum, class of codes to approach Shannon capacity is the class of orthogonal (convolutional) codes [3] or better yet, its extension, the super-orthogonal class [5].

The second point bears expanding upon. Whereas, for bandwidth constrained channels, existence of codes which approach the Shannon limit has long been established, the choice of the best code of a given complexity requires a lengthy search procedure (although so many nearly optimum codes are known to exist that the procedure to find one is far shorter than that required for finding the optimum). Furthermore, the performance of the best code is only known to within a moderately accurate bound. Conversely, when the bandwidth is allowed to expand arbitrarily relative to a single user's rate, a condition approximated by spread spectrum modulation, a near-optimum strategy is to employ orthogonal codes and a decoding process which successively cancels all previously decoded interfering users [3]. This procedure can support a total rate R_T which asymptotically approaches the overall Shannon capacity of (3). For a bandwidth expansion ratio, W/R, a constraint length K code from this class can be used as long as $2^K \leq$ W/R. A more nearly optimum class, the super-orthogonal codes [5], easily derived from the orthogonal class, allow the use of a constraint length K+2 for the same W/R ratio, with a resulting improvement in error rate. The code's structure is regular and easily described and implemented, while its performance is bounded more accurately than the generic bounds used for higher code rate codes. Similar to bandlimited channels, however, such implementation and performance is the result of an integrated approach to modulation and coding.

The conditions required to achieve the multiple access channel Shannon capacity, as assumed in [2] and [3] are

(a)   additive white Gaussian noise (AWGN) interference;

(b)   constant amplitude and phase of each user's carrier signal over long periods, so that coherent demodulation can be performed and each user's signal parameters can be measured accurately.

While (a) is closely approximated in spread spectrum systems as long as the bandwidth expansion ratio is sufficiently high (on the order of one hundred or more), (b) may be difficult to justify, particularly when users are in rapid motion and multipath is severe. We shall continue with both assumptions for two purposes. First, primary interest is to determine the advantage of cancelling interference due to other users' transmission and this is most evident in a model which is uncluttered by the additional disturbances due to multipath. Secondly, this model also serves to determine the limiting coded performance with and without cancellation.

Multi-Cellular Interference Limits on Performance

We proceed to apply the capacity formula (4) to a multi-cellular system wherein interference cancellation is performed on all same-cell users at the base station of each individual cell, but users of other-cell base stations can not be cancelled and thus introduce the background noise component; for simplicity and again to obtain the upper bound on performance, we ignore receiver thermal noise.

Then referring again to (4), S_T is the total signal power received at the base station contributed by all users of the given cell, and N_0W is the total power received at the base station from all users of all other cells. It is shown in [6] that when all cells are equally loaded, and all cell base stations employ power control on their populations of CDMA users, the total other-cell interference received by any given base station is proportional to S_T where the proportionality factor $f \approx 0.6$ for a fourth-power propagation law and

50

log-normal shadowing with standard deviation 8 dB. (f will be larger for lower-power propagation or greater standard deviations.) Thus letting

$$N_0 W = f \, S_T \tag{5}$$

we obtain from (4) that with ideal coding and with perfect cancellation of same-cell users, capacity is bounded by

$$M < \frac{W}{R} \log_2\left(1 + \frac{1}{f}\right) . \tag{6}$$

To assess the value of interference cancellation in a multi-cellular CDMA system, we next consider performance of ideally coded users without interference cancellation. In this case, there is no constraint on individual rates and received powers, so we take all to be equal: $S_j = S$, $S_T = MS$, $R_j = R$,
$R_T = MR$. Then reproducing the argument in [7], we have that, with perfect power control, the total interference to the base station receiver for each user from all the other users in the same cell and other cells is

$$I = (M - 1)\, S + M \, fS \approx M(1 + f)S \tag{7}$$

which is a slight overbound. Since all signals are spread over bandwidth W, we may define, as before, the total interference density by N_0 and hence $I = N_0 W$. Taking the ratio of I to S, the received signal power of each user, and noting that the bit energy $E_b = S/R$, we have

$$\frac{I}{S} = \frac{N_0 / W}{E_b / R} \tag{8}$$

and combining (7) and (8)

$$M \approx \frac{W / R}{\big((E_b / N_0)(1 + f)\big)} . \tag{9}$$

The value of E_b/N_0 is determined by the quality of the modem and the performance of the error correcting code in additive white Gaussian noise. It is lower bounded by

$$E_b / N_0 > \ell n \, 2 \tag{10}$$

which equals Shannon capacity for an infinitely wide (spread) bandwidth channel, approachable in the limit of infinite constraint length by an orthogonal convolutional code.

Thus, <u>without interference cancellation</u>, and with the same ideal error-correcting code class as assumed above with interference cancellation, we obtain from (9) and (10),

$$M < \frac{W/R}{(1+f)\ell n\, 2} \, . \tag{11}$$

Taking the ratio of the limiting rates (6) and (11), with and without cancellation, we have

$$\frac{\text{Limit on M With Perfect Cancellation}}{\text{Limit on M Without Perfect Cancellation}}$$

$$= (1+f)\,\ell n\,(1+1/f) \tag{12}$$

Table I provides numerical values for (6) (11) and (12) for various values of f.

f	M/(W/R) With Cancellation	M/(W/R) Without Cancellation	Ratio
0.2	2.58	1.20	2.15
0.4	1.81	1.03	1.75
0.6	1.42	0.90	1.57
0.8	1.17	0.80	1.46
1	1	0.72	1.39
2	0.58	0.48	1.21
3	0.42	0.36	1.15
4	0.32	0.29	1.12

Table I. Capacity Limits

Concluding Remarks

In the idealized case of error-control codes whose performance approaches the Shannon limit, it is shown that for a multi-cellular CDMA system, perfect interference cancellation of all users sharing a common cell increases capacity by less than 60% for fourth-power propagation and log-normal shadowing with 8 dB standard deviation, with even less improvement for lower-power laws. For uncoded transmission, or for much weaker codes, the improvements can be greater, but we may infer from the present results that the effect of interference cancellation diminishes as the error control power of the code approaches the ideal. Finally, it should be noted that, unlike the robust nature of improvements achieved through coding, interference

52

cancellation is a delicate operation. If not done extremely accurately (with some delay), as implied in this paper and described in Reference [3], it can even degrade performance. Also, for cancellation without decoding delay, low rate high-performance codes can not be used. An additional concern is the interaction between interference cancellation and power control, particularly since ideal cancellation performance can only be achieved with unequal received signal powers.

With these caveats and with the modest capacity increase achievable with interference cancellation, it would appear that there are simple, more robust and better proven methods [7] to achieve even greater improvements.

References

[1]   G. Ungerboeck, "Channel Coding with Multilevel/Phase Signals," <u>IEEE Trans. on Information Theory</u>, vol. IT-28, pp. 55-66, January, 1982.

[2]   A.D. Wyner, "Recent Results in the Shannon Theory," <u>IEEE Trans. on Information Theory</u>, vol. IT-20, pp. 2-10, January, 1974.

[3]   A.J. Viterbi, "Very Low Rate Convolutional Codes for Maximum Theoretical Performance of Spread-Spectrum Multiple-Access Channels," <u>IEEE Journal on Selected Areas in Communications</u>, vol. 8, pp. 641-649, May, 1990.

[4]   A.J. Viterbi, "Spread Spectrum Communications-Myths and Realities," <u>IEEE Communications Magazine,</u> pp. 11-18, 1979.

[5]   E. Zehavi and A.J. Viterbi, "On New Classes of Orthogonal Convolutional Codes," Bilkent International Conference in New Trends in Communication, Control, and Signal Processing, Ankara, Turkey, July, 1990.

[6]   A.J. Viterbi, A.M. Viterbi and E. Zehavi, "Other-Cell Interference in Cellular Power-Controlled CDMA" <u>IEEE Trans. on Communications</u>, vol. 42, pp. 1501-1504, April, 1994

[7]   K.S. Gilhousen, et al, "On the Capacity of a Cellular CDMA System," <u>IEEE Trans. on Vehicular Tech.</u>, vol. VT-40, pp. 303-312, May, 1991.

OPTIMUM PN SEQUENCES FOR CDMA SYSTEMS

Dilip V. Sarwate

Abstract

Most known methods for the design of hopping patterns for frequency-hopped code-division multiple-access (CDMA) systems and for the design of signature sequences for direct-sequence CDMA systems provide sequences that can be viewed as codewords (or the images of codewords) selected from low-rate Reed-Solomon codes. This chapter surveys sequence designs for CDMA systems from this viewpoint.

I.   Introduction

Code-division multiple-access (CDMA) systems can be classified as frequency-hopped CDMA systems (FH/CDMA) or direct-sequence (DS/CDMA) systems [3, 29, 35]. As the adjective *code-division* implies, a CDMA system allows several transmitters to share the available bandwidth by the use of different *codes* which are used to distinguish the signals at the receiver. In an FH/CDMA system, a transmitter changes (hops) its RF carrier frequency at regular intervals as prescribed by a *frequency-hopping pattern*. Careful choi-ce of the frequency-hopping patterns assigned to the transmitters allows the system designer to arrange matters so that the signals from two different transmitters hop to the same frequency only very infrequently. In contrast, a DS/CDMA transmitter phase-shift-keys its RF carrier with a *signature sequence* with very high pulse rate. Careful choice of the signature sequences assigned to the transmitters allows the system designer to arrange matters so that the signals from different transmitters interfere very little with each other in a receiver. The hopping patterns or signature sequences for CDMA systems are often referred to as *pseudonoise* (PN) sequences. This chapter contains a brief survey of various designs for such PN sequences from the viewpoint that most known design methods result in sequences that are code-words from low-rate Reed-Solomon codes, or the images of such codewords under some suitable linear mapping. Section II contains a survey of design methods for hopping patterns for FH/CDMA communication systems. Some of these methods explicitly use Reed-Solomon codes and the resulting frequency-hopping patterns are codewords from low-rate Reed-Solomon codes. Other methods were originally described using different terminology but the resulting frequency-hopping

The author is with the Coordinated Science Laboratory and the Department of Electrical and Computer Engineering, University of Illinois at Urbana-Champaign, Urbana, Illinois 61801 USA.

Preparation of this paper was supported by the Joint Services Electronics Program under Grant NOOO14–90-J–1270.

S.G. Glisic and P.A. Leppänen (eds.), Code Division Multiple Access Communications, 53-78.
© 1995 *Kluwer Academic Publishers. Printed in the Netherlands.*

patterns also can be viewed as code-words from low-rate Reed-Solomon codes. Section III shows that several well-known signature sequence sets for DS/CDMA communications (such as the Gold sequences and Kasami sequences [25]) also can be viewed in terms of codewords from low-rate Reed-Solomon codes. However, there seems to be no apparent connection between the properties of the Reed-Solomon codes and the resulting signature sequences,and viewing the signature sequences as Reed-Solomon codewords provides no new insights into possible new or improved designs.

II. Frequency-Hop Spread-Spectrum Systems

II.A Properties of Frequency-Hopping Patterns

Let q denote the number of frequency slots in a FH/CDMA system, and let f_i denote the center frequency of the i-th slot, $0 \leq i \leq q-1$. The center frequencies are usually chosen so that the slots are spaced uniformly across the frequency band allotted to the system. A frequency-hopping pattern is a sequence $x = (x_0, x_1, x_2, \ldots, x_{N-1})$ of N elements from the set $\{f_0, f_1, f_2, \ldots, f_{q-1}\}$ specifying the order in which the slots are to be used by a particular transmitter. Long messages are transmitted by repeating the frequency-hopping pattern as often as necessary. As in [25], let T denote the operator that shifts a sequence x cyclically to the left by one place, that is, $Tx = (x_1, x_2, \ldots, x_{N-1}, x_0)$. If T is applied k times to x, where $0 < k < N$, the result is $T^k x = (x_k, x_{k+1}, \ldots, x_{N-1}, x_0, x_1, \ldots x_{k-1})$, while $T^N x = x$. The period of x is the least positive integer M such that $T^M x = x$. Generally, M can be any divisor of N, but in most cases of interest, $M = N$. The sequences x, Tx, $T^2 x$, $\ldots$, $T^{M-1} x$ are distinct but *cyclically equivalent;* they are cyclic shifts of each other. The composition vector $N(x)$ of the sequence x is defined as

$$N(x) = (N_0(x), N_1(x) \ldots, N_{q-1}(x))$$

where, for $0 \leq i \leq q-1$, $N_i(x)$ denotes the number of times that frequency f_i occurs in x. Note that cyclically equivalent sequences have the same composition vector. Obviously,

$$N_i(x) \geq 0 \text{ for all i, and } \sum_{i=0}^{q-1} N_i(x) = N. \tag{1}$$

It can be shown [10] that (1) implies that

$$\sum_{i=0}^{q-1} N_i^2(x) = \|N(x)\|^2 \geq N \left\lfloor \frac{N}{q} \right\rfloor + \left\lceil \frac{N}{q} \right\rceil (N \bmod q) \tag{2}$$

where $\|\cdot\|$ denotes the norm of a vector and $(N \bmod q)$ denotes the least positive residue of N modulo q. Equality holds in (2) if and only if

$$N_i(x) = \left\lceil \frac{N}{q} \right\rceil \quad \text{for } (N \bmod q) \text{ values of } i. \tag{3a}$$

$$N_i(x) = \left\lfloor \frac{N}{q} \right\rfloor \quad \text{for } q - (N \bmod q) \text{ values of } i. \tag{3b}$$

Note that

$$N \left\lfloor \frac{N}{q} \right\rfloor + \left\lceil \frac{N}{q} \right\rceil (N \bmod q) = N \quad \text{if } N \le q, \tag{4}$$

and that, in general,

$$N \left\lfloor \frac{N}{q} \right\rfloor + \left\lceil \frac{N}{q} \right\rceil (N \bmod q) \ge \frac{N^2}{q} \tag{5}$$

with equality if and only if N is an integer multiple of q. As will be shown below, it is often possible to find frequency-hopping patterns that satisfy (3). Such hopping patterns spread the signal energy as uniformly as possible across the entire frequency band allotted to the FH/CDMA system. Uniform usage of the frequency slots can also be justified by a minimax approach to the design of frequency-hopping patterns for FH/CDMA systems operating in jamming or fading environments. If a signal visits one slot much more often than others, then it is more vulnerable to jamming or fading in that slot. Of course, such a signal is less vulnerable in those slots that it visits less frequently, but clearly the maximum vulnerability is minimized if the slots are utilized as uniformly as possible. Uniform usage of slots also minimizes the average multiple-access interference from other transmitters.

When several FH/CDMA transmitters are simultaneously active, it often happens that two or more transmitters hop to the *same* frequency slot at the same time. This event is called a *collision* or *hit* and it usually causes such severe signal degradation that it is best to detect collisions whenever possible, to delete the most severely degraded symbols, and then to reconstruct these deleted symbols by use of an erasures-and-errors correcting channel code [1]. Collisions sometimes are less of a problem in a *fast* FH/CDMA system in which a data symbol is transmitted over several dwells, than in a *slow* FH/CDMA system in which one or more data symbols is transmitted during a dwell. Note also that most FH/CDMA systems are not dwell-synchronous, that is, the relative delay between two signals need not be an integer multiple of the dwell duration. Thus, when the desired signal hops into a slot, a receiver tracking the signal may find that an interfering signal is already present in that slot, but that this interfering signal disappears before the end of the dwell interval. On the other hand, an interfering signal mays uddenly appear during a dwell and last until the end of the dwell interval. Such *partial hits* can be accounted for quite easily if one knows the number of (full) collisions that occur for each given delay between the two patterns in a dwell-synchronous system. Full collisions are counted by the *Hamming correlation functions* which are considered next.

II.B The Hamming Correlation Functions

Let x and y denote two frequency-hopping patterns with common period N. The number of hits that occur in one period of x due to interference from y is counted by the *Hamming crosscorrelation function* which was defined by Lempel and Greenberger [10] as

$$H_{x,y}(j) = \sum_{i=0}^{N-1} h(x_i, y_{i+j}), \quad 0 \le j \le N-1 \tag{6}$$

where j is the relative delay between the two frequency-hopping patterns, the sum $i+j$ is taken modulo N, and

$$h(a,b) = \begin{cases} 1, & \text{if } a = b, \\ 0, & \text{if } a \ne b. \end{cases}$$

Equivalently,

$$H_{x,y}(j) = N - d(x, T^j y) \tag{7}$$

where $d(x, y)$ denotes the Hamming distance between the sequences x and y. Naturally, one would like to have x and y such that $H_{x,y}(j)$ is as small as possible for all j, $0 \le j \le N-1$. The *Hamming autocorrelation function* $H_x(j)$ for the hopping pattern x is just $H_{x,x}(j)$ and has the obvious property that $H_x(0) = N$. One would like to choose x such that the *out-of-phase autocorrelation* values $H_x(j)$, $0 < j < N$, are as small as possible. Note that the out-of-phase autocorrelation is a measure of the self-interference suffered by a signal from a delayed replica of itself. Such replicas may be received because of specular multipath propagation or because a repeat-jammer is re-broadcasting the signal. The out-of-phase autocorrelation can also be used to estimate the likelihood of false lock in the initial acquisition and synchronization process that aligns the hopping pattern of the receiver's local oscillator with the received signal.

How small can the Hamming crosscorrelation and out-of-phase Hamming autocorrelation be? First, note that $0 \le H_{x,y}(j) \le N$ for all j. Let $\overline{H}_{x,y}$ denote the *average Hamming crosscorrelation value* and $\langle \cdot, \cdot \rangle$ denote the usual inner product of two vectors. Then

$$N \cdot \overline{H}_{x,y} = \sum_{j=0}^{N-1} H_{x,y}(j) = \sum_{i=0}^{q-1} N_i(x) N_i(y) = \langle \mathrm{N}(x), \mathrm{N}(y) \rangle. \tag{8}$$

Similarly, $\overline{H}_x$, the *average out-of-phase Hamming autocorrelation* for the sequence x is given by

$$(N-1) \cdot \overline{H}_x = \sum_{j=1}^{N-1} H_x(j) = \langle \mathrm{N}(x), \mathrm{N}(x) \rangle - N = \|\mathrm{N}(x)\|^2 - N. \tag{9}$$

If $N \leq q$, then (2), (4), and (9) lead to the trivially obvious conclusion that $\overline{H}_x \geq 0$. Furthermore, $\overline{H}_x = 0$ if and only if the hopping pattern satisfies (3), that is, if and only if each $N_i(x)$ is either 0 or 1. Such sequences are called *nonrepeating* hopping patterns because each frequency is used at most once in each period [30], and they are of interest for the purposes of initial acquisition and synchronization. Note that $\overline{H}_x = 0$ implies that $H_x(j) = 0$ *for all* j, $0 < j < N$. More generally, (2) and (5) show that the right side of (9) is at least $(N^2/q) - N$ and hence, for *any* sequence x,

$$\overline{H}_x > \frac{N}{N-1}\left(\frac{N}{q} - 1\right). \tag{10}$$

The right side of (10) is nonpositive for $N \leq q$, and in this case, $\overline{H}_x \geq 0$. as discussed above. On the other hand, if $N > q$, the right side of (10) exceeds $N/q - 1$. Since the Hamming correlation functions are integer-valued, it follows that for *any* sequence x,

$$\max_{0<j<N} H_x(j) \geq \left\lceil \frac{N}{q} \right\rceil - 1. \tag{11}$$

In comparison, for any j, $0 < j < N$, the expected value of $H_x(j)$ averaged over all possible q^N N-tuples in the set $\{f_0, f_1, f_2, \ldots, f_{q-1}\}^N$ is N/q. In other words, it is not possible to design hopping patterns whose autocorrelation values are significantly smaller than the average value for a randomly chosen hopping pattern.[*]

---

[*] A more careful analysis [10] leads to a more precise estimate for the right side of (11) but these results generally do not improve on the right side of (12). It is also worth noting that all these results can also be deduced by applying the well-known Plotkin upper bound on the average minimum distance of a block code [11] to a length N q-ary block code whose N codewords are $\{x, Tx, T^2x, \ldots, T^{N-1}x\}$.

58

Next, suppose that x and y are chosen randomly and independently from the set of q^N N-tuples in $\{f_0, f_1, f_2, \ldots, f_{q-1}\}^N$. Then, for all j, $0 \le j < N$, the expected value of $H_{x,y}(j)$ is N/q. Once again, it is not possible to do much better than this with nonrandom designs except in a few special cases which essentially correspond to frequency-division multiple-access (FDMA) schemes. For applications to FH/CDMA systems, consider X, a set of K hopping patterns of period N. From (8) it follows that

$$\sum_{x \in X} \sum_{y \in X} \sum_{j=0}^{N-1} H_{x,y}(j) = \left\| \sum_{x \in X} N(x) \right\|^2 . \tag{12}$$

Since $\displaystyle\sum_{x \in X} N_i(x) \ge 0$ for all i, and $\displaystyle\sum_{i=0}^{q-1} \sum_{x \in X} N_i(x) = KN$, it follows from analogy with (1) that the right side of (12) is smallest when

$$\sum_{x \in X} N_i(x) = \left\lceil \frac{KN}{q} \right\rceil \text{ for } (KN \bmod q) \text{ values of } i, \tag{13a}$$

$$\sum_{x \in X} N_i(x) = \left\lfloor \frac{KN}{q} \right\rfloor \text{ for } q - (KN \bmod q) \text{ values of } i, \tag{13b}$$

Let $\overline{H}_c(X)$ and $H_c(X)$ respectively denote the average and maximum values of $H_{x,y}(j)$ over all $K(K-1)$ pairs of distinct hopping patterns x and $y \in$ X and all j, $0 \le j < N$, Similarly, let $\overline{H}_a(X)$ and $H_a(X)$ respectively denote the average and maximum values of $H_x(j)$ over all K patterns $x \in$ X and all j, $0 < j < N$. Define $\overline{H}_{\max}(X) = \max\{\overline{H}_c(X), \overline{H}_a(X)\}$ and $H_{\max}(X) = \max\{H_c(X), H_a(X)\}$. Then, it follows from (5) and (12) that

$$K(K-1)NH_c(X) + K(N-1)H_a(X) \ge K(K-1)N\overline{H}_c(X) + K(N-1)\overline{H}_a(X)$$

$$= \left\| \sum_{x \in X} N(x) \right\|^2 - KN \ge \frac{(KN)^2}{q} - KN \tag{14}$$

and hence, if $KN > q$,

$$H_{\max}(X) \ge \overline{H}_{\max}(X) > \frac{N}{q} - \frac{1}{K}. \tag{15}$$

Thus, one or both of the maximum (and average) correlation values are bounded from below by a quantity which is almost the same as the expected value when random patterns are used.

The bound (15) on $H_{\max}(X)$ is useful only if $N \gg q$. However, if N is no larger than q, the bound is not only small but it hardly increases at all as K increases. In such instances, a larger lower bound on $H_{\max}(X)$ can be obtained as follows. Consider all the KN cyclic shifts of the hopping patterns in X as a q-ary nonlinear cyclic code of length N. The Singleton bound on the mini-mum distance of this code [11] together with (7) implies that

$$H_{\max}(X) \geq \left\lceil \log_q(KN) \right\rceil - 1. \tag{16}$$

For example, according to (16), for a set of q^s hopping patterns of length $N \leq q$, the maximum Hamming correlation value is at least s. However, bound (16) increases only logarithmically with N whereas bound (15) increases linearly with N, and thus the latter is tighter when applied to small sets of long hopping patterns.

Returning to (14), note that this can be written as

$$H_c(X) + \frac{1}{K-1} H_a(X) \geq \overline{H}_c(X) + \frac{1}{K-1} \overline{H}_a(X) \geq \left(\frac{KN}{q} - 1 \right) \tag{17}$$

showing that there is a tradeoff between the maximum (or average) crosscor-relation and maximum (or average) autocorrelation values – if careful design of hopping patterns reduces one maximum (or average) correlation value substantially below N/q, then the other maximum (or average) correlation value will be larger than N/q. However, since different weights are attached to the quantities in (17), a set of patterns with very small maximum (or average) crosscorrelation will necessarily have large maximum (or average) autocorrelation but a set of patterns with very small maximum (or average) autocorrelation need not necessarily have very large maximum (or average) crosscorrelation value. As an example, it can be shown that if $\overline{H}_c(X) = 0$ (which implies that $H_c(X) = 0$), then

$$H_a(X) \geq \overline{H}_a(X) > KN/q - 1 \gg N/q - 1.$$

In contrast, if $\overline{H}_a(X) = 0$ (which implies that $H_a(X) = 0$ and all the patterns are nonrepeating), (17) shows that

$$H_c(X) \geq \overline{H}_c(X) \geq \frac{K}{K-1} \frac{N}{q} - \frac{1}{K-1}$$

where the right side is only slightly larger than the right side of (15).

60

II.C Runs and Bursts

A hopping pattern x of length N is said to contain a *run* of length r of the frequency f_i beginning at position j if

$$(x_{j-1}, x_j, x_{j+1}, \ldots x_{j+r-1}, x_{j+r}) = (f_k, f_i, f_i, \ldots f_i, f_l)$$

where $f_k \neq f_i$, $f_l \neq f_i$ and the subscripts on x are taken modulo N. Long runs are undesirable for several reasons. First, the signal stays in the same slot for r successive dwells and is thus more vulnerable if that slot is being jammed or is in a deep fade. Second, long occupancy of a slot also makes the signal more vulnerable to interception by an unauthorized receiver. Finally,if long runs of the same frequency occur in two different patterns, then long *bursts* of full hits will occur whenever the relative delay is such that the runs arrive at the same time at a receiver. Although hits often can be detected and the corresponding symbols erased (and later restored by an errors-and-erasures correcting code,) a burst of hits causes a large number of erasures in a short period of time, and this may well lead to a decoding failure in the error-control system. This is because the blocklength of the error-control code is usually much smaller than the period of the hopping patterns, and thus a code that can handle H hits scattered over N dwells may well fail if many of these H hits occur in a burst and affect symbols belonging to the same codeword. Bursts of hits are caused not only by runs of the same frequency, but also by the occurence of identical subsequences in two hopping patterns, that is, if $(x_i, x_{i+1}, \ldots x_{i+r-1}) = (y_j, y_{j+1}, \ldots y_{j+r-1})$, where, as before, the sub-scripts are taken modulo N, then a burst of r hits occurs for a relative delay of $(j-i)$ dwells. Generally, these hits become partial hits if the relative delay is $(j-i+\varepsilon)$ dwells unless the subsequences contain repetitions of the same frequency. Now, let $B(X)$ denote the length of the longest burst of hits between any two hopping patterns in X and note that $H_{\max}(X) \geq B(X)$. The $K\,N$ subsequences of the form $(x_i, x_{i+1}, \ldots x_{i+r-1})$, cannot be distinct if $q^r < KN$, and therefore bursts of length r must occur for all r satisfying this inequality. It follows that

$$H_{\max}(X) \geq B(X) \geq \left\lceil \log_q(KN) \right\rceil - 1. \tag{18}$$

which not only provides a direct proof of (16) but also shows that one or more *bursts* of hits of length at least $\left\lceil \log_q(KN) \right\rceil - 1$ must occur.

II.D The Design of Hopping Patterns using Reed-Solomon Codes

Hopping patterns were defined earlier as sequences of elements from the set $\{f_0, f_1, f_2, \ldots, f_{q-1}\}$. However, it is not necessary that the elements of the set be the center frequencies of the slots. All the various properties of hopping patterns discussed above hold provided only that the set contains q

distinct elements. In short, one can also regard a hopping pattern as a sequence of elements from some arbitrary set of q distinct elements, and the pattern can always be transformed into a sequence of frequencies by a suitable one-one mapping from this set to $\{f_0, f_1, f_2, ..., f_{q-1}\}$. This is the viewpoint that will be taken in the remainder of this Section. In particular, hopping patterns will be viewed as sequences of elements from the finite field $GF(q) = GF(p^k)$ where p denotes a prime. Thus, it should not be too surprising that Reed-Solomon codes over $GF(q)$ are a source of excellent de-signs of hopping patterns. The concept is as follows.

Let N be a divisor of $q - 1$, and let α denote a primitive N-th root of unity in $GF(q)$. Let $C(N,\ t+1;\ i)$ denote the cyclic $(N,\ t+1)$ Reed-Solomon code over $GF(q)$ with parity-check polynomial

$$h(z) = \prod_{j=i}^{i+t}(z - \alpha^{-j})$$

generator matrix

$$\mathbf{G} = \begin{bmatrix} 1 & \alpha^i & \alpha^{2i} & \cdots & \alpha^{(N-1)i} \\ 1 & \alpha^{(i+1)} & \alpha^{2(i+1)} & \cdots & \alpha^{(N-1)(i+1)} \\ \vdots & \vdots & \vdots & \ddots & \vdots \\ 1 & \alpha^{(i+t)} & \alpha^{2(i+t)} & \cdots & \alpha^{(N-1)(i+t)} \end{bmatrix} \tag{19}$$

and minimum distance $N - t$. Suppose that two *cyclically inequivalent* codewords x and y are chosen to be hopping patterns. Since the code is cyclic, $T^j y$ is a codeword for any j. Furthermore, since the sequences are cyclically inequivalent, $T^j y \neq x$ for any j. Hence, $H_{x,y}(j) = N - d(x, T^j y) \leq t$. Similarly, if x is of period M where M divides N, then $H_x(j) = N - d(x, T^j x) \leq t$ for $j /, \equiv 0 \bmod M$. Thus, hopping patterns can be constructed by choosing one codeword from each cyclic equivalence class. However, for reasons noted earlier, it is usually desirable to use only those patterns that have full period N. Furthermore, since a pattern of period N may leave as many as $q - N$ slots unused, this construction is usually applied only to Reed-Solomon codes of length $q - 1$. Interestingly, in the early seventies, the above idea was dis-cussed by both Reed [22] and Solomon [30] in separate papers. Their solutions to the problem were somewhat different and these solutions are considered next.

II.E Reed's Construction

In [22], Reed considered hopping patterns obtained by choosing one code-word from each cyclic equivalence class of the codewords belonging to the code $C(q-1,\ t+1;\ 0)$ with generator matrix

$$\mathbf{G} = \begin{bmatrix} 1 & 1 & 1 & \cdots & 1 \\ 1 & \alpha & \alpha^2 & \cdots & \alpha^{(q-2)} \\ \vdots & \vdots & \vdots & \ddots & \vdots \\ 1 & \alpha^t & \alpha^{2t} & \cdots & \alpha^{(q-2)t} \end{bmatrix} \tag{20}$$

Note that α is a primitive element of the field $GF(q)$. The number of cyclic equivalence classes (i.e., the number of different hopping patterns) is given by [20]

$$\frac{1}{q-1} = \sum_{d|(q-1)} \varphi(d) q^{1+\lfloor t/d \rfloor}$$

where $\varphi(\cdot)$ is Euler's totient function. Of course, many of these equivalence classes contain codewords of period less than $q - 1$ (referred to as *nonprimitive* codewords henceforth) and thus are usually not of interest. Reed also gave a lower bound on the number of hopping patterns of period $q - 1$ (these are *primitive* codewords) that can be obtained from this code. This bound is based on the following simple argument. If α^j is a nonprimitive element of $GF(q)$, then the j-th row $(1, \alpha^j, \alpha^{2j}, \dots \alpha^{(q-2)j})$ of the generator matrix in (20) is a nonprimitive codeword. Suppose that s rows of $\mathbf{G}$ are nonprimitive codewords. The q^s linear combinations of these rows *usually* are nonprimitive codewords, but can be primitive codewords in some cases. On the other hand, nonzero linear combinations of the other $t + 1 - s$ rows are always primitive codewords. Since the sum of a primitive codeword and a nonprimitive codeword is always a primitive codeword, it follows that there are *at least* $q^s(q^{t+1-s} - 1) = q^{t+1} - q^s$ primitive codewords in the code. Hence, the set X of hopping patterns of period $q - 1$ obtained from $C(q-1,\ t+1;\ 0)$ contains at least

$$(q^{t+1} - q^s)/(q-1) = \sum_{j=s}^{t} q^j \ \geq q^t$$

hopping patterns. Furthermore $H_{\max}(X) = t$, and since $KN = K(q - 1) < q^{t+1}$ (which implies that $\lceil \log_q(KN) \rceil - 1 = t$), it follows that both (16) and

(18) are satisfied with equality. In other words, these hopping patterns are optimal with respect to these bounds.

A detailed description of the construction of these hopping patterns has been published recently [29]. As a specific simple example of this construction process, consider $C(q-1,2; 0)$ whose generator matrix has rows $(1, 1,..., 1)$ and $(1, \alpha, \alpha^2,...\alpha^{q-2})$ of period 1 and $q-1$ respectively. Then, the q hopping patterns are given by $(\beta,1)\mathbf{G}$ where $\beta \in GF(q)$. Put another way, the q sequences obtained via Reed's construction can be expressed as

$$x^{(j)} = (\beta_j, \beta_j,...\beta_j) + (1, \alpha, \alpha^2,...\alpha^{q-2}), \ 0 \le j \le q-1, \tag{21}$$

where the $\beta_j, 0 \le j \le q-1$ denote the elements of $GF(q)$. Since the i-th element of the j-th sequence is

$$x_i^{(j)} = \beta_j + \alpha^i, \quad 0 \le i < q-1, \quad 0 \le j \le q-1, \tag{22}$$

and since $\alpha^i \ne 0$, the element β_j does not occur in $x^{(j)}$ while any other β_l occurs exactly once. Thus, these hopping patterns are nonrepeating and they also satisfy (13) with each element of $GF(q)$ occurring a total of $q-1$ times in the q hopping patterns. Note also that for $i \ne j$, $\langle N(x^{(i)}), N(x^{(j)}) \rangle = q-2$ and that $H_{x^{(i)}, x^{(j)}}(l)$ has value 0 if $l = 0$ and value 1 otherwise.

Applying Reed's construction to $C(q-1, t+1; i)$ where $i \ne 0$ sometimes provides *minor* improvements on the above results. If $q = 2^k$, Reed's construction applied to $C(2^k-1,2; 0)$ provides 2^k hopping patterns as described by (21) with $H_{\max} = 1$. In comparison, the construction provides $2^k + 1$ hopping patterns with $H_{\max} = 1$ when it is applied to $C(2^k-1, 2; 1)$. This is because both α and α^2 are primitive elements, and hence the codewords of the form $(1,0)\mathbf{G}$ and $(\beta,1)\mathbf{G}$ are cyclically inequivalent primitive codewords. Letting x and y denote the two rows of $\mathbf{G}$ and noting that $T^i x = \alpha^i x$, this set of $2^k + 1$ hopping patterns can be expressed as

$$\{x, y, x+y, Tx+y, T^2 x+y,...T^{2^k-2} x+y\}$$

which is very similar to the representation of Gold sequences in Eq. (4.4) of [25]. Similarly, applying Reed's construction to $C(2^k-1, 3; 0)$ provides $2^{2k}+2^k$ hopping patterns with $H_{\max} = 2$, whereas it provides $2^{2k}+2^k + 1$ hopping patterns with $H_{\max} = 2$ when it is applied to $C(2^k-1, 3; 1)$ where k

is odd. More generally, if 2^k-1, is a Mersenne prime, then all the rows of the generator matrix of $C(2^k-1, t+1; 0)$ are primitive codewords, and hence Reed's construction provides

$$(2^{k(t+1)}-1)/(2^k-1) = \sum_{j=0}^{t} 2^{jk}$$

hopping patterns of period 2^k-1 with $H_{\max} = t$.

Although the hopping patterns obtained via Reed's construction are optimal with respect to the bounds (16) and (18), they can have some undesirable properties when $t > 1$. For example, consider the q^2 hopping patterns with $H_{\max} = 2$ obtained from $C(q-1, 3; 0)$ where q is assumed to be odd. It can be shown that some of the hopping patterns use only half of the available fre-quency slots, which is usually undesirable.

II.F. Solomon's Construction

In [30], Solomon considered the construction of hopping patterns based on *cosets* of Reed-Solomon codes. $C(q-1, t; 0)$ is a subcode of $C(q-1, t+1; 0)$, and hence the latter can bebe partitioned into q *cosets* of $C(q-1, t; 0)$. The coset representatives can be taken to be multiples of the last row of **G** in (20), i.e. $\beta_j \cdot (1, \alpha^t, \alpha^{2t}, \ldots \alpha^{(q-2)t})$ where, as before, the β_j denote the elements of $GF(q)$. Now, cyclically shifting all the codewords in one such coset results in another such coset (usually distinct from the first.) Solomon suggested using the coset with representative $(1, \alpha^t, \alpha^{2t}, \ldots \alpha^{(q-2)t})$ as a set of hopping patterns. Each coset contains q^t hopping patterns, and since these are all codewords in $C(q-1, t+1; 0)$, $H_{\max} = t$, for this set. In comparison, recall that Reed's set of hopping patterns also has $H_{\max} = t$, but contains *at least* q^t hopping patterns. As an example of Solomon's construc-tion technique, consider the repetition code $C(q-1, 1; 0)$ which is a subcode of $C(q-1, 2; 0)$,. The codewords in the coset are all of the form

$$x^{(j)} = (\beta_j, \beta_j, \ldots \beta_j) + (1, \alpha, \alpha^2, \ldots \alpha^{q-2}), 0 \le j \le q-1,$$

and these are *identical* to the hopping patterns exhibited in (21)! On the other hand, Reed's construction provides 2^k+1 hopping patterns with $H_{\max} = 1$ when it is applied to $C(2^k-1, 2; 1)$ over $GF(2^k)$ while Solomon's con-struction can provide only 2^k hopping patterns. There is also a minor prob-lem (that did not show up in the above simple examples) with the general version of Solomon's construction. If α^t is not a primitive element, then the coset representative $(1, \alpha^t, \alpha^{2t}, \dots \alpha^{(q-2)t})$ is not a primitive codeword, and hence the hopping patterns do not all have period $q-1$ (even though $H_{max} = t$ for the set.) Fortunately, there is a simple way around this difficulty. The code $C(q-1, t; 0)$ is also a subcode of $C(q-1, t+1; -1)$ and Solomon's set can be taken to be the coset with representative $(1, \alpha^{-1}, \alpha^{-2}, \dots \alpha^{-(q-2)})$. Since α^{-1} is a primitive element, all the hopping patterns in the coset are of period $q-1$, and the problem is thus eliminated.

II.G Titlebaum's Construction

Let $k = 1$ so that $GF(q) = GF(p)$ and consider the code generated by the matrix

$$\mathbf{G} = \begin{bmatrix} 1 & 1 & 1 & \cdots & 1 \\ 0 & 1 & 2 & \cdots & p-1 \end{bmatrix}$$

It was shown by Roth and Seroussi [23] that this is a maximum-distance-separable (MDS) *cyclic* code that is *equivalent* under column permutations to an *extended cyclic* Reed-Solomon code over $GF(p)$. The minimum distance of this code is $p - 1$, and hence the representatives of the $p-1$ equivalence cla-sses of period p form a set of nonrepeating hopping patterns with $H_a(X) = 0$ and $H_c(X) = 1$. The hopping patterns thus satisfy (3) and (13) with equa-lity and are optimal with respect to the bounds of Section II.B. It is easily shown that the representatives of these equivalence classes may be taken to be

$$(0, 1i, 2i, \dots (p-1)i), \quad 1 \le i \le p-1.$$

This form of the construction is due to Titlebaum [32] and Shaar and Davies [26]. Extensions to larger sets of hopping patterns can be found in [14] and [33].

A related construction, due to Shaar and Davies [27], is as follows. Suppose that 2^k-1 is a Mersenne prime, and consider the generator matrix of the Reed-Solomon code $C(2^k-1, 2^k-2; 1)$ over $GF(2^k)$. The rows of this generator matrix are primitive codewords of the form $(1, \alpha^t, \alpha^{2t}, \dots \alpha^{(2^k-2)t})$. Shaar and Davies proposed that these rows be considered as hopping patterns of period 2^k-1 over the alphabet $GF(2^k) - \{0\}$ of size 2^k-1. Clearly, $N(x) = (1, 1, \dots, 1)$ for all these hopping patterns, and hence $H_a(X) = 0$ and $\overline{H}_c(X) = 1$. In fact, $H_c(X) = 1$ for this set. This is because if the i-th and the j-th sequence collide in the m-th and the n-th dwells when the relative time delay is l dwells, then it must be that $\alpha^{im} = \alpha^{j(l+m)}$ and $\alpha^{in} = \alpha^{j(l+n)}$. However, these two equations cannot be satisfied simultaneously when $i \ne j$. Note that the Shaar-Davies construction cannot be used with fields of characteristic greater than 2 because $p^k - 1$ is always composite when $p > 2$ (except in the trivial case $p = 3$, $k = 1$.) Another construction which also requires an alphabet of prime cardinality is given in [4]. However, this is not directly related to Reed-Solomon codes and will not be discussed further here.

II.H   The Lempel-Greenberger Construction

The Reed and Solomon constructions use codewords from cyclic Reed-Solomon codes and hence the hopping patterns have period at most $q-1$. However, the constructions can be used to create large sets of hopping patterns of these relatively short periods. In contrast, in their seminal paper [10], Lempel and Greenberger showed how to create small sets of hopping patterns of very large period. In hindsight, the Lempel-Greenberger construction can also be viewed as a construction based on Reed-Solomon codes, and thus it fits in perfectly with the theme of this chapter.

Let Y denote an m-sequence (that is, a maximal-length linear feedback shift register sequence) of period $N = p^n - 1$ over $GF(p)$. Two important properties of Y are as follows:

1.   For any i, $0 < i < N$, there is a j, $0 < j < N$, such that $Y - T^i Y = T^j Y$.
2.   The n-tuples $(Y_i, Y_{i+1}, \ldots Y_{i+n-1})$, $0 \le i < N$, (with subscripts on Y being taken modulo N) are distinct and nonzero, that is, each of the $p^n - 1$ nonzero n-tuples occurs exactly once in a period of Y.

Suppose that $k \le n$. Then, since each nonzero k-tuple can be extended into p^{n-k} different nonzero n-tuples, the nonzero k-tuples all occur p^{n-k} times in a period of Y. Since the zero k-tuple can be extended into $p^{n-k}-1$ different *nonzero* n-tuples, it occurs $p^{n-k}-1$ times in a period of Y.

Sequences of k-tuples are the basis of the Lempel-Greenberger construction. Let y denote a sequence whose *elements* are successive overlapping k-tuples from Y. Thus,

$$y_i = (Y_i, Y_{i+1}, \ldots Y_{i+k-1}), \quad 0 \le i < N. \tag{23}$$

Since $GF(q) = GF(p^k)$ can be regarded as a k-dimensional vector space over $GF(p)$, one can also think of the y_i as elements of $GF(q)$. Property 2 of Y above thus implies that each nonzero element of $GF(q)$ occurs p^{n-k} times in a period of y while the zero element occurs $p^{n-k}-1$ times. Hence y satisfies (3). The Lempel-Greenberger set of hopping patterns can then be defined as the set of q sequences $\{y^{(j)} : 0 \le j \le q-1\}$ where

$$y_i^{(j)} = \beta_j + y_i, \quad 0 \le i < N, \quad 0 \le j \le q-1, \tag{24}.$$

and, as before, the $\beta_j, 0 \le j \le q-1$ denote the elements of $GF(q)$. Note that y itself is one of the members of this set of hopping patterns. Note also that each $y^{(j)}$ satisfies (3) with the element β_j occurring $p^{n-k}-1$ times and all the other elements occurring p^{n-k} times. Thus, the Lempel-Greenberger set also satisfies (13) with each element of $GF(q)$ occurring a total of N times in the q hopping patterns.

It was shown in [10] that for $1 < l < N$,

$$H_{y^{(j)}}(l) = p^{n-k} = \left\lceil \frac{N}{q} \right\rceil - 1 \tag{25}$$

and that if $i \ne j$, then

$$H_{y^{(i)}, y^{(j)}}(l) = \begin{cases} 0, & l = 0, \\ p^{n-k} = \left\lceil \dfrac{N}{q} \right\rceil, & 1 < l < N. \end{cases} \tag{26}$$

Thus, the hopping patterns are optimal with respect to the bounds (11) and (15). These properties can be proved straightforwardly using the properties of Y. These properties can also be deduced from the fact that the Lempel-Greenberger hopping patterns can be obtained by mapping a set of Reed and Solomon sequences over $GF(p^n)$ of the form exhibited in (21) to $GF(q)$. This idea is considered next.

Suppose without loss of generality that the m-sequence Y is in its *characteristic phase* so that there is a primitive element $\xi \in GF(p^n)$ such that $Y_i = Tr(\xi^i), 0 \le i < N$ where

$$Tr(z) = z + z^p + z^{p^2} + \cdots + z^{p^{n-1}}$$

is the *trace function* from $GF(p^n)$ to $GF(p)$ [16]. Now, with respect to the standard polynomial basis $\{1, \xi, \xi^2, \ldots, \xi^{n-1}\}$ for $GF(p^n)$ over $GF(p)$, the n-tuple $(a_0, a_1, \ldots, a_{n-1})$ represents the element $a_0 \cdot 1 + a_1 \xi^1 + \cdots + a_{n-1} \xi^{n-1}$. Let $\{\zeta_0, \zeta_1, \zeta_2, \ldots, \zeta_{n-1}\}$, denote the *dual basis* of the polynomial basis. If the n-tuple $(b_0, b_1, \ldots, b_{n-1})$ represents $Z = b_0 \zeta_0 + b_1 \zeta_1 + \cdots + b_{n-1} \zeta_{n-1}$. With respect to this dual basis, then according to the properties of dual bases, $b_j = Tr(Z\xi^j)$ for $0 \le j \le n-1$ [16]. Thus the n-tuple

$$(Y_i, Y_{i+1}, Y_{i+2}, ..., Y_{i+n-1}) = (Tr(\xi^i), Tr(\xi^{i+1}), Tr(\xi^{i+2}), ..., Tr(\xi^{i+n-1}))$$

$$= (Tr(\xi^i \xi^0), Tr(\xi^i \xi^1), Tr(\xi^i \xi^2), ..., Tr(\xi^i \xi^{n-1}))$$

is just the representation of ξ^i, with respect to the *dual* of the standard polynomial basis. Now consider the sequence Z of *successive overlapping n-tuples* from the *m*-sequence Y. The *i*-th element of Z is given by

$$Z_i = (Y_i, Y_{i+1}, Y_{i+2}, ..., Y_{i+n-1}) \tag{27}$$

and thus Z is a representation of the sequence $(1, \xi, \xi^2, ..., \xi^{N-1})$ over $GF(p^n)$ with respect to the dual of the standard polynomial basis.

Consider the Reed and Solomon set of sequences from the Reed-Solomon code $C(p^n-1, 2; 0)$ over $GF(p^n)$. Denoting the elements of $GF(p^n) = GF(N+1)$ as γ_j, $0 \le j \le N$, the *j*-th Reed and Solomon sequence can be expressed as

$$Z^{(j)} = (\gamma_j, \gamma_j, ... \gamma_j) + (1, \xi, ... \xi^{N-1}), 0 \le j \le N, \tag{28}$$

which is exactly of the form given in (21). Note that γ_j does not occur in $Z^{(j)}$ while all the other elements of $GF(p^n)$ occur exactly once. Now, suppose that the elements in the vectors in (28) are represented in the dual-polynomial basis of (27). View the first k entries of the representation of γ_j as an element of $GF(q)$, and note that p^{n-k} different γ_j give rise to the same element β_j (say) of $GF(q)$. Of course, the representation of ξ^i is just the right side of (27) and the first k entries are just $y_i = (Y_i, Y_{i+1}, ... Y_{i+k-1})$, as defined in (23). It follows that the sequence $y^{(j)}$ defined in (24) is the image of p^{n-k} Reed-Solomon codewords $Z^{(j)}$ under the *projection mapping* that maps an element of $GF(p^n)$ onto its first k coordinates in the dual-polynomial basis. Note also that only q different sequences over $GF(q)$ are ob-tained from this mapping. As a special case, the Lempel-Greenberger hopping patterns are identical to the Reed and Solomon patterns when $n = k$ and all field elements are represented with respect to the dual of the standard polynomial basis.

Let L denote a general linear mapping from the n-dimensional vector space $GF(p^n)$ over $GF(p)$ to the k-dimensional vector space $GF(q)$ over $GF(p)$,

University of Sunderland
St. Peter's Library
Tel: (0191) 515 3691

m: Code division multiple access
mmunications
ircode: 11110965150
ıe Date: 07/01/2008 23:55

m: Code division multiple access
mmunications
ircode: 11110965150
ıe Date: 07/01/2008 23:55

Please retain this receipt
to ensure you return items on time

Need to borrow a laptop?

Reserve one using our new 'Book-IT'
service.

For further information regarding all of our

services see the web pages at
http://my.sunderland.ac.uk

Plb

and let $L(Z^{(j)})$ denote the image of the Reed-Solomon hopping pattern $Z^{(j)}$ under L. The Lempel-Greenberger hopping patterns are obtained in the special case when L is the projection mapping that just deletes the last $n{-}k$ coordinates of the representation in the dual of the standard polynomial basis. All the Hamming correlation properties of the Lempel-Greenberger hopping patterns also apply to the more general sets. To see this, note that $Z^{(i)}$ and $Z^{(j)}$ are codewords in a linear cyclic code and hence $Z^{(i)} - T^l Z^{(j)}$ is also a codeword in the same code. Furthermore, if $l = 0$, this codeword consists of $\gamma_i - \gamma_j$ repeated N times, whereas if $0 < l < N$, then every element of $GF(p^n) except$ $\gamma_i - \gamma_j$ appears exactly once in this codeword. Now, for a relative time delay of l dwells, the Hamming correlation between $L(Z^{(i)})$ and $L(Z^{(j)})$ is just the number of times that 0 appears in the sequence $L(Z^{(i)}) - L(T^l Z^{(j)}) = L(Z^{(i)} - T^l Z^{(j)})$. But this sequence is just the image of $Z^{(i)} - T^l Z^{(j)}$ under L. Thus, if $\gamma_i - \gamma_j \in \ker(L)$, where $\ker(L)$ is the kernel of the linear transformation L, then $L(Z^{(i)}) = L(Z^{(j)})$, the correlation function under consideration is an autocorrelation, and 0 appears $p^{n-k}-1$ times for any l, $0 < l < N$. On the other hand, if $\gamma_i - \gamma_j \notin \ker(L)$, then $L(Z^{(i)}) \neq L(Z^{(j)})$, the correlation function under consideration is a crosscorrelation, and 0 appears $p^{n-k}-$ times if $l \neq 0$ and but it never appears if $l = 0$. Thus, the Hamming correlation properties (25) and (26) of the Lempel-Greenberger hopping patterns also hold in the more general case.

Another interesting property of the Lempel-Greenberger hopping patterns arises from the fact that the underlying sequence Y is an m-sequence of period p^n-1. According to Property 2 above, Y contains a run of length n of each nonzero element of $GF(p)$. It follows that the sequence y defined in (23) has a run of length $n-k+1$ corresponding to each nonzero element of $GF(p)$. Thus, at some time during each period, the signal will hop to the frequency slot corresponding to $(i,i,...,i)$ in $GF(q)$ and stay there for $n-k+1$ successive dwells. Since Y also contains runs of lengths $n-1$, $n-2,...$ the signal will return to the same slot many more times in each period though it will stay there for fewer successive dwells. Since $y^{(j)} = \beta_j + y$, the other Lempel-Greenberger patterns hop to slots corresponding to $\beta_j +(i,i,...,i)$ and stay there for $n-k+1$ successive dwells. In fact, given any frequency slot, there are $p-1$ Lempel-Greenberger patterns that will hop to that slot and stay there for $n-k+1$ successive dwells. All this is clearly undesirable for the reasons dis-cussed in Section II.C. Note that the maximum length of a burst of hits for Lempel-Greenberger hopping patterns is *at least* $n-k+1$ which is approx-imately k times larger than the lower bound of (18).

II.I   Generalizations of the Lempel-Greenberger Construction

The Lempel-Greenberger construction is optimal with respect to various correlation bounds but has the defect that long runs occur in the sequences and these runs are undesirable for the reasons outlined in Section II.C. Various generalizations of this construction are possible. For example, one can use the extension field $GF(p^s)$ instead of the prime field $GF(p)$. Thus, Y is an m-sequence of period $(p^s)^n - 1$ over $GF(p^s)$ and these generalized Lempel-Greenberger hopping patterns are formed by considering the sequence y of successive overlapping k-tuples from Y and adding a fixed k-tuple over $GF(p^s)$ to each element of y The hopping patterns are thus elements of $GF(p^{sk})$ and so on. None of the proofs in [10] really require the base field $GF(p)$ to be a prime field – they all hold *mutatis mutandis* if the base field is the extension field $GF(p^s)$ instead. Unfortunately, this also means that all the results asserted in the previous subsection also hold. In particular, these generalized Lempel-Greenberger hopping patterns also have runs of the same frequency, and produce bursts of hits for certain relative time delays, etc.

Recently, a different solution to the runs problem, based on a combination of the Lempel-Greenberger and Reed-Solomon approaches, has been proposed by Burger [2]. This method can be viewed as a product code construction in which one code is a Lempel-Greenberger hopping pattern and the other code is a Reed-Solomon hopping pattern. The method begins by constructing a set of q Lempel-Greenberger hopping patterns of period p^n-1 over $GF(q)$ as usual. Each transmitter uses a different $(q-1,1)$ Reed-Solomon code (actually a coset of a $(q-1,1)$ Reed-Solomon code) over $GF(q)$ to encode each symbol in its Lempel-Greenberger pattern. In other words, each symbol in the Lempel-Greenberger pattern is replaced by a codeword of length $q-1$ from the Reed-Solomon coset. Thus, the resulting hopping pattern has period $(p^n-1)(q-1) = p^{n+k} - p^n - p^k + 1$ over $GF(q)$. Now, the Reed-Solomon coset is a coset of $C(q-1, 1; 0)$, and the q different cosets assigned to the q transmitters are all subsets of $C(q-1, 3; q-2)$. Hence, any two codewords in the coset assigned to a transmitter collide in at most one position, while codewords from the cosets assigned to two different transmitters collide in at most two positions. The Hamming correlations of these patterns are not quite optimal but are none the less very good. More important however, is the fact that the runs of the same frequency that occur in the Lempel-Greenberger patterns have been broken up. It is also easy to show that at most 4 suc-cessive collisions can occur between two patterns which alleviates the burden on the error control system as well. Further details of this construction can be found in [2].

II.J Vajda's Construction

The use of a product code construction of hopping patterns has also been explored by Vajda [34] who took a cyclic product of two codes. The first code is obtained from $C(N, t+1; 0)$ over $GF(q)$ where N is a prime. As described in Section II.E, q^t hopping patterns of period N can be obtained from this code. Let r denote the largest integer such that $N^{r-1} \leq q^t$. Now consider a (nonlinear) cyclic code consisting of a set of N^{r-1} hopping patterns of period N *and all their cyclic shifts*. Note that the minimum distance of this nonlinear cyclic code is $N{-}t$. The other code is a coset of $C(M, K{-}1; 2)$ over $GF(N^r)$ that is a subcode of $C(M, K{+}1; 0)$ over $GF(N^r)$.where M is a divisor of $N^r - 1$. Thus, this code has $N^{r(K-1)}$ codewords over an alphabet of size N^r. and since the codewords belong to $C(M, K{+}1; 0)$, the minimum distance beween cyclic shifts of two codewords is at least $M{-}K$. Vajda has proposed using the cyclic product of these codes instead of the direct product discussed in Section II.I. Thus, each of the M N^r-valued symbols in a code-word of the second code is replaced by a column vector of length N con-sisting of a codeword of the first code. This creates a $N{\times}M$ matrix $\mathbf{Q} = Q_{i,j}$. Now,M and N are relatively prime, and hence the entries in $\mathbf{Q}$ can be read off in cyclic fashion to form a hopping pattern of length MN whose i-th symbol is $Q_{i \bmod N, j \bmod M}$. Since the cyclic productof a (n_1, k_1, d_1) cyclic code with a (n_2, k_2, d_2) cyclic code is a $(n_1 n_2, k_1 k_2, d_1 d_2)$ cyclic code [13], this hopping pattern is actually a codeword in a $(MN, (k{+}1)(K{+}1), (M{-}K)(N{-}t))$ cyclic code over $GF(q)$. There are $N^{r(K-1)}$ such hopping patterns, and it follows from(7) that, as shown in [34],

$$H_{\max} \leq MN - (M - K)(N - t) = Mt + KN - Kt.$$

As an example of this construction, let $q = 32$, $N = 31$, $t = 2$, $r = 3$,and $M = (31^3 - 1)/(31 - 1) = 993$. Let $K = 4$. Then, a set of 887,503,681 hopping patterns of period 30,783 over GF(32) is obtained. The maximum Hamming correlation is 2,102, so that there is, on the average, one hit every 14.64 symbols. In contrast, the Reed and Solomon sets of hopping patterns from over GF(32) provide 1,024 hopping patterns of period 31 with a maximum Hamming correlation of 2, that is, one hit every 15.5 symbols, which is very slightly better.

II.K Einarsson's Construction

Because of technological limitations, the frequency synthesizers used to produce frequency-hopped signals can change the carrier frequency no more than a few thousand times in one second. In a fast FH/CDMA system, the transmission of a symbol occurs over several hops, and it is necessary to use

72

M-ary signaling in order to achieve a reasonably large data rate. Einarsson [5] proposed a combined design of hopping patterns and *M*-ary modulation for use in such systems. In systems using this design,each transmitter is assigned a collection of *M* hopping patterns of length *N*, and transmits one *M*-ary data symbol per *N* dwells by choosing and transmitting one of the hopping patterns. The data rate is thus $N^{-1} \log_2(M)$ bits per dwell. Note however, that the receiver is now more complicated since it must track all *M* possible hopping patterns in order to determine which one is being transmitted. Thus, *M* different frequency synthesizers might be needed in each receiver. The Einarsson design uses *all* the nonzero codewords in the Reed-Solomon code C(q–1, 2; 0). Each transmitter is assigned all the codewords in a cyclic equivalence class. Thus, $M = N = q-1$. Since all these sequences are from *C(q*–1, 2; 0*)*, the number of hits between two patterns assigned to diff-erent transmitters is at most 1 regardless of the time delay between the two patterns. However,the number of hits per period can be guaranteed to be one only if the two transmitters are *frame-synchronous*. If the transmitters are only dwell-synchronous, then the tail end and the front end of two *possibly different* hopping patterns from an interfering transmitter can cause coll-isions,* and thus the number of hits per period can be two in some cases. There is also the question of the initial acquisition of synchronization in the receivers in such systems since the hopping patterns assigned to a transmitter are not cyclically inequivalent. In fact, the Hamming crosscorrelation bet-ween two hopping patterns assigned to the same transmitter can have values as large as $N - 1$.

III. Direct-Sequence Spread-Spectrum Systems

The desirable properties of signature sequences for DS/CDMA systems and a detailed tutorial survey of various constructions for these signature sequences can be found in [25]. Here, it is shown that most of these sequences can be viewed as images of codewords from Reed-Solomon codes.

III.A Binary *m*-sequences and Their Decimations

Let $N = 2^n - 1$ and let $x^{(1)}$ denote a binary *m*-sequence of period *N* in its characteristic phase, that is, $x_{2i}^{(1)} = x_i^{(1)}$. Then there is a primitive element α such that $x_i^{(1)} = Tr(\alpha^i)$, $0 \le i < N$, where the *trace function* was defined in Section II.H. Thus, $x^{(1)}$ is the binary image of the vector

$$X^{(1)} = (1, \alpha, \alpha^2, ..., \alpha^{N-1})$$

---

* A similar phenomenon in DS/CDMA systems gives rise to the *odd crosscorrelation* function of binary sequences (cf.}[25].)

over $GF(2^n)$ via the trace function. But, $X^{(1)}$ is a primitive Reed-Solomon codeword belonging to the cyclic Reed-Solomon code $C(N, 1; 1)$ with parity-check polynomial $h(z) = (z - \alpha^{-1})$. Thus, the binary m-sequence $x^{(1)}$ is the $GF(2)$ image of the Reed-Solomon codeword $X^{(1)}$.

More generally, for $0 < s < N$, the s-th decimation of $x^{(1)}$ is the sequence $x^{(s)}$ defined by $x_i^{(s)} = x_{si}^{(1)}$ where the subscripts are taken modulo N. Then, $x_i^{(s)} = x_{si}^{(1)} = Tr(\alpha^{si}) = Tr((\alpha^s)^i)$ and hence $x^{(s)}$ is the image of

$$X^{(s)} = (1, \alpha^s, (\alpha^s)^2, \ldots, (\alpha^s)^{N-1})$$

which is a codeword in $C(N, 1; s)$. Thus, the decimations of $x^{(1)}$ are also images of Reed-Solomon codewords. If $gcd(N, s) = 1$, α^s is primitive, $X^{(s)}$ is a primitive codeword and $x^{(s)}$ is an m-sequence. Otherwise, α^s is nonprimitive and both $X^{(s)}$ and $x^{(s)}$ have period $N/gcd(N, s) < N$. Note that since $x^{(1)}$ is in its characteristic phase, so is $x^{(s)}$, that is,

$$x_{2i}^{(s)} = x_{2si}^{(1)} = x_{si}^{(1)} = x_i^{(s)}.$$

It follows that

$$x^{(s \cdot 2^k)} = x^{(s)} \text{ for all } s, \ 0 < s < N, \text{ and all } k, \ 0 \leq k < n.$$

Since $x^{(s \cdot 2^k)}$ is the image of $X^{(s \cdot 2^k)}$, it follows that $X^{(s)}$ is not the only Reed-Solomon codeword that is mapped to $x^{(s)}$ by the trace function . Note that $X^{(s \cdot 2^k)}$ is a codeword in the cyclic Reed-Solomon code $C(N, 1; s \cdot 2^k)$ with parity-check polynomial $h(z) = (z - \alpha^{-s \cdot 2^k})$. This idea is used repeatedly in the remainder of this section.

III.B Dual-BCH Sequences

Consider the Reed-Solomon code $C(N, 2; 2)$ over $GF(2^n)$ whose generator matrix $\mathbf{G}$ has two rows: $X^{(2)}$ and $X^{(3)}$. Both rows are primitive codewords if n is odd, while if n is even, $X^{(3)}$ has period $N/3$. Thus, this Reed-Solomon code has $2^n + 1$ or 2^n cyclically inequivalent codewords of period N according as n is odd or even. A typical primitive codeword is of the form $X^{(2)} + \beta X^{(3)}$ where β is some element of $GF(2^n)$. Now, suppose that n is odd and that $\beta \neq 0$. Then $\beta X^{(3)}$ is just a cyclic shift of $X^{(3)}$, and hence the trace function (which is a *linear* map from $GF(2^n)$ to $GF(2)$) maps $X^{(2)} + \beta X^{(3)}$ to the binary sequence $x^{(2)} + T^i x^{(3)} = x^{(1)} + T^i x^{(3)}$ for some i, $0 \leq i < N$. If $\beta = 0$, the sequence is just $x^{(1)}$. One other sequence that can be added to the list of binary sequences obtained from $C(N, 2; 2)$ is $x^{(3)}$ itself. Thus, the collection of $2^n + 1$ sequences can be written as

$$\{x^{(1)}, x^{(3)}, x^{(1)} + T^i x^{(3)} : 0 \leq i < N\}. \tag{29}$$

When n is even, matters are slightly more complicated. Now, the image of $\beta X^{(3)}$ can be a cyclic shift of one of three different binary sequences $y^{(0)} = x^{(3)}$, $y^{(1)}$, or $y^{(2)}$ of period $N/3$. Thus, the 2^n binary sequences obtained can be written as

$$\{x^{(1)}, x^{(1)} + T^i y^{(0)}, x^{(1)} + T^i y^{(1)}, x^{(1)} + T^i y^{(2)} : 0 \leq i < N/3\}. \tag{30}$$

The sets exhibited in (29) and (30) are exactly those in Eq. (4.4) and (4.12) of [25]. They are called *dual-BCH sequences* because they can be obtained from the dual codes of double-error correcting binary BCH codes. Here, we have exhibited them as the images of codewords from the Reed-Solomon code $C(N, 2; 2)$ over $GF(2^n)$.

III.C Gold Sequences and Gold-like Sequences

The Gold sequences are a set of $2^n + 1$ binary sequences of the form

$$\{x, y, x + T^i y : 0 \leq i < N\} \tag{31}$$

where x and y are m-sequences of period N, and y is a *preferred decimation* of x. Some preferred decimations of the form $2^k + 1$ and $2^{2k} - 2^k + 1$ are given in Theorem 1 of [25]. Clearly, if the decimation $2^k + 1$ is being used, then these sequences can be obtained from the $2^n + 1$ cyclically inequivalent codewords of period N from $C(N, 2; 2^k)$. A similar construction does not always work with the decimation $2^{2k} - 2^k + 1$. Similarly, there are several other preferred decimations that have been discovered experimentally and the Gold sequences corresponding to these cannot be generated by this method.

The classical Gold decimation $2^{\lfloor (n+2)/2 \rfloor} + 1$ does not produce an m-sequence when $n \equiv 0 \bmod 4$ and thus the set of sequences obtained is this case cannot be expressed in the form (31). In this case, the set of *Gold-like sequences* contains 2^n sequences of period N of the form exhibited in (30) or Eq. (4.12) of [25]. The reason for this is that these sequences are the images of codewords from $C(N,2;2^{(n+2)/2})$, but $\alpha^{2^{(n+2)/2}}$ is an element of order N/3, and hence, only 2^n different cyclically inequivalent Reed-Solomon codewords can be formed. As in the case of Gold sequences, all sets of Gold-like sequences need not be obtainable from Reed-Solomon codewords. However, the best known cases of Gold sequences and Gold-like sequences are all images of Reed-Solomon codewords.

III.D   Kasami Sequences

Let n be even and let $w = x^{2^{n/2}+1}$. Then, w is an m-sequence of period $2^{n/2}-1$, and *the small set of Kasami sequences* [25] is given by

$$\{x, x + T^i w: 0 \le i < 2^{n/2}\}$$

The reader who has followed the above discussion will have no difficulty in recognizing that the small set of Kasami sequences is obtained as the images of codewords from the Reed-Solomon code $C(N,2;2^{n/2})$. The case of *the large set of Kasami sequences* is slightly complicated because it is necessary to consider the cases $n \equiv 0 \bmod 4$ and $n \equiv 2 \bmod 4$ separately. In either case, the sequences are the images of codewords from the Reed-Solomon code $C(N,2;2^{(n+2)/2})$. but the number of sequences obtained depends on whether $n \equiv 0 \bmod 4$ or $n \equiv 2 \bmod 4$ The details can be found in [25].

IV. Concluding Remarks

This chapter has discussed how Reed-Solomon codes can be used in the designof frequency-hopping patterns for FH/CDMA systems and signature sequences for DS/CDMA systems. Although signature sequences for DS/CDMA systems can be viewed as the images of Reed-Solomon codewords, there seems to be no obvious connection between the properties of the Reed-Solomon code and the correlation propertiesof the resulting signature sequences. Thus, the primary application of Reed-Solomon codes in sequence design for CDMA systems remains the design of hopping patterns for FH/CDMA systems.

V.   Acknowledgments

Much of the material in Section II of this chapter is an updated and considerably expanded version of [24] while some of the material in Section III is drawn from [25]. The author wishes to thank his co-author on those papers,

Professor Michael B. Pursley of Clemson University for many helpful discussions on this topic. The author also wishes to thank Professors Pentti Leppänen and Savo Glisic of the University of Oulu for the invitation to write this paper.

VI. References and Selected Bibliography

1. C. A. Baum and M. B. Pursley, "Bayesian methods for erasure insertion in frequency-hop communication systems with partial-band interference," *IEEE Transactions on Communications,* vol. 40, pp. 1231-1238, July 1992.

2. N. Burger, "The design of frequency hopping patterns for multiple-access communications," M. S. thesis, Department of Electrical and Computer Engineering, University of Illinois at Urbana-Champaign, Urbana, Illinois, January 1994.

3. G. R. Cooper and C. D. McGillem, *Modern Communications and Spread Spectrum,* New York: McGraw-Hill, 1984.

4. G. R. Cooper and R. W. Nettleton, "A spread spectrum technique for high-capacity mobile communications," *IEEE Trans. Vehicular Tech.,* vol. VT-27, pp. 264–275, 1978.

5. G. Einarsson, "Address assignment for a time-frequency-coded, spread spectrum system," *Bell Syst. Tech. J.,* vol. 59, pp. 1241–1255, 1980.

6. E. A. Geraniotis and M. B. Pursley, "Error probabilities for slow frequency-hopped spread-spectrum multiple-access communication over fading channels," *IEEE Trans. Commun.,* vol. COM-30, pp. 996–1009, 1982.

7. S. W. Golomb, *Shift Register Sequences.* San Francisco: Holden-Day, 1967.

8. J. B. Jevtic and H. B. Alkhatib, "Frequency-hopping codes for multiple-access channels: A geometric approach," *IEEE Trans. Inform. Theory,* vol. 35, pp. 477–481, March 1989.

9. P. V. Kumar, "Frequency-hopping code sequence designs having large linear span," *IEEE Trans. Inform. Theory,* vol. 34, pp. 146–151, January 1988.

10. A. Lempel and H. Greenberger, "Families of sequences with optimal Hamming correlation properties," *IEEE Trans. Inform. Theory,* vol. IT-20, pp. 90–94, January 1974.

11. J. H. van Lint, *A Course in Coding Theory,* Berlin: Springer-Verlag, 1989.

12. F. J. MacWilliams and N. J. A. Sloane, "Pseudorandom sequences and arrays," *Proc. IEEE,,* vol. 64, pp. 1715-1729, December 1976.

13. F. J. MacWilliams and N. J. A. Sloane, *The Theory of Error-Correcting Codes,* Amsterdam: North-Holland, 1977.

14. S. V. Maric and E. L. Titlebaum, "Frequency hop multiple access codes based upon the theory of cubic congruences," *IEEE Trans. Aerospace and Electronic Systems,* vol. 25, pp. 1035–1039, November 1990.

15. R. J. McEliece, "Some combinatorial aspects of spread spectrum communication systems," in *New Concepts in Multi-User Communications,* J. K. Skwirzynski (Ed.), NASI Sirjhoff and Noordhoff, 1981.

16. R. J. McEliece, *Finite Fields for Computer Scientists and Engineers,* Kluwer Press: Boston MA 1989.

17. R. M. Mersereau and T. S. Seay, "Multiple access frequency hopping patterns with low ambiguity," *IEEE Trans. Aerospace and Electronic Systems,* vol. AES-17, pp. 571–578, 1981.

18. B. Popovic, "New sequences for frequency hopping multiplex," *Electronics Letters,* vol. 24, June 1986.

19. B. Popovic, "Comment on'Code acquisition for a frequency hopping system'," *IEEE Trans. Commun.* May 1989.

20. M. B. Pursley, "Frequency-hop transmission for satellite packet switching and terrestrial packet radio networks," *IEEE Trans. Inform. Theory,* vol. IT-32, pp. 652–667, September 1986.

21. M. B. Pursley, "Reed-Solomon codes for frequency-hop communications," in *Reed-Solomon Codes and Their Applications,* S. B. Wicker andV. K. Bhargava (eds.), Piscataway, NJ: IEEE Press, 1994.

22. I. S. Reed, "k-th order near-orthogonal codes," *IEEE Trans. Inform. Theory,* vol. IT-15, pp. 116–117, January 1971.

23. G. M. Roth and G. Seroussi, "On cyclic MDS codes of length q over $GF(q)$," *IEEE Trans. Inform. Theory,* vol. IT-32, pp. 284-285, March 1986.

24. D. V. Sarwate and M. B. Pursley, "Hopping patterns for frequency hopped multiple access communications," *Conf. Record: IEEE Int. Conf. on Commun.,* (Toronto, Ontario: June 4-8, 1978), vol. 1, pp. 7.4.1–7.4.3.

25. D. V. Sarwate and M. B. Pursley, "Crosscorrelation properties of pseudorandom and related sequences," *Proc. IEEE,* vol. 68, pp. 593–619, May 1980.

26. A. A. Shaar and P. A. Davies, "Prime sequences: quasi-optimal sequences for OR channel code division multiplexing," *Electronics Letters,* vol. 19, pp. 888-890, 1981.

27. A. A. Shaar and P. A. Davies, "A survey of one-coincidence sequences for frequency-hopped spread-spectrum systems," *IEE Proc. – Part F: Commun., Radar, and Signal Proc.,* vol. 131, pp. 719-724, December 1984.

28. A. A. Shaar, C. F. Woodcock, and P. A. Davies, "Number of one-coincidence sequence sets for frequency-hopping multiple access communication systems," *IEE Proc. – Part F: Commun., Radar, and Signal Proc.,* vol. 131, pp. 725-728, December 1984.

29. M. K. Simon, J. K. Omura, R. A. Scholtz, and B. K. Levitt, *Spread Spectrum Communications.* Rockville, MD: Computer Science Press, 1985.

30. G. Solomon, "Optimal frequency hopping sequences for multiple access," *Proc. Symp. on Spread Spectrum Commun.,* Vol. 1, AD–915852, pp. 33–35, 1973.

31. H. Y. Song, I. S. Reed, and S. W. Golomb, "On the nonperiodic cyclic equivalence classes of Reed-Solomon codes," *IEEE Trans. Inform. Theory,* vol. 39, pp. 1431-1434, July 1993.

32. E. L. Titlebaum, "Time frequency hop signals, Part I: Coding based upon the theory of linear congruences," *IEEE Trans. Aerospace and Electronic Systems,* vol. AES-17, pp. 490–494, July 1981.

33. E. L. Titlebaum and L. H. Sibul,"Time frequency hop signals, Part II: Coding based upon the theory of quadratic congruences," *IEEE Trans. Aerospace and Electronic Systems,* vol. AES-17, pp. 494–501, July 1981.

34. I. Vajda,"Code sequences for frequency-hopping multiple-access systems," *IEEE Trans. Commun.* to appear.

35. R. E. Ziemer and R. L. Peterson, *Digital Communications and Spread Spectrum Systems.* New York: Macmillan, 1985.

Block demodulation – An overview

Ezio Biglieri,   Elena Bogani,   Monica Visintin

Abstract — We consider the use of block demodulation, or multi-symbol detection, in receivers for digital transmission. After reviewing the rationale behind it, we examine its performance for differential and double-differential detection of PSK and intersymbol-interference channels.

I. Introduction

It is well known that in the presence of disturbances that do not act independently from symbol to symbol, operation of a symbol-by-symbol demodulator is not optimum, and a maximum-likelihood sequence detector (MLSD) should be used instead. When the latter is too complex, or we have only a partial knowledge of the disturbance statistics, we may introduce in such channel an interleaver/deinterleaver pair, so that the correlation effects are removed. However, this practice may turn out to be harmful, as a better performance would be obtained by taking advantage of (rather than destroying) the existing correlations. When a MLSD cannot be implemented, suboptimum solutions should be sought that have a reduced performance penalty. One of these is based on the concept of block detection. This is performed by detecting independently blocks of more than one symbol, so that the detection process will be improved by a longer observation time that smooths out the disturbances.

In this paper we discuss the use of block detection for a number of channels. We first review the available literature, mainly devoted to detection of differentially-encoded signals affected by a random phase rotation which remains constant over a block. Further, we show some new results on double-differentially-encoded signals. Finally, some considerations on block detection of signals affected by intersymbol interference are presented.

II. Block demodulation

We assume that the received signal has the form [3]

$$r_k = s_k(\theta) + n_k,$$

The authors are with Dipartimento di Elettronica, Politecnico di Torino, Corso Duca degli Abruzzi, 24 – I-10129 Torino, Italy.
e-mail: <name>@polito.it
This research was supported by the Italian Space Agency (ASI).

S.G. Glisic and P.A. Leppänen (eds.), Code Division Multiple Access Communications, 79-93.
© 1995 *Kluwer Academic Publishers. Printed in the Netherlands.*

80

where s_k is the sample at discrete time k of the information-bearing (complex) transmitted signal, n_k is a sample taken from a stationary complex white Gaussian noise process with zero mean and variance σ^2, and θ is a random variable. The value of θ is not observed, and we assume its value to remain constant during an N-symbol interval.

We observe an N-tuple of received samples, denoted $\mathbf{r}$. We may write, with obvious notations,

$$\mathbf{r} = \mathbf{s}(\theta) + \mathbf{n}.$$

With the noise model assumed, the conditional probability of $\mathbf{r}$ given $\mathbf{s}$ and θ is

$$p(\mathbf{r} \mid \mathbf{s}, \theta) = \frac{1}{(2\pi\sigma^2)^N} \exp\left\{ -\frac{\|\mathbf{r} - \mathbf{s}(\theta)\|^2}{2\sigma^2} \right\} \tag{1}$$

The conditional probability of $\mathbf{r}$ given $\mathbf{s}$, to be maximized by the demodulator for optimum detection, is obtained by averaging (1) over θ. If we assume that $\|\mathbf{s}(\theta)\|$ does not depend on θ and $\mathbf{s}$, maximization of $p(\mathbf{r} \mid \mathbf{s})$ is tantamount to maximizing the value of the integral

$$\int_\Theta \exp\left\{ \frac{\Re[\mathbf{r} \cdot \mathbf{s}^*(\theta)]}{\sigma^2} \right\} p(\theta)\, d\theta,$$

where $\cdot$ denotes scalar product, Θ denotes the range of θ, and $p(\theta)$ is its probability density function. In the absence of side information, θ is assumed to be uniformly distributed over Θ. For example, if

$$\mathbf{s}(\theta) = \mathbf{s}e^{j\theta}$$

and

$$s_k = e^{j\phi_k},$$

where ϕ_k are discrete phases, then optimum detection is achieved [3] by finding

$$\max_{\phi_1,\ldots,\phi_N} |\mathbf{r} \cdot \mathbf{s}^*|^2. \tag{2}$$

This detection rule has an inherent phase ambiguity associated with it, since rotation of all the components of $\mathbf{s}$ by one and the same phase ϕ does not change the quantity to be maximized. To resolve this ambiguity, one should differentially encode the phase information at the transmitter. Letting

$$\phi_k = \phi_{k-1} + \Delta\phi_k$$

upon observation of the received signal over N symbol intervals we make a simultaneous decision about $N - 1$ information phases. The first sample of the block is used to provide a phase reference for the entire block. The minimum allowable value of N is thus $N = 2$, for which (2) corresponds to the standard decision rule for differentially detected PSK. The larger the value of N, the better the detector performance, at the price of a corresponding increase of its complexity.

A. Previous work on block demodulation

The first reference known to the authors in which the concept of block demodulation is applied to detection of M-ary PSK is [7]. Subsequently Leib and Pasupathy [8] observe that a reason for the performance loss of differentially coherent relative to coherent detection of PSK is that its phase reference is impaired by channel noise in the same way as the information phase. Thus, a receiver is proposed where the phase reference is obtained by averaging the past phase reference. These are obtained by removing the information phases from past symbols, and hence requires past decisions to be fed back. The performance of this new receiver for M-PSK and for MSK is found to bridge the gap between coherent and differentially coherent detection.

In [3], analysis of block demodulation (in the form described in the previous section) is carried out for uncoded M-PSK over the additive white Gaussian noise (AWGN) channel. It is proved that as $N \to \infty$ the bit-error-rate (BER) performance of block-detected M-PSK approaches that of ideal coherent-detection with differential encoding, and that this limiting performance is approached with small values of N. In particular, for binary PSK, extending N from 2 to 3 recovers more than half the power loss of differential versus coherent detection with differential encoding. For 4-PSK, the power gain achieved by increasing N from 2 to 3 is more than 1 dB. Similar analyses may be found in [9, 7, 15]. The issue of complexity in DPSK block demodulation is treated in [12].

In [4], the analysis of block demodulation over the AWGN channel is extended to trellis-coded M-PSK. Use of multiple trellis-coded modulation (TCM) [2, Chap. 7] with multiplicity (i.e., number of trellis-coded symbols per input symbol) equal to $N - 1$ in combination with block detection yields significant performance improvement over conventional detection even for small N.

Ho and Fung [6] study uncoded QPSK with a correlated Rayleigh fading channel model. Here the signal s_k is multiplied by a complex random variable u_k, a sample of a zero-mean correlated Gaussian process. They find that block detection is a very effective strategy for eliminating the irreducible error floor associated with a conventional differential detector. In all the cases investigated in [6] a detector with N as small as 3 is sufficient for this purpose. Moreover, it is discovered that block detection is not sensitive to the mismatch between the decoding metric and the channel fading statistics.

Divsalar and Simon [5] consider a wide variety of situations where block detection may be successfully used to improve performance over conventional differential detection. In particular, they examine uncoded and trellis-coded M-PSK and QAM transmitted over Rayleigh and Rice fading channels, with unknown slowly and fast-varying phase, and with or without channel-state information. The channel is endowed with a perfect (i.e., infinite-depth) interleaver.

Block detection may also be applied to modulation schemes other than PSK. In [14], uncoded full-response continuous-phase modulation (CPM) is examined. Again, it is shown that in the limit as $N \to \infty$ the BER of the block detector approaches that of the coherent receiver. The special case of GMSK is dealt with in [1]. Here it is shown that for $N \geq 5$ the BER performance of block detection comes close to ideal (i.e., maximum-likelihood sequence) detection by a few tenths of a dB. In [10] it is shown that the performance of MSK with block detection tends to that of binary noncoherently orthogonal FSK, and a modification is introduced in the demodulator to further improve the system performance.

A generalization of block demodulation is presented in [11]. The receiver here is a sliding-block detector. It processes a block of N consecutive received samples and makes a decision about the symbol in the middle of the block. Once a decision has been made, the block is moved ahead by one symbol interval. The complexity of this receiver may be reduced if the previous K decisions are fed back to the decoder. Analysis of this structure for differentially encoded PSK over the 2-ray frequency-selective Rayleigh fading channel shows that with $N = 5$ its performance is superior to conventional differential detection, with no error floor in nearly all the cases examined in [11].

III. Block detection of double-differentially encoded PSK

In this section we focus our attention to a technique known as Double Differential Phase Shift Keying (DDPSK). This was first introduced in the early 1970's, and recently resurrected [13] because it provides insensitivity to frequency offsets, due for example to Doppler effect. While with DPSK the transmitted phases are associated with differences of information phases $\theta_i - \theta_{i-1}$, with standard DDPSK they are associated with differences of differences, i.e., $\theta_i - 2\theta_{i-1} + \theta_{i-2}$.

A problem arising with symbol-by-symbol double differential detection is that in the presence of noise it causes a high noise correlation in the receiver output [13], which results in a considerable performance degradation (on the order of 3–4 dB) with respect to conventional differential detection. A significant portion of the performance degradation (about a couple of dB) can be bought back [13] by modifying the encoding scheme so that the transmitted phases are associated with the modified differences

$$\theta_i - \theta_{i-1} - \theta_{i-2} + \theta_{i-3}.$$

Simon and Divsalar [13] describe a block detection technique for double-differentially encoded PSK (DDPSK) which is actually based on that for DPSK. The first stage in the demodulator is the same used for the symbol-by-symbol detection; it evaluates the difference between the current and the previously received phases and converts the frequency offset into a phase offset; the second stage is a block detection scheme designed for unknown phase offset (DPSK). The authors realize that the demodulator is suboptimum.

Another suboptimum block demodulator, derived in the following, gives a better performance with respect to the one proposed in [13].

The transmitted signal complex envelope is

$$x(t) = x_k = Ae^{j\phi_k} \quad t \in [kT, (k+1)T)$$

and the corresponding received sample is assumed to be

$$r_k = x_k e^{j(\theta_k + \omega_k' kT)} + n_k = s_k + n_k$$

where θ_k and ω'_k are unknown phase and radian frequency offsets introduced by the channel. We assume, as in [13], that the phase and frequency shifts are constant over a period of N consecutive symbols and that $\theta_k = \theta$ and $\alpha = \omega'_k T$ are independent random variables uniformly distributed between $-\pi$ and π.

Let $\mathbf{r}_k$ represent the vector of the N consecutive samples $r_{k-(N-1)}, \ldots, r_k$, and $\mathbf{s}_k$ the vector of the received samples in the absence of noise. Then the probability density function of $\mathbf{r}_k$ given $\mathbf{s}_k$ is

$$p(\mathbf{r}_k|\mathbf{s}_k) = \frac{1}{(2\pi\sigma^2)^N} \exp\left\{ -\frac{\|\mathbf{r}_k - \mathbf{s}_k\|^2}{2\sigma^2} \right\}$$

The argument of the exponential function may be rewritten as

$$\|\mathbf{r}_k - \mathbf{s}_k\|^2 = \|\mathbf{r}_k\|^2 + \|\mathbf{s}_k\|^2 - $$
$$- 2A\Re\left[\sum_{i=0}^{N-1} r_{k-i} e^{-j[\phi_{k-i}+\theta+(k-i)\alpha]} \right]$$

so that the probability density function of $\mathbf{r}_k$ given the knowledge of the sequence of absolute phases ϕ_k, θ and α becomes

$$p(\mathbf{r}_k|\phi_k, \theta, \alpha) = \frac{1}{(2\pi\sigma^2)^N} \exp\left\{ -\frac{\|\mathbf{r}_k\|^2 + \|\mathbf{s}_k\|^2}{2\sigma^2} \right\} \times$$
$$\times \exp\left\{ \frac{A}{\sigma^2} \Re \sum_{i=0}^{N-1} r_{k-i} e^{-j[\phi_{k-i}+\theta+(k-i)\alpha]} \right\}$$
$$= C \exp\left\{ \frac{A}{\sigma^2} \Re \sum_{i=0}^{N-1} r_{k-i} e^{-j[\phi_{k-i}+\theta+(k-i)\alpha]} \right\}$$

where C is a constant, irrelevant here. By averaging out θ and α, we obtain the probability density function of $\mathbf{r}_k$ given only the knowledge of ϕ_k:

$$p(\mathbf{r}_k|\phi_k) = \int\int p(\mathbf{r}_k|\phi_k, \theta, \alpha)p(\theta)p(\alpha)\, d\theta\, d\alpha$$
$$= \frac{1}{4\pi^2} \int_{-\pi}^{\pi} \int_{-\pi}^{\pi} p(\mathbf{r}_k|\phi_k, \theta, \alpha)\, d\theta\, d\alpha.$$

The optimum demodulator chooses the vector $\widehat{\phi}_k$ which maximizes the conditional density function $p(\mathbf{r}_k|\phi_k)$, i.e.

$$\widehat{\phi}_k = \max_{\phi_k} {}^{-1} p(\mathbf{r}_k|\phi_k).$$

Since the double integral cannot be easily computed, it is approximated by using a truncated power series for the exponential function inside the integrand $p(\mathbf{r}_k|\phi_k, \theta, \alpha)$.

$$p(\mathbf{r}_k|\phi_k) \simeq \frac{C}{4\pi^2} \sum_{m=0}^{L} \frac{1}{m!} \int_{-\pi}^{\pi} \int_{-\pi}^{\pi} \left(\frac{A}{\sigma^2}\right)^m \times$$

$$\times \quad \left\{\Re \sum_{i=0}^{N-1} r_{k-i} e^{-j[\phi_{k-i}+\theta+(k-i)\alpha]}\right\}^{m} d\alpha \, d\theta$$

$$= \quad \frac{C}{4\pi^2} \sum_{m=0}^{L} \frac{1}{m!} I_m(\phi_k)$$

It can be easily shown that

$$I_0(\phi_k) = 4\pi^2$$
$$I_1(\phi_k) = I_3(\phi_k) = \cdots = I_{2n+1}(\phi_k) = 0$$
$$I_2(\phi_k) = \left(\frac{A}{2\sigma^2}\right)^2 8\pi^2 \sum_{i=0}^{N-1} |r_{k-i}|^2$$

all independent of ϕ_k and thus useless for taking a decision. The first significant term is $I_4(\phi_k)$, which, after some manipulations, may be written as

$$I_4(\phi_k) = \left(\frac{A}{2\sigma^2}\right)^4 12\pi\Re\left\{\sum_{i1=0}^{N-1}\sum_{i2=0}^{N-1}\sum_{i3=0}^{N-1}\sum_{i4=0}^{N-1} r_{k-i1} r_{k-i2} r^*_{k-i3} r^*_{k-i4}\times\right.$$
$$\left. \times \quad e^{-j[\phi_{k-i1}+\phi_{k-i2}-\phi_{k-i3}-\phi_{k-i4}]} \int_{-\pi}^{\pi} e^{j[i1+i2-i3-i4]\alpha} d\alpha\right\}$$

The integral appearing in I_4 is null if $i1 + i2 - i3 - i4 \neq 0$, otherwise it is equal to 2π. Moreover, if $i1 + i2 = i3 + i4$ and $i1 = i3$ or $i1 = i4$, the integral is nonzero, but the resulting term does not give any information on ϕ_k, since $[\phi_{k-i1} + \phi_{k-i2} - \phi_{k-i3} - \phi_{k-i4}] = 0$. Therefore the significant part of I_4 may be written as

$$I_4(\phi_k) = \left(\frac{A}{2\sigma^2}\right)^4 24\pi^2\Re \sum_{i1=0}^{N-1}\sum_{i2=0}^{N-1}\sum_{i3=0,i3\neq i1}^{N-1}\sum_{i4=0,i4\neq i1}^{N-1} r_{k-i1} r_{k-i2} r^*_{k-i3} r^*_{k-i4} \times$$
$$\times \quad e^{-j[\phi_{k-i1}+\phi_{k-i2}-\phi_{k-i3}-\phi_{k-i4}]} \delta(i1 + i2 - i3 - i4)$$

where $\delta(\cdot)$ denotes the Kronecker function which is equal to 1 if its argument is zero, otherwise is equal to 0.

In a similar way it can be shown that the next term in the summation, I_6, may be expressed as

$$I_6(\phi_k) = \left(\frac{A}{2\sigma^2}\right)^6 40\pi\Re\left\{\sum_{i1=0}^{N-1}\sum_{i2=0}^{N-1}\sum_{i3=0}^{N-1}\sum_{i4=0}^{N-1}\sum_{i5=0}^{N-1}\sum_{i6=0}^{N-1}\right.$$
$$r_{k-i1} r_{k-i2} r_{k-i3} r^*_{k-i4} r^*_{k-i5} r^*_{k-i6} \times$$
$$\times \quad e^{-j[\phi_{k-i1}+\phi_{k-i2}+\phi_{k-i3}-\phi_{k-i4}-\phi_{k-i5}-\phi_{k-i6}]} \times$$
$$\left. \times \quad \int_{-\pi}^{\pi} e^{j[i1+i2+i3-i4-i5-i6]\alpha} d\alpha\right\}$$

Notice that I_6 depends on $\left(\frac{A}{2\sigma^2}\right)^6$, while I_4 depends on $\left(\frac{A}{2\sigma^2}\right)^4$. Therefore, to include also I_6 in the approximate expression of $p(\mathbf{r}_k|\phi_k)$, we need the value of the signal-to-noise ratio A/σ^2.

For each value of N the term I_4 takes on a particular simplified expression, which is evaluated in the following for $N = 3$ and $N = 4$. In the case $N = 3$, the only useful terms in the summation are those for which $i1 + i2 = i3 + i4 = 2$, i.e.

$i1$	$i2$	$i3$	$i4$
0	2	1	1
2	0	1	1
1	1	0	2
1	1	2	0

and they all give the same contribution to I_4, which becomes

$$
\begin{aligned}
I_4(\phi_k) &= 4 \times \left(\frac{A}{2\sigma^2}\right)^4 24\pi^2 \Re\left\{ r_k r_{k-2} r_{k-1}^* r_{k-1}^* e^{-j[\phi_k + \phi_{k-2} - \phi_{k-1} - \phi_{k-1}]} \right\} \\
&\propto \Re\left\{ r_k r_{k-1}^* r_{k-1}^* r_{k-2} e^{-j[\Delta\phi_k - \Delta\phi_{k-1}]} \right\} \\
&= \Re\left\{ r_k r_{k-1}^* r_{k-1}^* r_{k-2} e^{-j\Delta^2\phi_k} \right\}.
\end{aligned}
$$

This corresponds to the standard detection rule for DDPSK. Notice that in this case exactly the same contributions arise from I_6 and all the following terms in the power series expansion, and that the decision rule is actually optimum.

In the case $N = 4$ the useful terms correspond to the indices listed in the following table:

$i1$	$i2$	$i3$	$i4$
0	2	1	1
2	0	1	1
1	1	0	2
1	1	2	0
1	3	2	2
3	1	2	2
2	2	1	3
2	2	3	1
0	3	1	2
0	3	2	1
3	0	1	2
3	0	2	1
1	2	0	3
2	1	0	3
1	2	3	0
2	1	3	0

and we get

$$
\begin{aligned}
I_4(\phi_k) \;\propto\; &\; \Re\left\{ r_k r_{k-2} r_{k-1}^* r_{k-1}^* e^{-j[\Delta\phi_k - \Delta\phi_{k-1}]} +\right. \\
&+ \; r_{k-1} r_{k-3} r_{k-2}^* r_{k-2}^* e^{-j[\Delta\phi_{k-1} - \Delta\phi_{k-2}]} + \\
&+ \; \left. 2 r_k r_{k-3} r_{k-1}^* r_{k-2}^* e^{-j[\Delta\phi_k - \Delta\phi_{k-2}]} \right\} \\
=\; &\; \Re\left\{ r_k r_{k-1}^* r_{k-1}^* r_{k-2} e^{-j\Delta^2\phi_k} + \right. \\
&+ \; r_{k-1} r_{k-2}^* r_{k-2}^* r_{k-3} e^{-j\Delta^2\phi_{k-1}} + \\
&+ \; \left. 2 r_k r_{k-1}^* r_{k-2}^* r_{k-3} e^{-j[\Delta^2\phi_k + \Delta^2\phi_{k-1}]} \right\}.
\end{aligned}
$$

In this case the decision rule is suboptimum, since I_6 generates other combinations of received samples. Notice that the decision rule may again be expressed in terms of the double phase difference $\Delta^2\phi_k$, as for the case $N = 3$.

Simulation results obtained over an AWGN channel (with $\theta = \alpha = 0$) for the case of a DD-BPSK modulation are shown in Fig. 1.a, for values of N ranging from 3 (standard demodulation) to 10; the results obtained with a standard demodulator for a D-BPSK system are shown for comparison. Notice that the block demodulation of DDPSK signals dramatically decreases the bit error rate, and the case $N = 10$ gives results that are only 0.4 dB worse than coherent BPSK at a bit error rate equal to 10^{-5}.

The comparison with the two-stage detection schemes [13] over an AWGN channel is given in Fig. 1.b. These results show that the two-stage demodulator with block length $N = 10$ is outperformed by the block detection scheme with the same observation interval. Then it may be noted that the two-stage block detection scheme gives very similar results for the standard DDPSK modulator with delay combination T, T and the modified one with delay combination $T, 2T$.

Simulations have been carried out also for the cases of Rayleigh and Rician channels: here θ is uniformly distributed, but α is not, and both vary during the N-symbol observation interval. The autocorrelation function of the fading process was assumed to be

$$
R(\tau) = J_0(2\pi f_d \tau),
$$

which is typical for radio mobile systems in urban environments [6]. Fig. 2 shows the results obtained with a DD-BPSK modulation scheme in the case of a Rician channel with normalized fading rate $f_d T$ equal to 0.05, direct-to-faded path power ratio K ranging from 10 to 30 dB, and block decoding with $N = 10$. The bit error rate of the same system over an AWGN channel is shown as reference.

Fig. 3 shows the results obtained with a Rayleigh channel with normalized fading rate $f_d T$ equal to 0.01 for various values of N. As it may be seen, the block demodulator ($N > 3$) works better than the symbol-by-symbol demodulator ($N = 3$) only when the signal to noise rate E_b/N_0 is low, while the asymptotic bit error rate grows if N is increased. Actually, for low values of E_b/N_0 the channel is approximately AWGN (additive noise is the most significant impairment), and the behavior is thus similar to that shown in Fig. 1. On the contrary, when E_b/N_0 is high, the main cause of errors is fading, but the assumption of a constant frequency

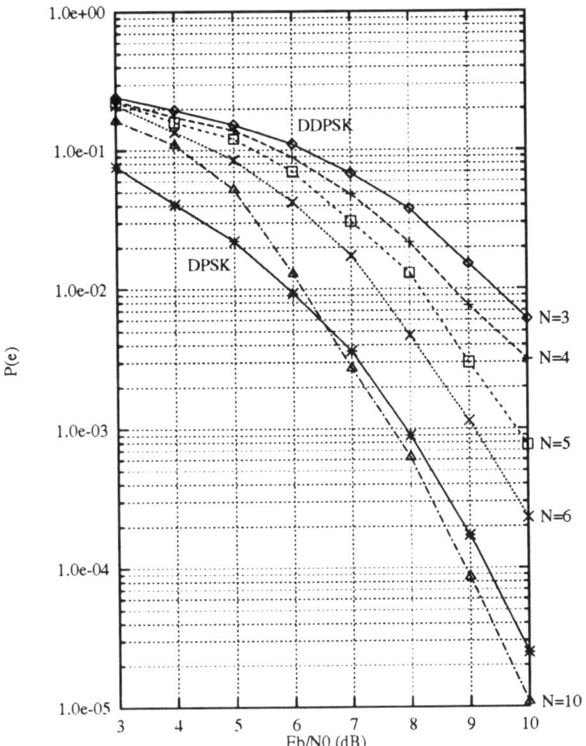

Fig. 1.a Simulation results obtained over an AWGN channel. Solid lines correspond to standard symbol-by-symbol decision for D-BPSK ($N = 2$) and DD-BPSK, dotted lines correspond to block detection with DD-BPSK for different values of the observation length N.

offset is not valid: the variations of the instantaneous frequency with respect to an average constant value inside the observed block become more and more relevant as N grows, and consequently the error probability increases. The same phenomenon may be observed for block detection of DPSK signals. Other relevant results are that, over a Rayleigh channel, DPSK performs always better than DDPSK for any value of the observation length, that the block detection scheme described here is not able to suppress the error floor, and that the two-stage detector [13] presents a lower BER at high E_b/N_0 for the same observation length.

IV. Block detection in an interference environment

Let us consider the following interference model. The signal vector observed at the output of the channel is

$$\mathbf{r} = \mathbf{x} + \mathbf{i} + \mathbf{n},$$

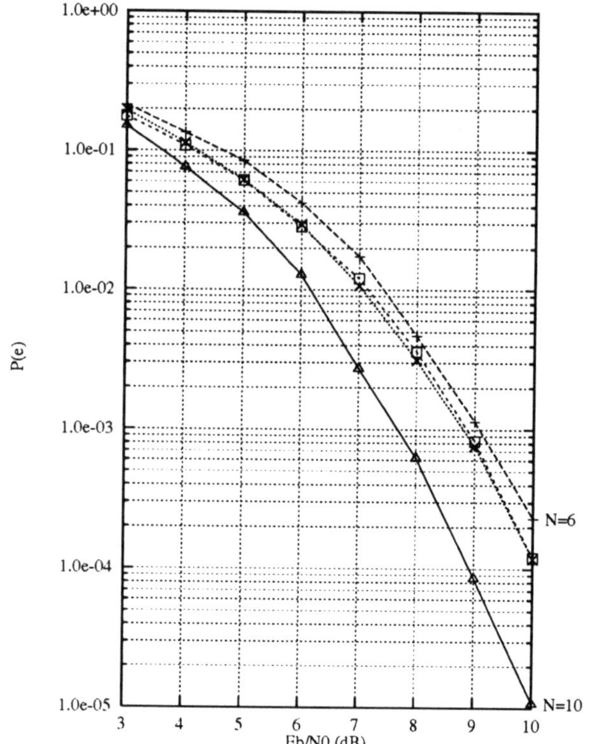

Fig. 1.b Simulation results for a DD-BPSK system obtained over an AWGN channel. The upper and lower curves correspond to the block detection scheme for $N = 6$ and $N = 10$ respectively. The curves in between correspond to the two-stage block demodulator proposed in [13] with delay combination T, T and $N = 10$ ($\times$), and with delay combination $T, 2T$ and $N = 11$ ($\square$).

where the noise vector **n** is an N-tuple of Gaussian independent random variables with mean zero and variance $N_0/2$, while **i** represents a disturbance independent of the signal and the noise.

We further assume that N-dimensional interleaving takes place in the transmission, i.e., that adjacent blocks are independent. Under this assumption, the N-dimensional channel is memoryless. The receiver uses a Gaussian metric, i.e., it chooses $\hat{\mathbf{x}}$ if

$$\|\mathbf{r} - \hat{\mathbf{x}}\| \leq \|\mathbf{r} - \mathbf{x}\|$$

for all **x**.

Some insight can be gained by calculating the pairwise error probability $P[\mathbf{x} \rightarrow \hat{\mathbf{x}}]$. We have

$$P[\mathbf{x} \rightarrow \hat{\mathbf{x}}] \quad = \quad P[\|\mathbf{r} - \hat{\mathbf{x}}\| \leq \|\mathbf{r} - \mathbf{x}\| \mid \mathbf{x}]$$

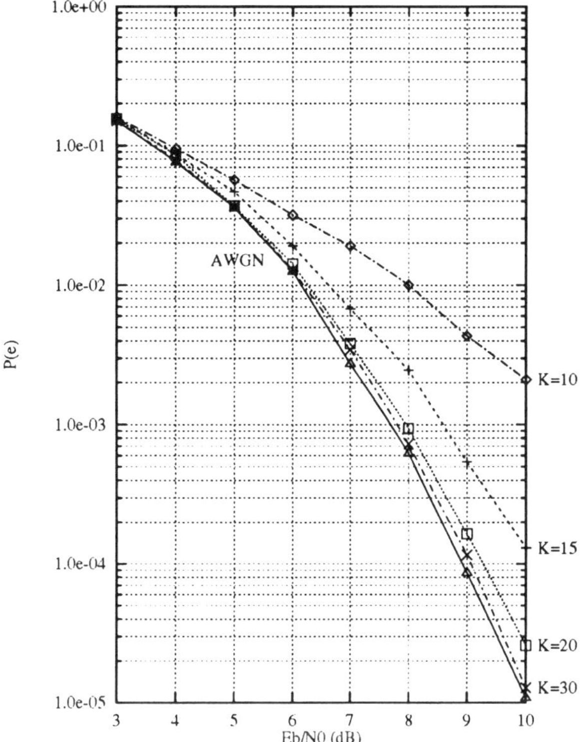

Fig. 2. Simulation results for a DD-BPSK system obtained over a Rician channel. The solid line corresponds to an AWGN channel, while the dotted ones correspond to a Rician channel with normalized fading rate $f_d T = 0.05$. K indicates the power ratio between the direct and the faded path expressed in dB. The decisions are taken over blocks of $N = 10$ symbols.

$$= \mathbf{E}\left\{Q\left(\frac{\|\widehat{\mathbf{x}} - \mathbf{x}\| - 2(\mathbf{i}, (\widehat{\mathbf{x}} - \mathbf{x})/\|\widehat{\mathbf{x}} - \mathbf{x}\|)}{\sqrt{2N_0}}\right)\right\}$$

where the expectation is taken with respect to the random vector $\mathbf{i}$.

It is seen that with respect to the case of no interference, in which the pairwise error probability depends only on the distance $\|\widehat{\mathbf{x}} - \mathbf{x}\|$, here this distance is reduced by twice the projection of the interference over a unit-norm vector parallel to $\widehat{\mathbf{x}} - \mathbf{x}$. If $\mathbf{i}$ affects only a few dimensions, increasing the dimensionality of $\mathbf{x}$, i.e., the observation length, will cause a larger number of dimensions to be unaffected by the interference, thus improving the detector performance. Some quantitative considerations can be derived by examining the error exponent of the channel created by block detection.

A. Computing the error exponent

Consider then a K-tuple $(\mathbf{x}_1, \cdots, \mathbf{x}_K)$ of transmitted signal vectors. The pairwise error probability, i.e., the probability that the likelihood of $(\widehat{\mathbf{x}}_1, \cdots, \widehat{\mathbf{x}}_K)$ be

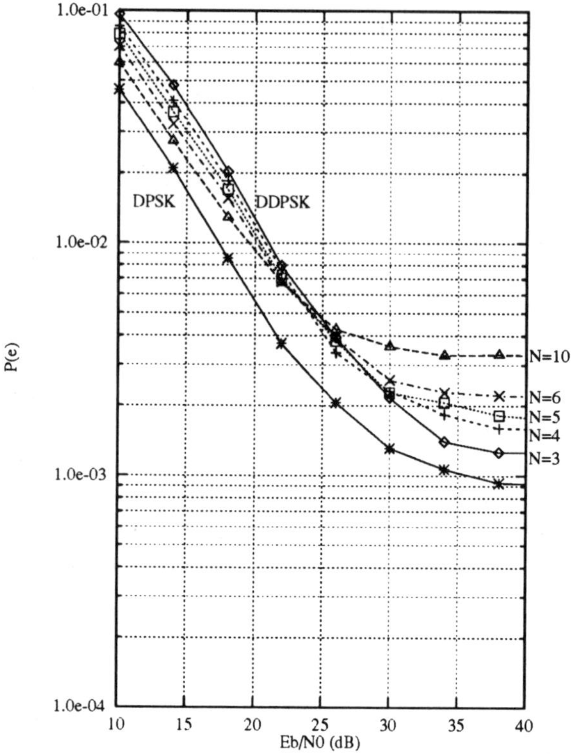

Fig. 3. Simulation results for a DD-BPSK system obtained over a Rayleigh channel with normalized fading rate $f_dT = 0.01$. The continuous line corresponds to standard demodulation of D-BPSK ($N = 2$), while the dotted lines correspond to block detection of DD-BPSK for different values of the observation length N.

greater than that of the transmitted K-tuple can be bounded from above by using the Chernoff bound, which yields

$$P[(\mathbf{x}_1, \cdots, \mathbf{x}_K) \rightarrow (\widehat{\mathbf{x}}_1, \cdots, \widehat{\mathbf{x}}_K)] \leq$$
$$\leq \prod_{k=1}^{K} \mathbf{E}\left\{\exp\left[-\lambda[\|\widehat{\mathbf{x}}_k - \mathbf{x}_k\|^2 - 2(\mathbf{i}_k, \widehat{\mathbf{x}}_k - \mathbf{x}_k) - 2(\mathbf{n}_k, \widehat{\mathbf{x}}_k - \mathbf{x}_k)]\right]\right\}$$

where $\mathbf{E}$ is to be taken with respect to $\mathbf{n}$ and to $\mathbf{i}$, and $\lambda \geq 0$, the Chernoff parameter, should be optimized. Since the optimum value of λ here would depend on k, we rather choose $\lambda = 1/2N_0$, which provides the upper bound:

$$P[(\mathbf{x}_1, \cdots, \mathbf{x}_k) \rightarrow (\widehat{\mathbf{x}}_1, \cdots, \widehat{\mathbf{x}}_k)] \leq \prod_{k=1}^{K} \mathbf{E}\left\{\exp\left[-\frac{1}{4N_0}[\|\widehat{\mathbf{x}}_k - \mathbf{x}_k\|^2 - 4(\mathbf{i}_k, \widehat{\mathbf{x}}_k - \mathbf{x}_k)]\right]\right\},$$

where the expectation is now with respect to $\mathbf{i}$ alone.

Next, apply a random coding argument. With the assumption that the code alphabet consists of M distinct signals, randomly select the $2K$ vectors $\mathbf{x}_k$ and $\hat{\mathbf{x}}_k$, where each vector is independently selected with uniform probability. The average pairwise error bound is then

$$\overline{P[(\mathbf{x}_1, \cdots, \mathbf{x}_k) \rightarrow (\hat{\mathbf{x}}_1, \cdots, \hat{\mathbf{x}}_k)]} \le 2^{-\rho_0 K}$$

where we have defined

$$\rho_0 = -\log_2\left\{ \frac{1}{M^2} \sum_{\mathbf{x}} \sum_{\hat{\mathbf{x}}} \mathbf{E} \exp\left[-\frac{1}{4N_0}[\|\hat{\mathbf{x}} - \mathbf{x}\|^2 - 4(\mathbf{i}, \hat{\mathbf{x}} - \mathbf{x})] \right] \right\}.$$

B. Intersymbol interference

A case in which the model described above can be applied is offered by intersymbol interference channel. Let $\{u_k\}$ denote a sequence of independent, identically distributed data symbols, $\{h_k\}_{k=0}^{L}$ the impulse response of the discrete channel with memory L, and $\{x_k\}$ the sequence of samples received at the output of the discrete channel before the addition of the noise. We have

$$x_k = \sum_{i=0}^{L} h_i u_{k-i}.$$

Take now blocks of observed data with length N (here we assume $N > L$ for simplicity). Then we may write the received signal blocks in the form

$$\mathbf{r} = \mathbf{x} + \mathbf{i} + \mathbf{n}$$

where

$$\mathbf{x} = f(\mathbf{u}),$$

with $f(\cdot)$ a deterministic function that we assume to be invertible, and $\mathbf{i}$ is independent of $\mathbf{x}$. Specifically, $\mathbf{x}$ depends only on the data symbols in vector $\mathbf{u}$, and constitutes a *distorted* version of $\mathbf{u}$. If the channel has been identified exactly, then $\mathbf{u}$ can be recovered exactly from the knowledge of $\mathbf{x}$. The vector $\mathbf{i}$ represents the *dispersion* term: it depends on data that are outside the observed block.

For illustration's sake, consider a simple example obtained by taking $L = 2$ and $N = 5$. Under our assumptions we may write

$$\mathbf{x} = \begin{bmatrix} h_0 & 0 & 0 & 0 & 0 \\ h_1 & h_0 & 0 & 0 & 0 \\ h_2 & h_1 & h_0 & 0 & 0 \\ 0 & h_2 & h_1 & h_0 & 0 \\ 0 & 0 & h_2 & h_1 & h_0 \end{bmatrix} \mathbf{u},$$

where

$$\mathbf{x} = [x_k, x_{k+1}, x_{k+2}, x_{k+3}, x_{k+4}]'$$

and

$$\mathbf{u} = [u_k, u_{k+1}, u_{k+2}, u_{k+3}, u_{k+4}]'.$$

Moreover,

$$\mathbf{i} = \begin{bmatrix} h_1 u_{k-1} + h_2 u_{k-2} \\ h_2 u_{k-1} \\ 0 \\ 0 \\ 0 \end{bmatrix}$$

Notice that for optimum block-detection the dispersion term $\mathbf{i}$ should be averaged out to provide the metric. However, since this would imply a considerable complexity in metrics calculation, we use instead the simple "Gaussian" metric $\|\mathbf{r} - \mathbf{x}\|$. Some quantitative considerations on the behavior of block detection in this situation can be obtained by examining the simple case of $L = 2$ above, along with 4-PSK, various values of N, and a channel with impulse response $h_0 = 1$, $h_1 = 2\epsilon$, and $h_2 = \epsilon$. Here the first two dimensions of the received signal are affected by dispersion due to ISI, while the remaining dimensions are only subject to distortion, which can be compensated for under the assumption of a known channel impulse response. In other words, ISI acts *randomly* on the first two component of the signals, and *deterministically* on the remaining components.

With $\epsilon = 0$ (no intersymbol interference) a value $\rho_0 = 1.9$ bits per symbol requires a ratio energy per bit over noise power spectral density $E_b/N_0 = 5.1$ dB. With $\epsilon = 0.1$ the increment in signal to noise ratio needed to achieve the same value of ρ_0 is 2 dB with $N = 1$, (symbol-by-symbol detection), 1.3 dB with $N = 2$, 0.9 dB with $N = 3$, 0.7 dB with $N = 4$, and only 0.4 dB with $N = 5$. Thus, block detection with $N = 5$ improves the detector performance by about 1.6 dB versus conventional demodulation.

V. Conclusions

We have examined block detection, as a suboptimum receiver scheme that improves upon symbol-by-symbol detection while maintaining a limited complexity. A number of channels has been considered, namely, those generated by differential and double-differential encoding with and without fading, and intersymbol interference channels. The results available from the literature and those presented here show the attractiveness of this detection scheme as well as its intrinsic limitations.

References

[1] A. Abrardo, G. Benelli, and G. Cau, "Multiple-symbols differential detection of GMSK," *Electronics Letters*, Vol. 29, No. 25, pp. 2167–2168, 9th December 1993.

[2] E. Biglieri, D. Divsalar, P. J. McLane, and M. K. Simon, *Introduction to Trellis-Coded Modulation with Applications*. New York: Macmillan, 1991.

[3] D. Divsalar and M. K. Simon, "Multiple-symbol differential detection of MPSK," *IEEE Trans. Commun.*, Vol. 38, No. 3, pp. 300–308, March 1990.

[4] D. Divsalar, M. K. Simon, and M. Shahshahani, "The performance of trellis-coded MDPSK with multiple symbol detection," *IEEE Trans. Commun.*, Vol. 38, No. 9, pp. 1391–1403, September 1990.

[5] D. Divsalar and M. K. Simon, "Maximum-likelihood differential detection of uncoded and trellis coded amplitude-phase modulation over AWGN and fading channels – Metrics and performance," *IEEE Trans. Commun.*, Vol. 42, No. 1, January 1994.

[6] P. Ho and D. Fung, "Error performance of multiple-symbol differential detection of PSK signals transmitted over correlated Rayleigh fading channel," *IEEE Trans. Commun.*, Vol. 40, No. 10, pp. 1566–1569, October 1992.

[7] P. Y. Kam, "Maximum-likelihood digital sequence estimation over the Gaussian channel with unknown carrier phase" *IEEE Trans. Commun.*, Vol. 35, No. 7, pp. 764-7, July 1987.

[8] H. Leib and S. Pasupathy, "The phase of a vector perturbed by Gaussian noise and differentially coherent receivers," *IEEE Trans. Inform. Theory*, Vol. 34, No. 6, pp. 1491–1501, November 1988.

[9] H. Leib and S. Pasupathy, "Optimal noncoherent block demodulation of Differential Phase Shift Keying (DPSK)", *AEÜ*, Vol. 45, No. 5, pp. 299-305, 1991.

[10] H. Leib and S. Pasupathy, "Block-coded noncoherent MSK", *Proc. Intern. Conf. on Commun., ICC'90*, pp. 448-452, Apr. 1990.

[11] W. Liu and P. Ho, "On multiple-symbol differential detection of PSK signals in frequency selective fading channels," *Unpublished manuscript*, 1993.

[12] K. Mackenthun, "A fast algorithm for maximum likelihood detection of QPSK or $\pi/4$-QPSK sequences with unknown phase", *Proc. of the Third IEEE Intern. Symp. on Personal, Indoor and Mobile Radio Commun.*, Boston, MA, pp. 240-4, Oct. 1992.

[13] M. K. Simon, and D. Divsalar, "On the implementation and performance of single and double differential detection schemes", *IEEE Trans. Commun.*, Vol. 40, No. 2, pp. 278–291, Febr. 1992

[14] M. K. Simon and D. Divsalar, "Maximum-likelihood block detection of noncoherent continuous phase modulation," *IEEE Trans. Commun.*, Vol. 41, No. 1, pp. 90–98, January 1993.

[15] S. G. Wilson, J. Freebersyser, and C. Marshall, "Multisymbol detection of M-DPSK", *Proc. Globecom '89*, Dallas, TX, pp. 1692-7, Nov. 1989.

Chapter 3

Interference Suppression

ADAPTIVE MULTIUSER DETECTION

Sergio Verdú

1. Introduction

Spurred by its applications in Code Division Multiple Access, multiuser detection has grown from its origins more than ten years ago to a vibrant research and development activity in industry and academia. As the needs to increase capacity in multiuser radio channels become more pressing, it is safe to expect that the interest in the subject will grow in the near future.

The development of multiuser detection has proceeded along a path which is typical of other areas in communications. Initially, optimum solutions were obtained along with the best possible performance achievable in Gaussian noise channels [1]. Those results showed a huge gap between the optimum performance and the performance of the conventional single-user detector (which neglects the presence of multiaccess interference). In particular, they showed that the near-far problem is not a flaw of CDMA, as widely believed, but of the inability of the conventional receiver to exploit the structure of the multiaccess interference. This feature of multiuser detection sidesteps the need for sophisticated high-precision power control in mobile communication systems. Thus, an increase in the complexity of the base station enables a considerable reduction in the complexity of the mobile transmitters. Equally important to the near-far resistant property of optimum multiuser detection, is the performance gain that it promises even in situations of exact power control (equal-power reception). This performance gain results in lower power consumption and processing gain requirements, which translate into increased battery lifes and lower bandwidth in order to support the same information rates.

The second stage in the development of multiuser detection was devoted to the analysis and design of multiuser detectors that could achieve significant performance gains over the conventional receiver without incurring in the exponential (in the number of users) complexity of the optimum multiuser detector. Notable among those efforts were the decorrelating detector of Lupas and Verdú [2], [3]; the multistage detector of Varanasi and Aazhang [4], [5]; the decision-feedback multiuser detectors of Duel-Hallen [6], [7]; and the suboptimum detectors of Xie, Rushforth and Short [8], [9].

Motivated by the channel environments encountered in many CDMA applications, the design of multiuser detectors for channels with fading, multipath, or noncoherent modulation has attracted considerable attention, as exemplified by the works of Varanasi [10]; Vasudevan and Varanasi [11], [12]; Zvonar and Brady [13], [14], [15]; Fawer and Aazhang [16].

97

S.G. Glisic and P.A. Leppänen (eds.), Code Division Multiple Access Communications, 97-116.
© 1995 *Kluwer Academic Publishers. Printed in the Netherlands.*

98

The foregoing multiuser detectors depend on various parameters such as received amplitudes and crosscorrelations which are usually not fixed and known beforehand. Therefore, another important thrust in research in multiuser detection is the design of *adaptive* detectors, which self-tune the detector parameters from the observation of the received waveform. The very recent, and already considerable, literature on this subject is surveyed in the present tutorial paper.

Sections 2 and 3 contain background material used throughout the paper on the multiaccess channel model, optimum multiuser detection and the decorrelating detector. A comprehensive tutorial exposition of these and other topics can be found in [17]. Section 4 deals with the MMSE linear multiuser detector and its adaptive implementations. Section 5 gives an overview of adaptive tentative-decision based detectors such as those that use successive cancellation and decision-feedback. Section 6 deals with blind multiuser detection, and in particular, it presents a multiuser detector which is optimally near-far resistant and requires no more knowledge than the conventional single-user detector. Section 7 is devoted to multiuser detection using learning neural-networks

2.   Optimum Detection

The asynchronous CDMA white Gaussian noise channel considered in this paper is

$$y(t) = \sum_{k=1}^{K} \sum_{i=-M}^{i=M} A_k b_k[i] s_k(t - iT - \tau_k) + \sigma n(t) \tag{1}$$

where

- K is the number of users.

- A_k is the received amplitude of the kth user.

- $b_k[i] \in \{-1,+1\}$ is the data stream modulated by the kth user.

- s_k is the unit-energy signature waveform of the kth user.

- T is the inverse of the data rate (duration of s_k).

- $\tau_k \in [0, T)$ is the kth user's offset.

- $n(t)$ is normalized white Gaussian noise.

- σ^2 is the background noise power spectral density.

The model in (1) can be generalized to take into account a number of features that are relevant in practice such as quadrature and nonbinary modulation, signature waveforms spanning more than one bit epoch, intersymbol interference, etc. For conceptual clarity it is best not to generalize the model in those directions; instead, it is usually conceptually advantageous to consider the special case of (1) where the users are symbol-synchronous:

$$y(t) = \sum_{k=1}^{K} A_k b_k s_k(t) + \sigma n(t) \quad t \in [0, T] \tag{2}$$

In most cases, the analysis of multiuser detectors for the synchronous channel (2) contains all the key ingredients necessary for the analysis of the more general channel (1).

In multiuser detection, it is frequently useful to examine performance in the situation of low background noise, $\sigma \to 0$. To that end, the *asymptotic multiuser efficiency* of the kth user (whose bit error rate is denoted by $P_k(\sigma)$ is defined as

$$\eta_k = \lim_{\sigma \to 0} \frac{\sigma^2}{A_1^2} \log 1/P_k(\sigma) \tag{3}$$

which is simply the degradation (measured in SNR) suffered by a user due to the presence of other users in the channel. The worstcase asymptotic multiuser efficiency over all received interfering amplitudes is called the *near-far resistance,* denoted by $\overline{\eta}_k$. For most multiuser detectors, near-far resistance is not an overly conservative performance measure because the worst-case usually does not occur at large interfering ratios. Therefore, it is an attractive performance measure even for receivers that employ some power control.

The optimum receiver for (2) processes the received waveform with a bank of matched filters, which produce a vector of observables:

$$\mathbf{y} = \mathbf{RAb} + \mathbf{n} \tag{4}$$

where $\mathbf{A} = diag\{A_1, \cdots A_K\}$, $\mathbf{b} = [b_1 \cdots B_K]^T$, $\mathbf{n}$ is a zero mean Gaussian vector with covariance matrix $\mathbf{R}$ and $\mathbf{R}$ is the crosscorrelation matrix whose ij coefficient is

$$\rho_{ij} = \int_0^T s_i(t) s_j dt \tag{5}$$

The optimum detector that selects the most likely data vector based on the observation of y must solve an NP-complete combinatorial optimization algorithm [18]. Thus, no algorithm polynomial in K is known for optimal multiuser detection. In the asynchronous case, the receiver consists of a matched filter front-end followed by a Viterbi algorithm [1]. The number of states of the Viterbi algorithm is exponential in K with metrics that are very simple to compute in terms of the matched filter outputs and crosscorrelations. The optimum kth user near-far resistance [19], [3] is equal to the minimum energy of any multiuser signal modulated by $\{-1,0.+1\}$, with fixed $A_1 b_1[0] = 1$. In terms of the crosscorrelation matrix, the near-far resistance of the kth user is equal to the reciprocal of the kk element of the inverse of the crosscorrelation matrix:

$$\overline{\eta}_k = \frac{1}{R_{kk}^{-1}} \tag{6}$$

In practical terms, this means that there is a huge performance gap between the conventional single-user matched filter and the optimum achievable performance. For example, while the near-far resistance of the conventional receiver is zero, the expected optimum near-far resistance using direct-sequence spread-spectrum signature waveforms with N chips per symbol is lower bounded by [20]

$$E\left[\overline{\eta}_1\right] \geq 1 - \frac{K-1}{N} \tag{7}$$

These notable performance gains are obtained at the expense of:

- the signature waveforms of all users must be known.

- the received amplitudes of all users must be known.

- the timing of all users must be acquired.

- exponential complexity in the number of users.

- a centralized structure which demodulates all transmitters.

Remarkably, as we will see in Section 6, recent progress in adaptive multiuser detection has resulted in a receiver which achieves optimally near-far resistant multiuser detection with none of the above shortcomings.

3.   Decorrelating Detector

The decorrelating detector outputs the signs of the matched filter outputs in (4) multiplied by the inverse crosscorrelation matrix R-1, i.e., it takes the sign of the vector

$$\mathbf{R}^{-1}\mathbf{y} = \mathbf{Ab} + \mathbf{R}^{-1}\mathbf{n}.$$

Thus, in the hypothetical absence of background noise, the decorrelating linear transformation recovers the transmitted bits without multiuser interference. In the asynchronous case, the decorrelating detector generalizes to an infinite impulse-response filter [3]. The decorrelating detector is the maximum likelihood solution in the absence of any knowledge about the received amplitudes. A major result of Lupas and Verdu [2], [3] is that the decorrelating detector achieves optimum near-far resistance. The bit error rate of the decorrelating detector is independent of the interfering amplitudes. This is because the decorrelating linear transformation projects the received waveform on a subspace which is orthogonal to the space spanned by the interfering signature waveforms. In comparison with the above requirements of the optimum detector, the decorrelating detector has the following features:

- the signature waveforms of all users must be known.

- the received amplitudes need not be known or estimated.

- the timing of all users must be acquired.

- the matrix inverse $\mathbf{R}^{-1}$ must be computed.

- it lends itself to decentralized implementation, demodulating only the desired user.

The optimal near-far resistance property of the decorrelating detector coupled with the fact that it does not require knowledge of the received amplitudes make the decorrelating detector attractive from the standpoint of implementation. The main disadvantage is the computation required to obtain the decorrelating coefficients from the crosscorrelations. In the case of synchronous direct sequence spread-spectrum, Chen and Roy [21] report a recursive least squares (RLS) computation of the decorrelating detector coefficients which requires knowledge of all signature sequences but sidesteps the need to perform computations with crosscorrelations. In the asynchronous case, the processing window can be truncated to the bit of interest as suggested in [22], [23]; or it can span a truncated sliding window as proposed in [24]. In the latter case, the dynamic updating of the decorrelating detector coefficients in response to variations in the crosscorrelations has been investigated in [25]. Mitra and Poor [26] advocate detecting the presence and identity of a new transmitter by processing the residual signal that results by subtracting from the received signal the multiuser signal modulated by the decorrelating detector decisions.

The optimality of the near-far resistance of the decorrelating detector with DPSK modulation has been established by Varanasi [10]. The decorrelating detector has also been used in the conjunction with DPSK and individual rake matched filters for each user (to combat multipath) in [27].

4. Linear MMSE Multiuser Detection

The decorrelating detector may have worse bit error rate than the conventional detector when all the interferers are very weak [3]. This means that it should indeed be possible to incorporate (exact or approximate) knowledge of the received amplitudes in order to obtain a linear multiuser detector that outperforms the decorrelating detector. Minimum mean-square error (MMSE) linear detection is one approach to this problem. According to this criterion, one chooses the $K \times K$ matrix $\mathbf{M}$ that achieves

$$\min_{\mathbf{M} \in R^{k \times k}} E\left[\|\mathbf{b} - \mathbf{M}\mathbf{y}\|^2\right] \tag{9}$$

where the expectation is with respect to the vector of transmitted bits $\mathbf{b}$ and the noise vector $\mathbf{n}$ which as we saw has zero mean and covariance matrix equal to $\sigma^2 \mathbf{R}$. Without invoking the Gaussian nature of $\mathbf{n}$ it is possible to show that the linear MMSE detector replaces the inverse crosscorrelation matrix $\mathbf{R}^l$ by the matrix

$$\left[\mathbf{R} + \sigma^2 \mathbf{A}^{-2}\right]^{-1}.$$

Thus the linear MMSE has the aforementioned features of the decorrelating detector, except that it requires knowledge of the received amplitudes. If either the background noise level or the kth user received energy dominates, then the MMSE detector approaches the conventional single user matched filter; on the other hand, as the background noise level vanishes $\sigma \to 0$, the MMSE detector approaches the decorrelating detector. Therefore, the asymptotic multiuser efficiency and the near-far resistance of the MMSE detector are the same as those of the decorrelating detector. In particular, it also achieves optimal near-far resistance. The linear MMSE multiuser detector was originally proposed by Xie, Short and Rushforth [9] in the asynchronous case, and much earlier in single-user dual-polarization channels [28], which can be viewed as two-user synchronous channels.

As long as the background noise is weak, there is little point in incurring in the additional complexity over the decorrelating detector required by the need to track received amplitudes. However, the great advantage of the linear MMSE detector is the ease with which it lends itself to adaptive implementation with training sequences.

The contribution of the kth user to the penalty function in (9) is equal to

$$E\left[(b_k - <c, y>)^2\right]. \tag{10}$$

where the linear transformation has been denoted by c. The gradient of the cost function inside the expectation in (10) is equal to

$$2(<c, y> - b_k)y. \tag{11}$$

Because of the convexity of (10) in c, the gradient descent adaptive algorithm is

$$c[j] = c[j-1] - \mu(<c[j-1], y[j]> - b_k[j])y[j] \tag{12}$$

will converge (with infinitesimally small step size ,u) to the argument that minimizes the penalty function in (10). The update of the impulse response in (12) has the following features:

- the data stream (training sequence) of the desired user must be known.

- the received amplitudes need not be known or estimated.

- the signature waveforms of the interferers need not be known.

- the timing of the desired user must be acquired.

- the timing of interfering users need not be acquired.

- knowledge of the signature waveform of the desired user is not necessary, but it facilitates the initialization of the algorithm.

- it can be implemented in an asynchronous channel, with the only requirement that the timing of the desired user be acquired. The longer the allowed impulse response, the better the performance will be, with a judicious truncation achieving almost the same performance as a doubly infinite filter response.

The gradient descent algorithm shown in (12) is the simplest adaptation law that minimizes (10). Other more complex, but faster, algorithms can be used instead, based, for example, on recursive least-squares or in lattice structures (e.g. [29]).

In addition to the aforementioned earlier reference [28], the adaptive linear MMSE detector was proposed by Madhow and Honig [30], Rapajic and Vucetic [31] and Miller [32], [33]. The implementation of (12) is carried out with finite-dimensional vectors whose dimensionality is equal to (or twice) the number of chips per symbol. Several methods have been proposed in order to lower complexity in systems with large processing gains; for example the cyclically shifted filter bank of [30], the replacement of simple tap delays by first-order low-pass filters in [34], and the symmetric

104

dimension reduction scheme in [35]. Lee [36] observes that the RLS algorithm is ill-conditioned in near-far environments with high SNR, and proposes a transformation of the chip matched filter outputs to overcome this problem. Significant speed-up is reported with both the gradient descent and RLS implementations of the MMSE criterion.

Joint adaptive multiuser detection and timing recovery is achieved with an RLS algorithm by Zvonar and Brady [37], [38] and with a steepest descent algorithm by Smith and Miller [39].

An interesting alternative to the minimization of mean-square error has been proposed by Mandayam and Aazhang [40]. It uses a stochastic gradient algorithm to minimize probability of error, which (for a linear detector) can be written as the sum of Q-functions. The gradient of this penalty function admits (via the chain rule) a closed-form expression. For low background noise, and assuming that at each step of the adaptation the detector can be guaranteed to have positive asymptotic efficiency (so that the adaptive law operates in the region where the cost function is convex), this detector should converge to the optimum linear multiuser detector obtained by Lupas and Verdu [2] which makes better use of the amplitudes than the MMSE detector.

5. Tentative-Decision Based Multiuser Detectors

One of the simplest ideas in multiuser detection is that of successive cancellation: detect the data of the strongest user with a conventional detector and then subtract the signal due to that user from the received waveform. The process can then be repeated with the resulting waveform which contains no trace of the signal due to the strongest user assuming no error was made in its demodulation. This technique has the disadvantage that it requires extremely accurate estimation of the received amplitudes, and unless the users can be ordered so that the received amplitudes satisfy

$$A_1 \gg A_2 \gg \cdots A_K$$

its performance is actually worse than that of the decorrelating detector which requires no knowledge of the received amplitudes. A related technique is the multistage detection of Varanasi and Aazhang where the first stage consists of a bank of conventional detectors [4] or a decorrelator [5]; the second stage assumes that the previous decisions are correct and simply cancels the corresponding signals from the received waveform, thereby resulting in a clear single-user channel in the event that previous decisions are indeed correct. The decorrelating decision-feedback detector of Duel-Hallen [41] (and its adaptive version in [21]) incorporates features common to both successive cancellation and multistage detection with a decorrelating front-end. Similarly, it is possible to assume that decisions made about earlier bits in an asynchronous system are correct and therefore they can be cancelled, as in conventional single-user decision-feedback equalization (DFE). The application of this idea to multiuser detection goes back to [42].

This philosophy has been adopted in the synchronous case by   and Falconer [43] and in a multipath QPSK multiaccess channel by Abdulrahman, Falconer and Sheikh [44], [45] which uses a fractionally-spaced DFE detector whose feedforward and feedback coefficients are adapted to minimize mean-square error using training sequences. Another adaptive multiuser detector based on DFE is experimentally demonstrated by Stojanovic and Zvonar for a channel with severe multipath [46]. Kohno, Imai, Hatori and Pasupathy [47] consider a CDMA channel with limited bandwidth for which they design an adaptive MMSE detector that uses decision-feedback to remove intersymbol interference. The first stage in that detector (which uses knowledge of all the signature waveforms) performs preliminary decisions which are then used in the adaptive stage. Rapajic and Vucetic [31], [48] find no improvement over the adaptive MMSE detector by incorporating the possibility of decision feedback. Adaptive versions of the multistage detectors of Varanasi and Aazhang have been proposed by Chen, Siveski and Bar-Ness [49] (in the case of conventional tentative decisions) and [50], [51] (in the case of a decorrelating first stage). In those detectors, the first stage is nonadaptive and requires knowledge of all the signature waveforms. However, the interference canceller is adaptive and does not require knowledge of amplitudes. The adaptation is carried out by gradient descent of the energy of the difference between y and the output of the linear adaptive canceller (or a different penalty function in [52]), and therefore, it does not require training sequences. Another adaptive two-stage multiuser detector based on soft tentative decisions is proposed by Brady and Catipovic [53], which uses knowledge of training sequences and signature waveforms in order to adapt to the channel parameters and refine a coarse initial estimate of timing and phase.

6.   Blind Multiuser Detection

The requirement of training sequences in the multiuser detectors surveyed above is a cumbersome one in multiuser communications. Since transmitters start and finish their transmissions asynchronously, the "birth" (or "death") of an interferer requires the recomputation of the adaptive receiver coefficients. Often, decision-directed operation of the adaptive detector is not robust enough to take care of those sudden changes, and the desired user must be asked to interrupt its data transmission so that a training sequence is transmitted.

In this section we will review a recent adaptive multiuser detector due to Honig, Madhow and Verdu [54] which has the following features:

- it achieves optimal near-far resistance.

106

- (approximate) knowledge of the signature waveform of the desired user is required.

- the timing of the desired user must be acquired.

- the received amplitudes need not be known or estimated. the signature waveforms of the interferers need not be known.

- the timing of interfering users need not be acquired.

- Training sequences are not required for any user.

Therefore, we will see that it is possible to attain the same near-far resistance as the optimum receiver, the same asymptotic efficiency as the decorrelating detector, and the same bit error rate as the linear MMSE detector with no more than the knowledge assumed by the conventional single-user detector. Although the approach of this detector is reminiscent of that of anchored minimum energy blind equalization proposed in [55], the solution of [54] does not have a counterpart in single-user communication, in contrast to the above multiuser detectors. With few changes, it is possible to generalize the design and analysis of the blind multiuser detector below to the asynchronous case.

The blind multiuser detector of [54] adapts a linear transformation of the observations whose impulse response is c_1 (assuming that the desired user is $k = 1$), and outputs the decision

$$\hat{b}_1 = \text{sgn}(<y, c_1>) \tag{13}$$

Any linear multiuser detector can be written in a *canonical orthogonal decomposition:*

$$c_1 = s_1 + x_1 \tag{14a}$$

where

$$<s_1 + x_1> = 0 \tag{14b}$$

The only c_1 that cannot be represented in this form are those for which

$$<c_1, s_1> \neq 0 \tag{15}$$

but the decisions are scale invariant, and if c_1 were orthogonal to s_1, the bit error rate would be 0.5. Thus, the freedom we lose in the decomposition (14), is a freedom we do not need to have.

Let us focus attention on adapting x_1, while preserving orthogonality to s_1. The energy (or more precisely, the second moment) of the output of the linear transformation

$$< y, s_1 + x_1 >$$

has three additive independent components: the first due to the desired user, the second due to the multiaccess interference, and the third due to the background noise. The first component is transparent to the choice of x_1. Thus, by varying x_1 can only change the energy of the second and third components. Accordingly, a very simple and sensible strategy is to choose x_1 that minimizes the output energy:

$$MOE(x_1) = E\left[\left(< y, s_1 + x_1 >\right)^2\right] \tag{16}$$

We would expect that if the background noise is comparatively small, the argument x_1 that minimizes (16) is such that it (almost) eliminates the contribution of the multiuser interferers to the output, in other words $s_1 + x_1$ would approach the decorrelating detector. For higher background noise, x_1 would try to attenuate the contribution of the multiaccess interference, but without becoming too large in norm, and thus contributing a large component due to the background noise. We need to speculate no further about the nature of the minimum output energy detector, because it is easy to check that the output energy in (16) is a translated version of the meansquare error:

$$MOE(x_1) = E\left[\left(A_1 b_1 - < y, s_1 + x_1 >\right)^2\right] + A_1^2 \tag{17}$$

Therefore, the x_1 that minimizes (16) is such that $s_1 + x_1$ is the MMSE linear detector of Section 4. If the minimum output energy detector is the MMSE detector, what is the point of this alternative derivation? The adaptive minimization of mean square error requires training sequences, whereas the minimization of output energy does not. Therefore, the minimization of the convex cost function (16) lends itself to blind adaptation. The simple method of projected gradient descent is adopted in [54] to show the following blind adaptation rule, which is guaranteed to converge globally:

$$x_1[i] = x_1[i-1] - \mu Z[i]\left(y[i] - Z_{MF}[i]s_1\right) \tag{18}$$

where Z_{MF} and Z are the outputs of the conventional single-user matched filter and of the proposed linear transformation:
$$Z_{MF}[i] = < y[i], s_1 >, \tag{19}$$
$$Z[i] = < y[i], s_1 + x_1[i-1] >, \tag{20}$$

The generalization of (18) to the asynchronous case is straightforward. In fact, in order to write the key equation (17) we did not invoke any structure of the multiaccess interference. In the asynchronous case, we can work with signals (or finite dimensional vectors) that span only one bit, or in order to improve performance, we can lengthen the duration of the linear transformation on both sides of the timelimited signal s_1. As usual, it should be possible to speed up convergence speed at the expense of computational complexity by adopting an RLS-based method.

The foregoing simple blind adaptive multiuser detector, which as we have seen, has no more requirements than the conventional detector, and yet, converges always to an optimally near-far resistant solution, is ideally suited to cope with transients due to initialization, powering on/off of interferers, or sudden changes in received power. The slower variations that occur due to offset drift, slow fading, etc. could be followed more closely (albeit, less robustly), by an MMSE adaptive detector operating in decisiondirected mode, in lieu of training sequences.

In practice, there will always be some mismatch between the received signature waveform s_1 of the desired user and the assumed (nominal) waveform $\hat{s}_1$. So the natural question to investigate is how robust will the blind multiuser detector be to mismatch? The answer depends on the background noise level. If $\hat{s}_1$ is different from s_1 as well as from the other interfering signature waveforms, one can always choose x_1 orthogonal to $\hat{s}_1$, so that $\hat{s}_1 + x_1$ will be orthogonal to the signals of all users: $s_1, \cdots s_K$. This may require an x_1 with huge norm, but if $\sigma \to 0$, then this will indeed be the solution that minimizes output energy. This means that in high SNR channels, the foregoing detector is not robust at all against mismatch in the nominal signal. In particular, as long as $\hat{s}_1$ is different from s_1, the asymptotic multiuser efficiency is equal to zero. An increase in background noise will have a robustifying effect. In that case a desired signal suffering a small mismatch will not be cancelled because that would require an x_1 with very large energy, and thus a correspondingly large contribution to the output energy due to the background noise. Fortunately, we can achieve the same robustifying effects of background noise even in high SNR situations by simply putting a constraint on the maximum allowable energy of x_1, referred to as *surplus energy* in [54]. The modified blind algorithm with constrained surplus energy is

$$x_1[i] = \beta x_1[i-1] - \mu Z[i] \left(y[i] - Z_{MF}[i] s_1 \right) \tag{21}$$

where $0 < \beta < 1$. Note that the conventional single-user receiver corresponds to a blind multiuser detector with zero surplus energy, while allowing unlimited surplus energy makes the detector nonrobust against desired signal mismatch in high SNR channels. A good choice for the surplus energy is the energy necessary to eliminate the interfering signals, which turns out to be -1 plus the reciprocal of the near-far resistance of a desired user with signal $\hat{s}_1$ and interfering signals $s_1, \ldots s_k$ In general, it is necessary that the nominal signal $\hat{s}_1$ be closer to s_1 than to the space spanned by the interfering signals. Provided this is satisfied and the blind detector has reached a stage in its convergence where the bit error rate is not too high, the assumed nominal can be refined by correlating the received waveform with the decisions of the user of interest:

$$\frac{1}{L}\sum_i \hat{b}_1[i]y[i].$$

(22)

The estimator in (22) converges, by the law of large numbers, to a scaled version of the received signature waveform of the desired user s_1.

There have been other efforts in blind multiuser detection. Oda and Sato [56] consider a multidimensional generalization of the conventional single-user blind equalization methods that attempt to minimize a nonconvex function of the output. The channel model can be specialized to synchronous CDMA; however since the equalizer in [56] does not use knowledge of any signature waveforms or data, bit error rate performance would be poor for weak users. Convergence is (as in the single-user case) not guaranteed using this method. Soon and Tong [57] develop a blind identification algorithm for a synchronous noiseless multiuser channel, which requires introducing a different amount of correlation in the data modulated by each user. The method is based on the singular value decomposition of the estimated covariance of the vector of observables (obtained by fractional sampling). Paris [58] proposes a blind selftuning maximum likelihood sequence estimator which, in principle, could be used for optimum asynchronous multiuser detection without prior knowledge of amplitudes and crosscorrelations.

Iltis and Mailaender [59] give a multiuser counterpart of the single-user algorithm due to Abend and Frichtman [60]. Their adaptive algorithm results in optimum symbol-by-symbol decisions without knowledge of amplitudes or chip-offsets. Complexity is exponential in the length of the processing window.

7.   Neural Network Multiuser Detectors

The first paper that considered the applicability of adaptive neural network receivers to multiuser detection is due to Aazhang, Paris and Orsak [61] where they study a multilayer perceptron. Each node in the first stage computes a nonlinear function of a linear transformation of the matched filter outputs. The number of neurons grows exponentially with the number of users. The signature waveforms are assumed known and training sequences are employed in order to adapt the linear transformation of the matched filter outputs. The adaptation is by gradient descent of mean square error (which in the context of neural networks is known as backpropagation), although in this problem this cost function is not convex

110

and has local minima. Two different configurations are simulated: with training sequences for the desired users, and with training sequences for all users. Simulations show that this difference turns out to have a very important effect on the nature of the detector to which the network converges. Assuming knowledge of the desired user's spreading code, Mitra and Poor [62] give a convergence analysis of a single layer perceptron, which can be viewed as a modified version of (12) where the update term is multiplied by a nonlinear function of the adaptive linear transformation. A so called radial-basis-function neural network is proposed in [63] for single-user equalization and investigated in [64] in synchronous multiuser detection. The number of nodes is exponential with the number of users, and the decision statistic is a linear combination of nonlinear transformations of the observables.

Miyajima, Hasegawa and Haneishi [65] propose a Hopfield neural network for synchronous multiuser detection using the likelihood function as the energy function to be minimized. The weights of the network are nonadaptive and equal to the crosscorrelations times the corresponding amplitudes, both of which are assumed known. When the true minimum of the function is found, the decisions are optimum. Although the network does not always converge to the global minimum, this approach has shown promise in the solution of other NP-complete combinatorial optimization problems. It is shown empirically in [65] that the probability of convergence to spurious local minima increases with the number of users, the background noise level, or when the interfering signals are weak. However, the achieved bit error rate is near-optimum.

Finally, we mention the application of Kohonen's Self-Organizing Map to synchronous multiuser detection [66]. This algorithm works with a matched filter bank front-end, and thus, it assumes knowledge of the signature waveforms; however it does not require the use of training sequences or knowledge of amplitudes in order to adapt the decision boundaries of the detector.

References

1. S. Verdu, "Minimum Probability of Error for Asynchronous Gaussian Multiple-Access Channels," *IEEE Trans. on Information Theory,* vol. IT-32, pp. 85-96, Jan. 1986.

2. R. Lupas and S. Verdu, "Linear Multiuser Detectors for Synchronous Code-Division Multiple-Access Channels," *IEEE Trans. Information Theory,* vol. IT-35, pp. 123-136, Jan. 1989.

3. R. Lupas and S. Verdu, "Near-Far Resistance of Multiuser Detectors in Asynchronous Channels," *IEEE Trans. Communications,* vol. COM-38, Mar. 1990.

4.   M. K. Varanasi and B. Aazhang, "Multistage Detection in Asynchronous Code-Division Multiple-Access Communications," *IEEE Trans. Communications,* vol. 38, pp. 509-519, Apr. 1990.

5.   M. K. Varanasi and B. Aazhang, "Near-Optimum Detection in Synchronous Code-Division Multiple-Access Systems," *IEEE Trans Communications,* pp. 725--736., May 1991

6.   A. Duel-Hallen, "Decision-Feedback Multiuser Detector for Synchronous Code-Division Multiple Access Channel," *IEEE Transactions on Communications,* vol. 41, pp. 285-290, Feb.1993.

7.   A. Duel-Hallen, "A Family of Multiuser Decision-Feedback Detectors for Asynchronous Code-Division Multiple Access Channels," *IEEE Transactions on Communications .* accepted for publication

8.   Z. Xie, C. K. Rushforth, and R. T. Short, "Multi-User Signal Detection Using Sequential Decoding," *IEEE Transactions on Communications,* pp. 578-583., May 1990.

9.   Z. Xie, R. T. Short, and C. K. Rushforth, "A Family of Suboptimum Detectors for Coherent Multi-User Communications," *IEEE Journal on Selected Areas in Communications,* pp. 683-690., May 1990.

10.   M. K. Varanasi, "Noncoherent Detection in Asynchronous Multiuser Channels," *IEEE Trans Inform Theory,* pp. 157-- 176, January 1993.

11.   S. Vasudevan and M. K. Varanasi, "Optimum Diversity Combiner Based Multiuser Detection for Time-Dispersive Rician Fading CDMA Channels," *IEEE Journal of Selected Areas in Communications,* 1994.

12.   M. K. Varanasi and S. Vasudevan, "Multiuser Detectors for Synchronous CDMA Communication over Nonselective Rician Fading Channels," *IEEE Trans Communications,* 1994. to appear

13.   Z. Zvonar and D. Brady, "Suboptimum Multiuser Detector for Synchronous CDMA Frequency-Selective Rayleigh Fading Channels," *IEEE Transactions on Communications,* 1995.

14.   Z. Zvonar and D. Brady, "Multiuser Detection in Single-Path Fading Channels," *IEEE Transactions on Communications,* March 1994.

15.   Z. Zvonar and D. Brady, "Differentially Coherent Multiuser Detection in Asynchronous CDMA Flat Rayleigh Fading Channels," *IEEE Transactions on Communications,* 1994.

16. U. Fawer and B. Aazhang, "A Multiuser Receiver for Code Division Multiple Access Communications over Multipath Fading Channels," *IEEE Transactions on Communications*. to appear

17. S. Verdu, "Multiuser Detection," *Advances in Detection and Estimation*, JAI Press, 1993.

18. S. Verdu, "Computational Complexity of Optimum Multiuser Detection," *Algorithmica*, vol. 4, pp. 303-312, 1989.

19. S. Verdu, "Optimum multiuser asymptotic efficiency," *IEEE Trans. Communications*, vol. COM-34, pp. 890-897, Sept. 1986.

20. U. Madhow and M. L. Honig, "MMSE Detection of Direct Sequence CDMA Signals: Performance Analysis for Random Signature Sequences, " *IEEE Transactions on Information Theory*. submitted

21. D.S. Chen and S. Roy, "An Adaptive Multi-user Receiver for CDMA Systems," *IEEE J. on Selected Areas in Communications, pp.* 808-816, June 1994.

22. S. Verdu, "Recent Progress in Multiuser Detection," *Proc. 1988 Int. Conf. Advances in Communications and Control Systems,* vol. 1, pp. 66-77, Baton Rouge, LA, Oct. 1988.

23. A. Kajiwara and M. Nakagawa, "Crosscorrelation Cancellation in SS/DS Block Demodulator," *IEICE Trans of Fundamentals of Electronics and Computer Sciences,* vol. E74, No. 9, pp. 2596-2602, September 1991.

24. S. S. H. Wijayasuriya, G. H. Norton, and J. P. McGeehan, "Sliding Window Decorrelating Algorithm for DS-CDMA Receivers," *Electronics Letters,* vol. 28, pp. 1596 - 98, Aug. 1992.

25. S. S. H. Wijayasuriya, G. H. Norton, and J. P. McGeehan, "A Novel Algorithm for Dynamic Updating of Decorrelator Coefficients in Mobile DS-CDMA," *PMIRC,* Yokohama, Japan, Sept. 1993.

26. U. Mitra and H.V. Poor, "A Generalized Adaptive Decorrelating Detector for Synchronous CDMA Systems," *IEEE Trans on Communications,* 1995.

27. S. S. H. Wijayasuriya, J. P. McGeehan, and G. H. Norton, "RAKE Decorrelating Receiver for DS-CDMA Mobile Radio Networks," *Electronics Letters,* vol. 29, pp. 395 - 96, Aug. 1993.

28. H.E. Nichols, A.A. Giordano, and J.G. Proakis, "MLD and mse Algorithms for Adaptive Detection of Digital Signals in the Presence of Interchannel Interference," *IEEE Trans. on Information Theory,* vol. IT-23, pp. 563-575, Sep 1977.

29. J.G. Proakis, *Digital Communications,* McGraw Hill, New York:, 1983.

30. U. Madhow and M. Honig, "MMSE Interference Suppression for Direct-Sequence Spread Spectrum CDMA," *IEEE Trans. Communications,* to appear.

31. P.B. Rapajic and B.S. Vucetic, "Adaptive receiver structures for asynchronous CDMA systems," *IEEE Journal on Selected Areas in Communications.*

32. S. L. Miller, "An Adaptive Direct-Sequence Code-Division Multiple-Access Receiver for Multiuser Interference Rejection," *IEEE Trans. Communications,* 1994.

33. S. L. Miller, "Transient Behavior of the Minimum Mean Squared Error Receiver for Direct-Sequence Code-Division Multiple-Access Systems," *MILCOM'94.*

34. E.G. Strom and S.L. Miller, "A Reduced Complexity Adaptive Near-Far Resistant Receiver for DS-CDMA," *Proc., GLOBECOM* '93, Houston, Texas.

35. E.G. Strom and S.L. Miller, "Optimum Complexity Reduction of Minimum Mean Square Error DS-CDMA Receivers," *VTC* '94.

36. E. K. B. Lee, "Rapid Converging Adaptive Interference Suppression for Direct-Sequence Code-Division Multiple Access Systems, " *Proc. 1993 IEEE Globecom Conf.,* pp. 1683-1687, Dec. 1993.

37. Z. Zvonar and D. Brady, "Adaptive Multiuser Receiver for Fading CDMA Channels with Severe ISI," *Proc. of the 1993 Conference on Information Sciences and Systems,* Baltimore, Maryland, March 1993.

38. Z. Zvonar and D. Brady, "Adaptive Multiuser Receiver for Shallow-Water Acoustic Telemetry Channels," J. *Acoust. Soc. Am.,* vol. 93, No. 4, Pt. 2, p. 2286, April 1993.

39. R.F. Smith and S.L. Miller, "Code Timing Estimation in a Near-Far Environment for Direct-Sequence Code-Division Multiple-Access," MILCOM '94.

40. N.B. Mandayam and B. Aazhang, "An Adaptive Multiuser Interference Rejection Algorithm for Direct-Sequence Code Division-Multiple-Access (DS-CDMA) Systems," *1994 IEEE Symp. Information Theory,* Trondheim, June 1994.

41. A. Duel-Hallen, "Decorrelating Decision-Feedback Multiuser Detector for Synchronous CDMA," *IEEE Trans. Communications,* submitted for publication.

42. R. Kohno, H. Imai, and M. Hatori, "Cancellation Techniques of Co-Channel Interference in Asynchronous Spread Spectrum Multiple Access Systems," *Trans. IECE (Electronics and Communications in Japan),* vol. 66, pp. 416-423, May 1983.

43. M. Abdulrahman and D.D. Falconer, "Cyclostationary Crosstalk Suppression by Decision Feedback Equalization on Digital Subscriber Loops," *IEEE J. Selected Areas in Communications,* pp. 640-649, April 1992.

44. M. Abdulrahman, D.D. Falconer, and A.U.H. Sheikh,, "Equalization for Interference Cancellation in Spread Spectrum Multiple Access Systems," *Proc. IEEE Vehic. Technol. Conf.,* Denver, Co, May, 1992.

45. M. Abdulrahman, A.U.H. Sheikh, and D.D. Falconer, "DFE Convergence for Interference Cancellation in Spread Spectrum Multiple Access Systems," *Proc. IEEE Vehic. Technol. Conf.,* New Jersey, May, 1993.

46. M. Stojanovic and Z. Zvonar, *Adaptive Spatial/Temporal Multiuser Receivers for Time-Varying Channels with Severe ISI.*

47. R. Kohno, H. Imai, M. Hatori, and S. Pasupathy, "An Adaptive Canceller of Cochannel Interference for Spread-Spectrum Multiple-Access Communication Networks in a Power Line," *IEEE Journal on Selected Areas in Communications,* vol. 8, No. 4, May 1990.

48. P.B. Rapajic and B.S. Vucetic, "Adaptive decision aided techniques for interference minimization in asynchronous CDMA systems," *Proc. of the 1994 IEEE Int. Symposium on Information Theory,* Trondheim, Norway, June 27-July 1, 1994.

49. Z. Siveski, Y. Bar-Ness, and D. W. Chen, "Adaptive Multiuser Detector for Synchronous Code Division Multiple Access Applications," *1994 International Zurich Seminar on Digital Communications,* Zurich, Switzerland, March 8- 11, 1994, .

50. D.W. Chen, Z. Siveski, and Y. Bar-Ness, "Synchronous Multiuser CDMA Detector with Soft Decision Adaptive Canceler," *Proc. of the 1994 Conference on Information Sciences and Systems,* Princeton, NJ, March 16-18, 1994.

51. B. Zhu, N. Ansari, Z. Siveski, and Y. Bar-Ness, "Convergence and Stability Analysis of a Synchronous Adaptive CDMA Receiver," *IEEE Trans. Communications,* submitted..

52. Y. Bar-Ness, Z. Siveski, and D. W. Chen, *Bootstrapped Decorrelating Algorithm for Adaptive Interference Cancellation in Synchronous CDMA Communication Systems.* preprint

53. D. Brady and J.A. Catipovic, "Adaptive Multiuser Detection for Underwater Acoustical Channels," *IEEE Journal of Oceanic Engineering.* to appear

54. M. Honig, U. Madhow, and S. Verdu, "Blind Adaptive Interference Supression for Near-Far Resistant CDMA," *1994 IEEE Globecom,* Dec. 1994. to appear. Also Patent Pending

55. S. Verdu, B. D. O. Anderson, and R. Kennedy, "Blind Equalization without Gain Identification," *IEEE Trans. Information Theory,* vol. 39, pp. 292-297, Jan. 1993.

56. H. Oda and Y. Sato, "A Method of Multidimensional Blind Equalization," *IEEE Int. Symp. on Information Theory, p.* 327, San Antonio, TX, Jan. 1993.

57. V. C. Soon and L. Tong, "Blind Equalization under Cochannel Interference," *Proc 27th Annual Conf. Informahon Sciences and Systems,* Johns Hopkins University, Baltimore, Md., March 1993.

58. B.P. Paris, "Asymptotic Properties of Self-Adaptive Maximum-Likelihood Sequence Estimation," *Proc 1993 Conf. Information Sciences and Systems,* pp. 161-166, Johns Hopkins University, Baltimore, Md., 1993.

59. R. A. Iltis and L. Mailaender, "An Adaptive Multiuser Detector with Joint Amplitude and Delay Estimation," *IEEE J. Selected Areas on Communications,* pp. 774-785, June 1994.

60. K. Abend and B. Fritchman, "Statistical detection of communication channels with intersymbol interference," *Proc. IEEE,* vol. 58, pp. 779-785, May 1970.

61. B. Aazhang, B.-P. Paris, and G.C. Orsak, "Neural Networks for Multiuser Detection in Code-Division Multiple-Access Communications," *IEEE Trans on Communications,* vol. 40, No. 7, July 1992.

62. U. Mitra and H.V. Poor, "Adaptive Receiver Algorithms for Near-Far Resistant CDMA," *IEEE Trans on Communications,* 1995.

63. S. Chen, B. Mulgrew, and S. McLaughlin, "Adaptive Bayesian Equalizer with Decision Feedback," *IEEE Signal Processing Magazine,* Aug. 1991.

64. U. Mitra and H.V. Poor, "Neural Network Techniques for Adaptive Multi-User Demodulation," *IEEE Journal on Selected Areas of Communications*. submitted for publication

65. T. Miyajima, T. Hasegawa, and M. Haneishi, "On the Multiuser Detection Using a Neural Network in Code-Division Multiple-Access Communications," *IEICE Trans Communications,* vol. E76-B, No. 8, August 1993.

66. A. Hottinen, "Self-Organizing Multiuser Detection," *Proc. IEEE ISSSTA '94, p. 152,* Oulu, Finland, July 1994.

Spatial and Temporal Filtering for Co-Channel Interference in CDMA

Ryuji Kohno

Abstract — This paper provides a brief review of the challenges and opportunities that arise for improving user capacity of spread spectrum multiple access (SSMA) or code division multiple access (CDMA) based on spread spectrum techniques. Co-channel interference (CCI) in CDMA is the most dominant factor in the limitation of user capacity, while capacities of FDMA and TDMA capacities are primarily bandwidth limited. From an information theoretical viewpoint, however, an optimum multi-user receiver for CDMA can be derived by utilizing CCI as redundant information distributed among multiple accessing users. In practical channels such as personal, indoor, mobile radio communication channels, more feasible receivers are required to combat with multipath fading that causes estimation errors of channel characteristics due to time variance. An overview of several CCI cancellation schemes to carry out sub-optimum but practical multi-user receivers for CDMA is provided mainly from our works by classifying them to the processing on temporal domain, that on spatial one, and that on both domains.

1   INTRODUCTION

The Japanese standard for wireless local area networks (LANs) [1] that was established December 1992 is accelerating research and development of spread spectrum (SS) systems for consumer and personal communications. Its main regulation is to use SS modulation such as direct-sequence (DS), frequency-hopping (FH) or their hybrid in 2.4 GHz band. Many advantages of SS techniques such as robustness against interference and noise, less interference to other systems and efficient use based on CDMA of frequency spectrum reveal wide wireless applications for commercial use [2]-[5]. FCC in USA opened the industrial, scientific,

S.G. Glisic and P.A. Leppänen (eds.), Code Division Multiple Access Communications, 117-146.
© 1995 *Kluwer Academic Publishers. Printed in the Netherlands.*

and medical (ISM) frequency band for unlicensed operation of SS systems in 1985 and modified its rules to permit wider bandwidth in 1990 [6].

Moreover, SS techniques have a shot at becoming a mass-market technology in the digital standard for cellular telephones. Both CDMA and TDMA have been proposed as the transmission standard for digital cellular telephone systems and products. Their battle is ongoing. CDMA offers up to about four to six times more capacity than first generation TDMA. That improvement of capacity depends on some additional techniques such as power control, voice activation, soft hand-off, and sectorization [7].

Efficient utilization of available frequency spectra is of major importance in not only wireless LANs and digital cellular telephones but also other consumer and personal communications. If mature techniques to improve capacity of CDMA more are established, CDMA will win the battle.

The purpose of this paper is to give a overview of several recently developed techniques to improve capacity of CDMA. In particular, the paper will focus on cancellation of co-channel interference (CCI) in CDMA which restricts the number of available users who can simultaneously access a channel or the capacity of CDMA [8][9]. Since optimum detection for synchronous CDMA systems or M-ary SS ones can be reduced to an optimum coding and decoding problem for a multiple channel in coding theory, this paper will discuss optimum and sub-optimum multi-user detection for asynchronous CDMA systems, in which transmission of every user is not synchronized one another.

In wireless personal, indoor and mobile communication channels, it is difficult to carry out optimum detection for asynchronous CDMA due to unknown and time-varying characteristics of the channels. Therefore, sub-optimum multi-user detection based on CCI cancellation must be more practical. Several schemes of CCI cancellation based on temporal or/and spatial filtering are introduced. In the last session, combined spatial and temporal filtering based on frequency spectral estimation is proposed and investigated.

2 CDMA SYSTEM MODEL

In this section, a model of CDMA is described in order to clarify a problem in CDMA. For the sake of simplicity, a DS/CDMA system

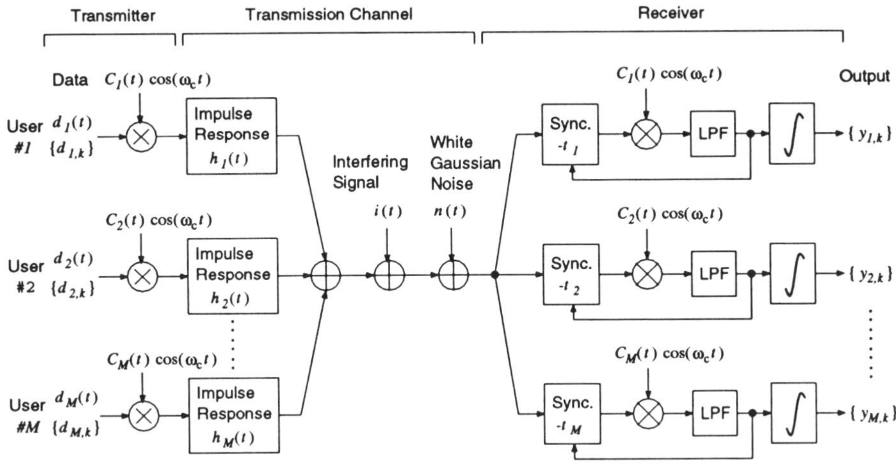

Figure 1: A Model of a DS/CDMA System

model is formalized.

Fig.1 shows a model of DS/CDMA system, where M users are transmitting individually spread spectrum signals $S_m(t), m = 1, 2, \cdots, M$ at the same time and in the same frequency band. This system can also be considered as the reverse link or uplink of a single cell in a cellular mobile communication system [10].

In order to spread a signal spectrum in a transmitter of the mth user, the message sequence signal $d_m(t)$ is directly multiplied by signature, spreading or pseudo-noise (PN) sequence signal $C_m(t)$ which has a much higher symbol rate than $d_m(t)$. $S_m(t)$ can be written as

$$S_m(t) = Re\{d_m(t)C_m(t)\exp[j\omega_c t + \theta_m]\} \tag{1}$$

where $j = \sqrt{-1}$, ω_c and θ_m are the carrier frequency and the phase offset, respectively. $d_m(t)$ and $C_m(t)$ are expressed by

$$d_m(t) = \sum_{k=-\infty}^{\infty} d_{m,k} U_{T_d}(t - kT_d), \tag{2}$$

$$C_m(t) = \sum_{j=-\infty}^{\infty} C_{m,j} U_T(t - jT). \tag{3}$$

where $U_T(t)$ is the unit rectangular pulse defined as $U_T(t) = 1$ for ($0 \leq t < T$) and $U_T(t) = 0$ otherwise. $d_{m,k}$ is a symbol of the message

120

sequence $\{d_{m,k}\}$ of the mth user at instant kT_d, $C_{m,j}$ is a symbol of the spreading sequence $\{C_{m,j}\}$ assigned to the mth user at instant jT. In the case of BPSK, $d_{m,k}, C_{m,j} \in \{+1, -1\}$. Here, T is the duration of a 'chip' in the spreading sequences, while T_d is that of a bit of message sequences. We assume that T_d is a multiple of T and define $N = T_d/T$.

In a receiver, multiplexed SS signals from all users are received and multiplied again by the same spreading sequence signal used in transmitter, in order to transform the spread spectrum of the desired SS signal into its original narrow band. The received signal $r(t)$ can be represented as

$$r(t) = \sum_{m=1}^{M} \int_{-\infty}^{\infty} h_m(\tau) S_m(t-\tau) d\tau + i(t) + n(t) \tag{4}$$

where $h_m(t)$ is an impulse response of the transmission channel from the mth user to the receiver, $i(t)$ is an interfering or jamming signal, $n(t)$ is channel noise. In multipath fading channels, the impulse response of the mth user's link takes on the form

$$h_m(t) = \sum_{\lambda=1}^{L_m} g_{m,\lambda} \delta(t - t_{m,\lambda}) e^{j\theta_{m,\lambda}} \tag{5}$$

Each path's excess path delay, $t_{m,\lambda}$, is organized in order of increasing magnitude with λ such that $t_{m,1} = 0$ and $T_c \leq t_{m,2} \leq \ldots \leq t_{m,L_m} < \Delta$. Moreover, each is uniformly distributed over the interval $[T_c, \Delta]$ where Δ represents the maximum excess path delay possible. $g_{m,\lambda}$ is the path amplitude fro the λ-th path and $\theta_{m,\lambda}$ is the path phase.

All the multiplexed SS signals except the desired SS signal interfere with the desired SS signal, due to the crosscorrelation of the different spreading sequences assigned to individual users, and is described as co-channel interference (CCI).

The output signal of the correlator for the nth user can be derived by integrating $r(t)C_n(t)$ over $T_d = NT$. When spreading sequence acquisition is performed, the sample value $y_{n,k}$ of the output at instant kT_d, can be represented in the form

$$
\begin{aligned}
y_{n,k} &= h_{n,0} d_{n,k} \\
&+ \sum_{i \neq 0} \sum_{j=0}^{N-1} h_{n,j+iN} \Theta_{n,n}(j) d_{n,k-i} \\
&+ \sum_{i} \sum_{j=0}^{N-1} \sum_{m=0}^{n-1} [\sum h_{m,j+iN} \{R_{n,m}(j+j_m) d_{m,k-i-1} + \hat{R}_{n,m}(j+j_m) d_{m,k-i}\}
\end{aligned}
$$

$$+ \quad \sum_{m=n+1}^{M} h_{m,j+iN}\{R_{n,m}(j+j_m)d_{m,k-i} + \hat{R}_{n,m}(j+j_m)d_{m,k-i+1}\}]$$

$$+ \quad i_{n,k}$$

$$+ \quad n_{n,k} \tag{6}$$

where $h_{m,j} = h_m(jT)$. It is assumed that $j_m = (t_{n,0} - t_{m,0})/T \bmod N$. $R_{n,m}(j)$ and $\hat{R}_{n,m}(j)$ are the partial crosscorrelation functions between the spreading sequences of the n-th and the m-th user [9].

In (6), the first term of the right hand side indicates the desired baseband data, the second term, the intersymbol interference (ISI) in the desired user's signal, the third, the co-channel interference (CCI) due to the undesired user, the fourth, the interfering signal component, and the fifth, the noise component. The magnitude of the co-channel interference is determined by the periodic or even crosscorrelation function $\Theta_{n,m}(j) = R_{n,m}(j) + \hat{R}_{n,m}(j)$ and $\hat{\Theta}_{n,m}(j) = R_{n,m}(j) - \hat{R}_{n,m}(j)$ depending on $d_{m,k} = d_{m,k-1}$ and $d_{m,k} = -d_{m,k-1}$, respectively.

In the presence of CCI, the crosscorrelation between the spreading sequences of desired and undesired users prevents acquisition and tracking of the spreading sequence and restricts the number of users simultaneously accessing the channel. This is because side lobes of the output signal of the correlator hamper the detection of the main lobe of autocorrelation and the side lobes increase with the number of available users.

3 MULTI-USER DETECTION FOR CDMA

Note from the above-discussion that CCI due to crosscorrelation of spreading sequences hampers establishment of acquisition and tracking, and limits the capacity of CDMA. CCI results in the "near-far " problem that relates to the problem of very strong signals from undesired users at a receiver swamping out the effects of weaker signals from desired user [2]. Therefore, CDMA receivers should be designed so as to improve the synchronization and the capacity against the near-far problem.

3.1 SINGLE-USER RECEIVERS AND MULTI-USER RECEIVERS

CDMA receiver structures can be classified as either multi-user or single-user structures in general. The conventional single-user detector assigned to each user is described in the previous section. Another example is the multipath combining receiver by Lehnert *et al.*[11]. Optimum single-user detector was proposed by Poor *et al.*[12] The conventional single-user receiver decides the correlator output for each user only in the interval corresponding to the data symbol individually among multiple-access users. In the decision, CCI is considered as an undesired component like noise.

From an information theoretical viewpoint, however, we can utilize CCI as redundant information which multiple-access users are sharing in a common channel or a multi-user channel, in order to afford performance gains over the conventional single-user receiver. The receiver by which all multiple-access users' signals can be correctly detected using CCI is termed the multi-user receiver. The multi-user receiver can be used at a cell site, that is a base station, to detect all multiple-access users' signals.

In contrast with the single-user receiver, for multi-user receivers, the decision of all the users' data is an inter-dependent process. Although their structures are much more complicated, their BER performance can greatly exceed that of the conventional correlator. These receivers vary in complexity from the optimum multi-user detector presented independently by Kohno *et al.* [13][14] and Verdu [15] to the suboptimum detectors developed independently by Kohno *et al.* [16][17], Masamura [18] Xie *et al.* [19], and Varanasi *et al.* [20].

Apart from [17], these detectors were all designed for the AWGN channel. On the other hand, the receiver of [17] which adaptively cancels the CCI was designed for time-varying baseband channels. Yoon *et al.* proposed multi-user receiver which can be considered as a continuation of the research of [17], and introduced a cascade of CCI cancellers and its application to multipath fading channels [21]-[23] independent of Grant *et al.* [24].

The above-mentioned receivers have been discussed mainly for usual DS/CDMA in which all simultaneously transmitted signals are received as the sum of their envelopes. It is termed an *ADDER channel*. On the otherhand, FH/CDMA based on multilevel FSK developed by Goodman *et al.* [25] is considered an *OR channel*, whose receiver has no information erasure due to frequency collision or hit because an energy detector outputs total energy of multiple access users' transmitted signals in individual frequency bands. For such an *OR channel*, we can improve the capacity of CDMA by applying appropriate error-correcting coding and decoding [26][27] and cancelling CCI [28] different from them for

ADDER channel. Kawahara *et al.* introduced similar *OR logic operation* into DS/CDMA so as to utilize efficient decoding for the multi-user channel [29].

3.2 OPTIMAL MULTI-USER RECEIVER

Without loss of generality and for the sake of notational simplicity, (6) can be rewritten if there is no intersymbol interference or multi-path distortion in a channel, no narrowband interference and only two users accessing the channel.

$$y_{1,k} = h_{2,0}R_{1,2}\mathbf{d_{2,k-1}} + h_{1,0}\mathbf{d_{1,k}} + h_{2,0}R_{2,1}\mathbf{d_{2,k}} + n_{1,k} \qquad (7)$$

$$y_{2,k} = h_{1,0}R_{2,1}\mathbf{d_{1,k}} + h_{2,0}\mathbf{d_{2,k}} + h_{1,0}R_{1,2}\mathbf{d_{1,k+1}} + n_{2,k} \qquad (8)$$

where $R_{1,2}$ and $R_{2,1}$ are partial crosscorrelation factors. It is noted from these equations that the correlator outputs $y_{1,k}$ and $y_{2,k}$ can be considered as multi-dimensional finite-state machine outputs corrupted with noise or multi-dimensional analogue convolutional encoder output with noise. Fig.2 illustrates its equivalent multi-dimensional tapped delay lines. Therefore, we can achieve optimum multi-user detection for

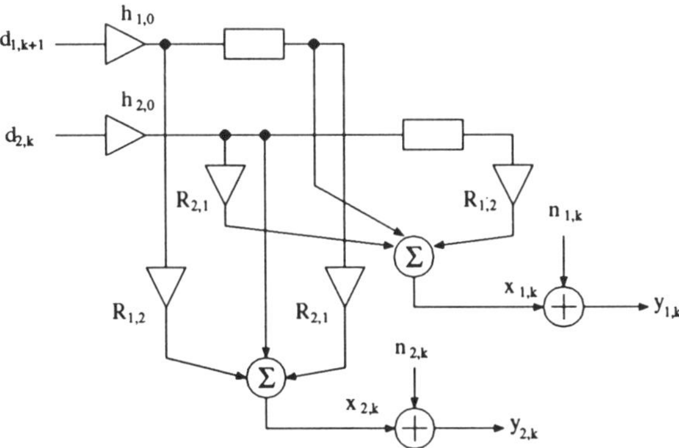

Figure 2: An Equivalent Tapped Delay Line Model of Asynchronous DS/CDMA Channel (two users)

CDMA by using a Viterbi algorithm that is well known as an efficient maximum likelihood sequence estimation (MLSE) for a convolutional code, a trellis coded modulation and so on. In general, the optimum

K-user receiver consists of a bank of single-user correlators in Fig.1 followed by a Viterbi algorithm whose complexity per binary decision is $O(2^K)$ [13]-[15].

4 CO-CHANNEL INTERFERENCE CANCELLATION BY TEMPORAL FILTERING

The optimum multi-user receiver affords the much performance gain over conventional single-user one, but expends some possible disadvantages such as higher complexity and the need to know the relative characteristics of the multi-user channel, e. g. amplitudes, path delays and path phases in (5). For indoor and mobile radio communications, implementation complexity should be reduced to a feasible level even if the performance degrades slightly from optimum one. Estimation of the channel characteristics is used to be more difficult due to its time-variance. In fact, presently developed CDMA cellular radio networks don't include the optimum muti-user receiver.

From this section, CCI cancellation techniques which carry out sub-optimum but more practical multi-user receivers are explained. There are several methods which strive to reduce or remove CCI. The design of spreading sequences attempts to produce sets of sequences with optimum correlation properties [9][30][31]. To further aid the receiver in cancelling CCI, temporal, spatial signal processing and their combination can be employed.

4.1 SINGLE STAGE CCI CANCELLER

The receiver must obtain initial data estimates for use in CCI cancellation. We simply perform a hard decision upon the k-th user's correlator output $y_{k,n}$ in (6) as shown in Fig.1; $\hat{b}_n^{(k)} = sgn(y_{k,n})$. Each initial data estimate, $\hat{b}_n^{(k)}$, of the k-th user's originally transmitted data $d_{k,n}$ at instant $nT_b(T_b = T)$, is first multiplied by its respective synchronized spreading sequence and carrier signal to create a replica of the originally transmitted signal as shown in Fig. 3. Then the signal is passed through a transversal filter(TF) which emulates channel $h_k(t)$. The structure of the TF is shown in Fig. 4. Each tap coefficient, at the appropriate excess path delay, $t_{k,\lambda}$, is $G_{k,\lambda} = g_{k,\lambda} \cos \phi_{k,\lambda}$. The estimation of the channel parameters may be feasible by an adaptive updating algorithm [17][22][23].

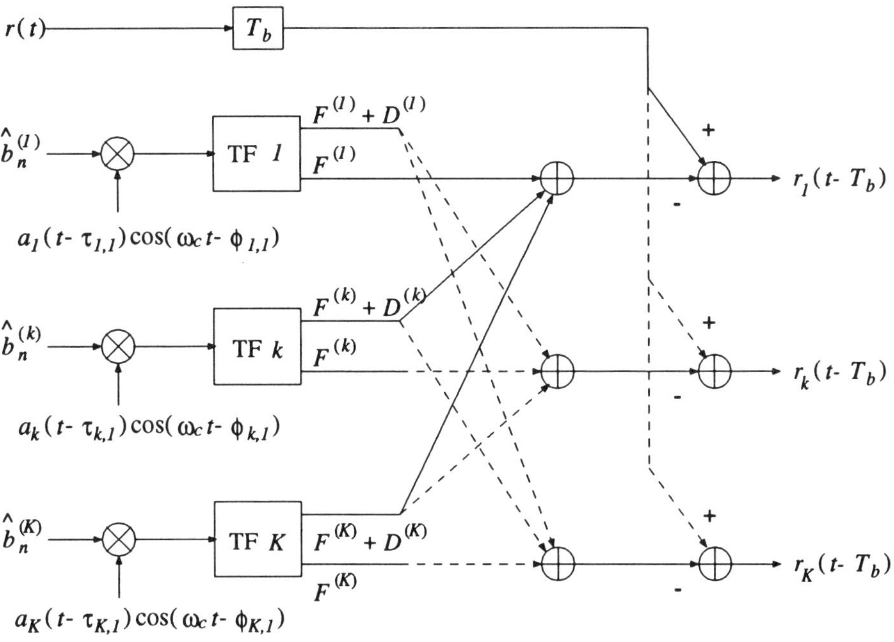

Figure 3: CCI Canceller

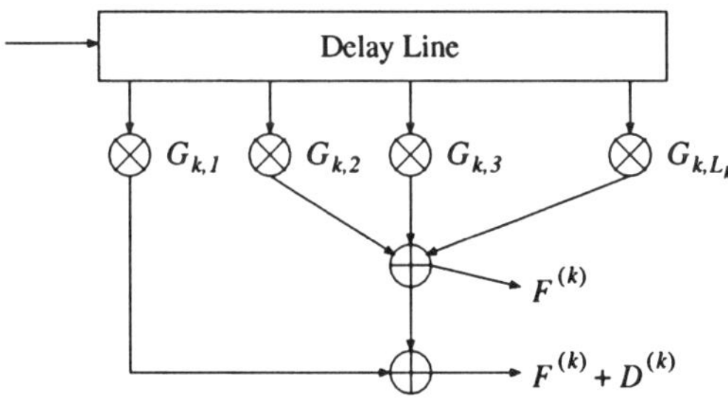

Figure 4: kth Transversal Filter (TF)

The output of the filter consists of two signals: the sum, $F^{(k)}$, of the kth user's faded terms ($\lambda \geq 2$) and the full replica of his received signal component which is the sum of $F^{(k)}$ and his first path's signal ($\lambda = 1$), $D^{(k)}$. By adding the m-th user's own faded terms, $F^{(m)}$, with all the other users' $F^{(k)} + D^{(k)}$, the canceller can create a replica of the CCI of the m-th user (including his intersymbol interference). How closely it resembles the actual CCI depends on two things: the number of correct data estimates and the accuracy of the channel estimates. The CCI replica of each user is subtracted from the original received signal, $r(t)$, to produce a "cleaned" received signal, $r_m(t)$, for the m-th user. This signal is passed to a second set of correlators as shown in Fig. 5 to obtain a second set of data estimates. The accuracy of the canceller depends on the ratio of the number of correct data estimates to the total number data estimates. As long as the ratio is high, more of the CCI can be removed instead of being added.

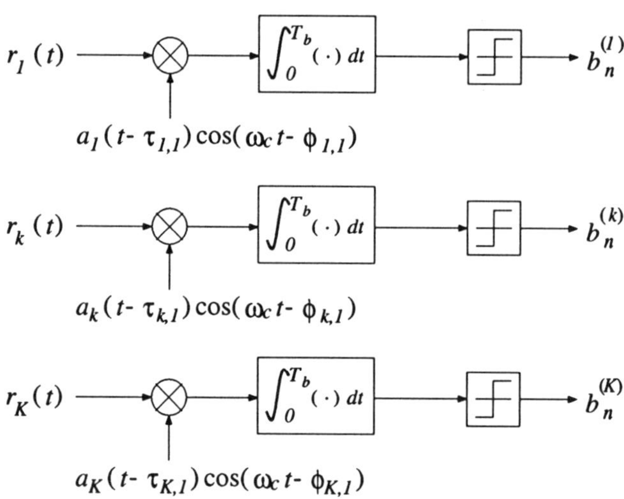

Figure 5: Data Decision

4.2 CASCADE OF CCI CANCELLER

Next, a cascade of such CCI canceller and data decision stages (Figs. 3 and 5) can be constructed as shown in Fig. 6. The decisions of the final stage give the final data decisions for the receiver. Intuitively, if the number of incorrect data decisions at the first stage in comparison to that at the initial data estimation has been reduced, more accurate replicas can be reconstructed at the CCI canceller of the second stage. In repeating this process, at a certain stage, the data decisions should converge to the actually transmitted data values.

The probabilities of bit error in the cascade of CCI cancellers which was calculated from the theoretical BER analysis and from computer simulations are shown in Figs. 7-9. The system, channel and simulation parameters are listed in [32].

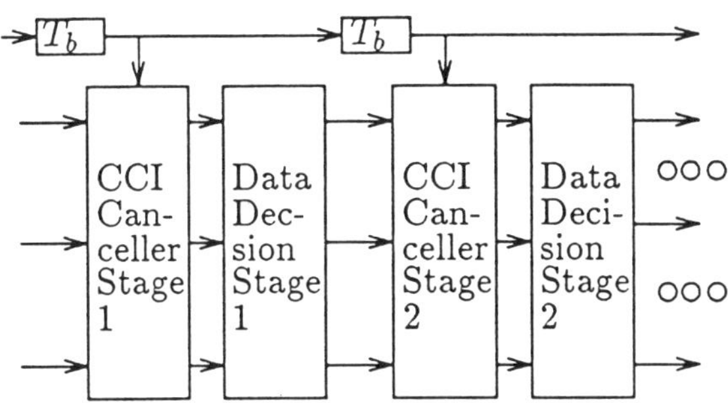

Figure 6: Cascade of CCI Cancellers

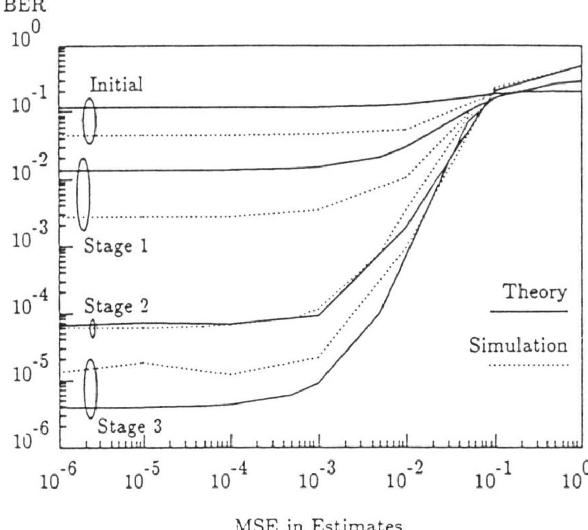

Figure 7: BER vs. MSE with $K = 64$ and $E_b/N_0 = 10$ dB for up to three canceller stages. $\bar{L} = 2$ and $N = 127$.

128

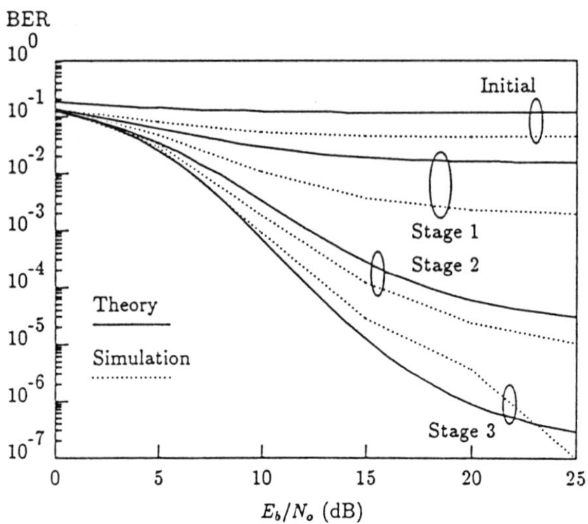

Figure 8: BER vs. E_b/N_0 with $K = 64$ and MSE $= 10^{-2}$ for up to three canceller stages. $\bar{L} = 2$ and $N = 127$.

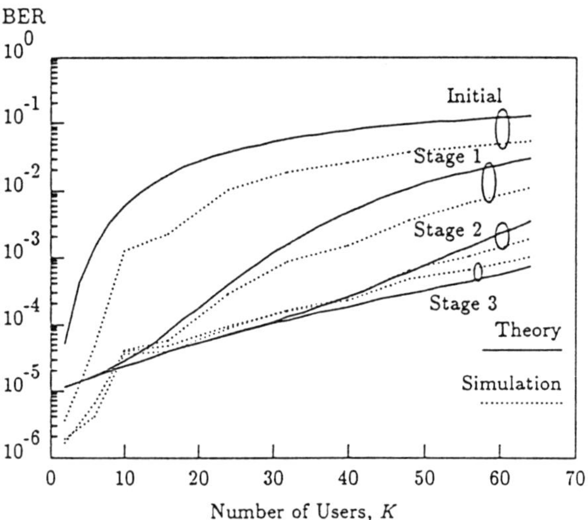

Figure 9: BER vs. K with $E_b/N_0 = 10$ dB and MSE $= 10^{-2}$ for up to three canceller stages. $\bar{L} = 2$ and $N = 127$.

4.3 COMBINATION OF CCI CANCELLATION AND DECODING

Performance of the cascaded CCI canceller depends upon both channel estimation errors and initial decision errors as Figs. 7 and 8 illustrate BER performance corresponding to such errors. In order to improve robustness against both kinds of errors, we have proposed to introduce decoding of an error-correcting code into the cascade of CCI cancellation [39].

Instead of using a hard decision device in Fig.5, we can decode the data operating on the stream of matched filter outputs to obtain more reliable initial estimation of data. Moreover, the more reliable decoded data are employed in updating the adaptive transversal filter of Fig.4 to achieve more accurate channel estimation. This idea originates in combination of cancelling and decoding for intersymbol interference(ISI) [35] and its cascade structure [40]. Fig.10 shows simulation results of BER vs. E_b/N_0 for the combined decoding and cancelling, cancelling without decoding and initial decision.

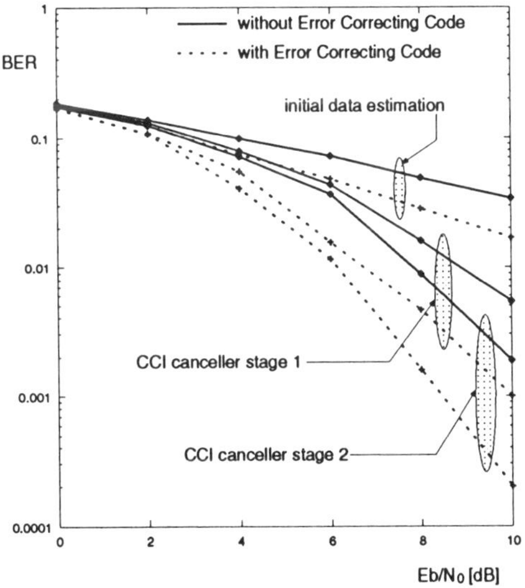

Figure 10: BER vs. E_b/N_0 with (7,4) Hamming Code, datarate=$9600bps$, $K = 30$, for up to two canceller stages. $\bar{L} = 2$ and $N = 127$.

5 CO-CHANNEL INTERFERENCE CANCELLA-TION BY SPATIAL FILTERING

An adaptive array antenna is useful in suppressing interfering signals because it can adaptively control directivity of the antenna by updating weights of element antennas even if the desired signal's arrival angle or directions-of-arrival(DOA) is unknown [33]. Fig.11 shows a basic structure of adaptive array antenna.

An adaptive array antenna has some differences in applications to CDMA and other narrowband systems. An array antenna is applied in CDMA mainly to distinguish a desired user's SS signal from undesired ones which have much less correlation with the desired user's SS signal, while it is employed in narrowband radio systems to reject multipath delayed signals which are transmitted from the same information source as a desired signal. This difference affects algorithms of updating weight coefficients. Moreover, an adaptive array antenna has different filtering performance for a wideband signal and narrowband one.

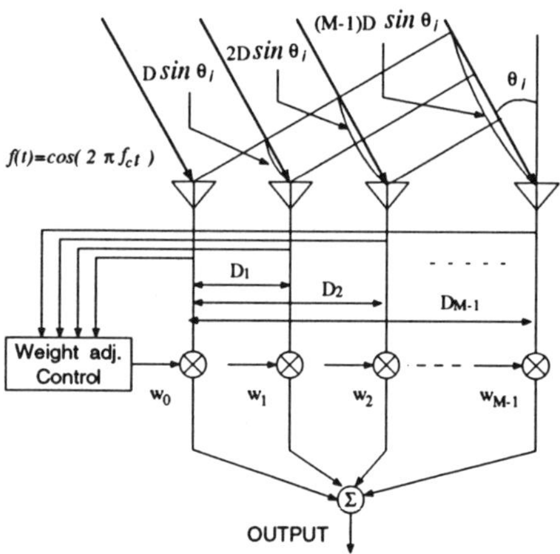

Figure 11: An Adaptive Array Antenna

5.1 SPATIAL AND TEMPORAL FREQUENCY PERFOR-MANCE OF ARRAY ANTENNAS

In general, an adaptive array antenna is controlled so as to have maximum gain to direction of desired signal and null to that of interfering signals. It is called a spatial filter. On the otherhand, since an array antenna such as LMS array can pass a signal possessing the same frequency component as the reference signal and reject that doing the different frequency one, it can also be considered as a temporal filter.

In fact, a temporal and spatial transfer function of array antenna can be represented by

$$H(\omega_t, \theta) = \sum_{n=1}^{M} W_n e^{-j(n\frac{2\pi\omega_t D sin(\theta)}{c})} \tag{9}$$

where ω_t and θ are temporal frequency and arrival angle(DOA) of signals, respectively. $W_n (n = 1, 2, \cdots, M)$ are weight coefficients, D is spatial interval among element antennas and c is a propagation speed of electric magnetic wave.

In an array antenna, there are different time delays among element antennas according to arrival angles of the signal waves. We can equivalently represent an array antenna as a model of tapped delay line in Fig.12, in which the delay intervals T_i of signal waves with arrival angles θ_i $(i = 0, 1, 2, ...I)$ can be written by

$$T_i = \frac{D}{c} \sin(\theta_i), \tag{10}$$

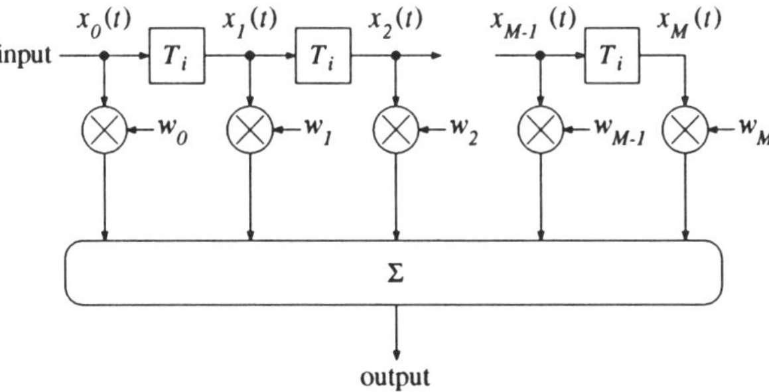

Figure 12: Equivalent Temporal Filter Model of a Line Array Antenna

If arrival angles and received power of the signal waves are known, optimum weights of the array can be obtained by solving the Wiener-Hopf equation. Then the frequency performance of the array can be derived. Figs.13 and 14 show the frequency performance and the antenna pattern of array with two elements when there are two waves with arrival angles of 0 and 60 degree.

132

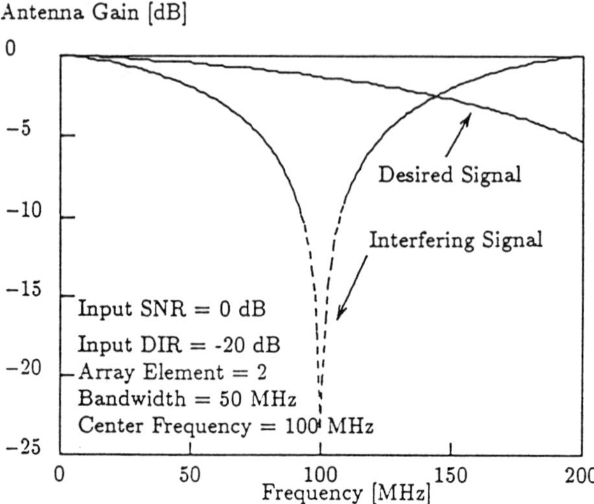

Figure 13: Temporal Frequency Performance of the Array Antenna with Two Elements

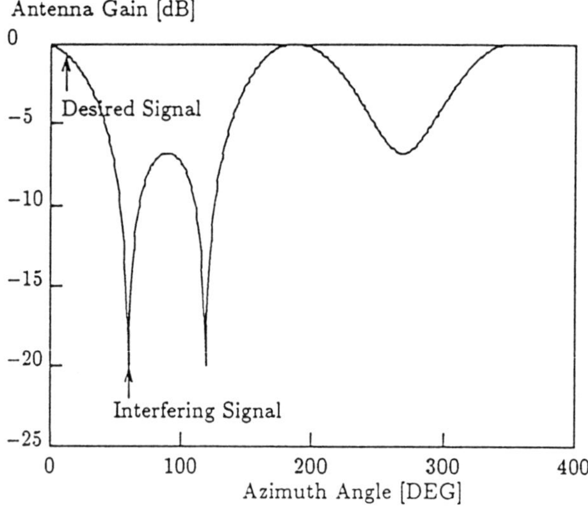

Figure 14: Antenna Pattern of the Array Antenna with Two Elements

5.2 TDL ARRAY ANTENNA FOR WIDEBAND SIGNALS

A tapped delay line(TDL) array antenna was proposed to reject wideband interference [33]. Fig.15 shows a structure of the TDL array antenna. Since TDL of each element antenna can more efficiently utilize correlation of received signals than a usual line array antenna, the frequency performance can be improved for wide-band signals such as interfering ss signals.

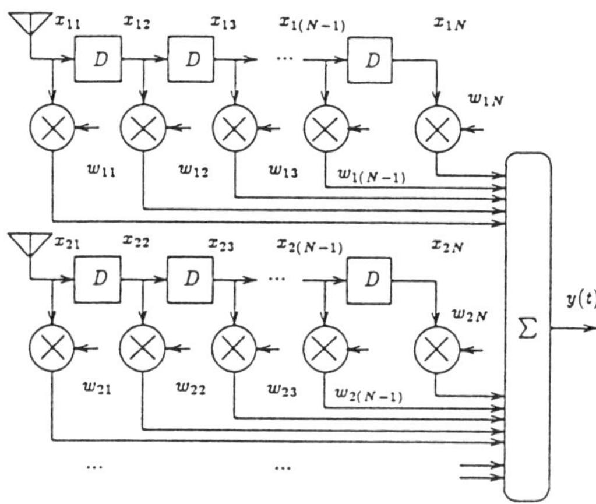

Figure 15: Structure of a Tapped Delay Line (TDL) Array Antenna

Fig.16 shows the frequency performance of the TDL array antenna with two elements and five taps in an element. It is noted that TDL array of Fig.15 can reject wider band interference than ordinary array of Fig.11.

5.3 ADAPTIVE ARRAY ANTENNA USING THE PROCESSING GAIN

In an adaptive array antenna, it is usually assumed that the reference signal is correct and the array can suppress interference effectively enough to obtain correct reference signal. However, since the weights are updated by the error signal or the difference between the reference signal and the antenna output which includes residual interference and noise, it is difficult to achieve stable convergence of the weights when power of interference and/or noise is much larger than that of desired

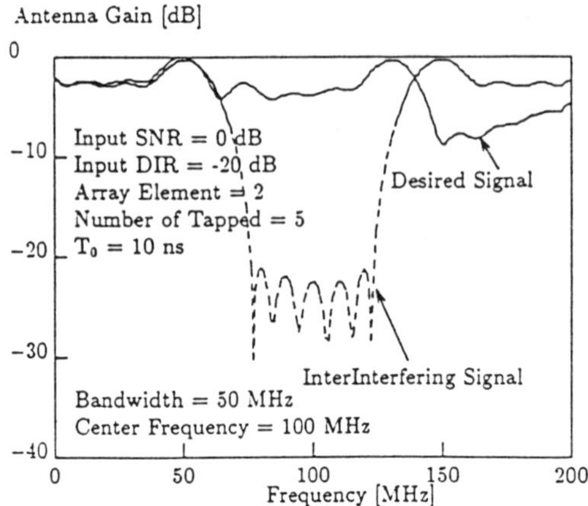

Figure 16: Temporal Frequency Performance of the TDL Array Antenna with Two Elements

signal, i.e. low DIR and DNR, which are usual in CDMA.

In order to solve the above-mentioned problem, we proposed a method to more correctly update the weights using the inherent processing gain of a single-user receiver for DS/CDMA [34]. Fig.17 shows a DS/CDMA receiver with the proposed method of updating array. As the error signal for updating the weights, we can use the difference between the input and output signals of threshold decision. The input signal includes less interference and noise than the output signal of array even in case of low DNR and DIR, because these undesired signal can be suppressed by the processing gain, that is defined as a ratio of RF bandwidth-to-baseband one.

However, convergence of the weights is disturbed by the delay in the controlling loop which is due to despreading or demodulation in the correlator. If duration of a data bit equals to a period of spreading sequence, i.e. N chips, the delay will be N. It is assumed that weights are updated chip by chip. In order to calculate the weights at instant k, we can use only the delayed error which is obtained at instant $k - N$. Therefore, stable convergence of the weights cannot be guaranteed in a time-varying channel such as a fading channel.

In order to improve the convergence, we can introduce the same way as an AEDEC, that is an automatic equalizer including a decoder of error-correcting code [35].

In particular, in order to reject co-channel interference with wide-

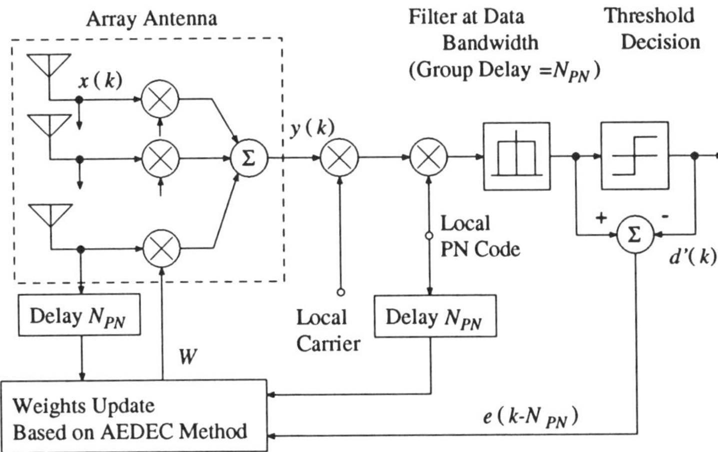

Figure 17: A DS/CDMA Receiver with the Proposed Adaptive Array Antenna Using the Processing Gain

band in DS/CDMA when DIR and DNR are low, we can introduce the proposed updating method using the processing gain to the TDL array antenna. A DS/CDMA receiver with the TDL adaptive array antenna using the processing gain can be constructed by combining Figs.15 and 17.

Computer simulations illustrate performance of the proposed TDL adaptive array antenna using the processing gain in comparison with conventional ones. In simulations, we assumed that there are desired and interfering SS signal waves with arrival angles of 20 and 60 degree, respectively, a carrier frequency of $100MHz$, spreading sequences with a chip rate of $25Mc/s$ and a period of $40nsec$, a data rate of $98Kb/s$, and a SS signal bandwidth of $50MHz$. The TDL array antenna has two elements and five taps in an element with a tap delay interval of $10nsec$ $(r = 2)$.

Fig.18 shows their output DIR corresponding to the bandwidth of SS signals. From this figure, we confirm that the proposed TDL array antenna can achieve the best DIR performance for wide-band SS signals.

Figs.19 and 20 show BER performances vs. the input DNR and the input DIR, respectively. We can confirm that the proposed TDL adaptive array antenna using the processing gain improves the BER performance

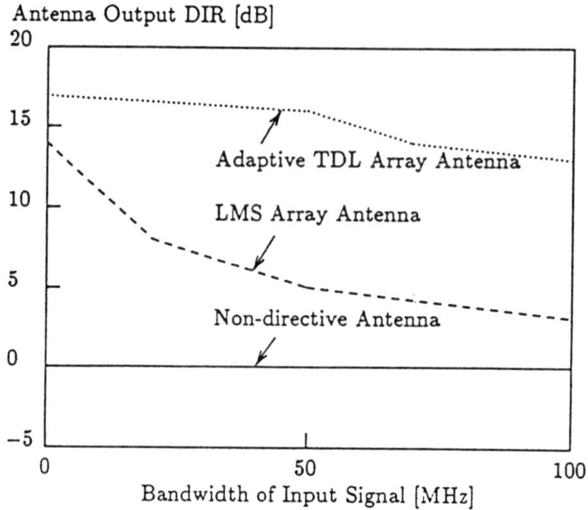

Figure 18: Output DIR Corresponding to Bandwidth of SS Signals

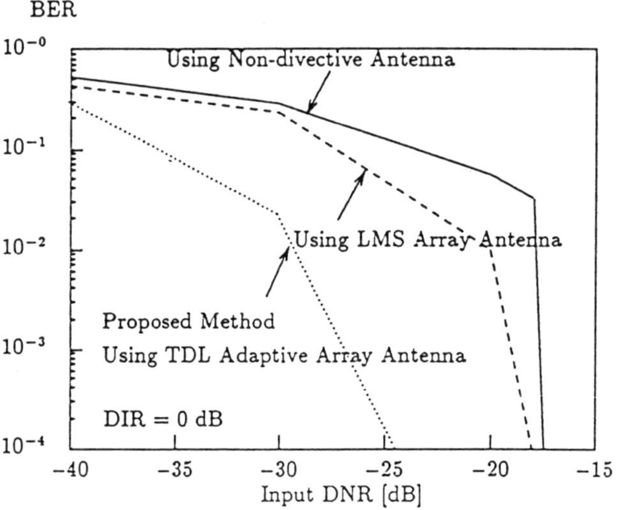

Figure 19: BER vs. Input DNR

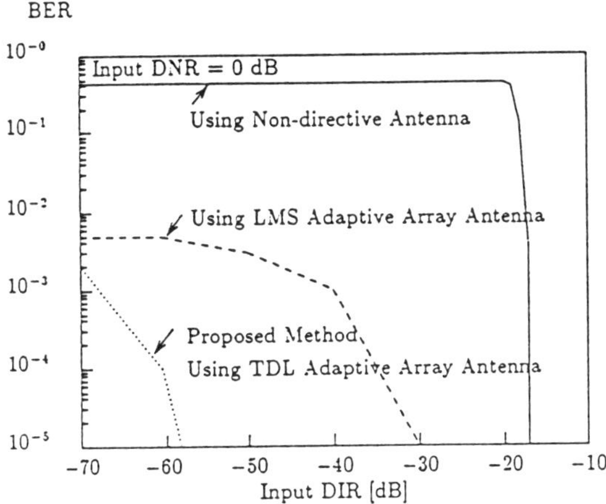

Figure 20: BER vs. Input DIR

in low DNR and DIR.

6 CO-CHANNEL INTERFERENCE CANCELLATION BY COMBINED SPATIAL AND TEMPORAL FILTERING

6.1 COMBINATION OF ARRAY ANTENNA AND DIGITAL FILTER

If there is a high-level interfering SS signal from an undesired user with the same arrival angle as that of a desired user, an array antenna cannot suppress it. In order to solve this problem, combination of an adaptive array antenna in the spatial domain and an adaptive interference canceller in the temporal domain was proposed [36]. The proposed system can suppress interfering SS signals,i.e. CCI, with arrival angles different from that of a desired user by using a null steering array antenna and eliminate by means of a canceller the residual interference and CCI having an arrival angle the same as that of the desired SS signal.

Fig.21 shows the structure of the DS/CDMA system based on combination of adaptive spatial and temporal filtering, which consists of array antenna and canceller using adaptive digital filters (ADFs). Fig.22 shows BER of demodulated data as a function of the D/I ratio for the same arrival angle. The proposed system can achieve stable demodulation and improve BER performance even in a heavy interference channel where a conventional array antenna system cannot achieve acquisition.

138

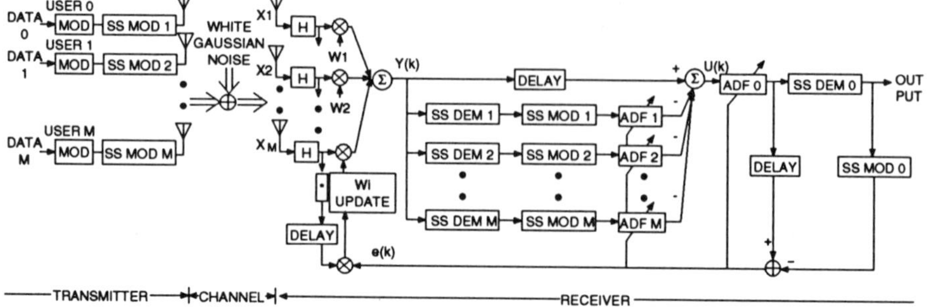

(a) Parallel Canceller structure

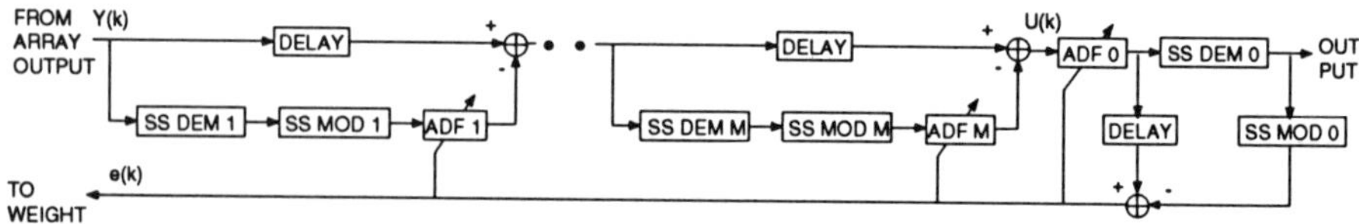

(b) Serial Canceller Structure

Figure 21: A DS/CDMA System with an adaptive array antenna including a canceller of CCI

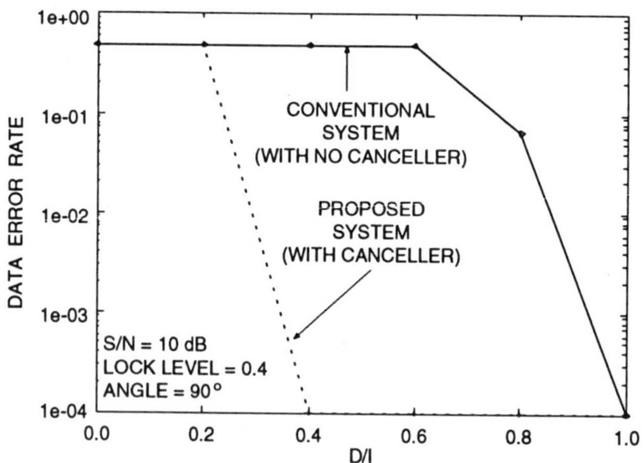

Figure 22: Demodulated BER vs. D/I Ratio

6.2 COMBINATION OF CCI CANCELLATION AND POWER CONTROL

In the serial structure of the canceller shown in Fig.21(b), interfering SS signals or CCIs from multiple access users are cancelled in the order of decreasing receiving power because it is easy to achieve acquisition, demodulation and cancellation of an interfering SS signal having greater power and its cancellation makes it possible or easy to cancel other interfering SS signals [36][37].

This discussion suggests that the performance of CCI cancellation or the capacity of CDMA depends upon receiving signal power distribution among multiple access users. In fact, when the receiving power distribution is exponential, the capacity can be improved to be 1.5 times as large as that in uniform power distribution.

Although transmission power control is considered as a key technology for increasing the capacity in a DS/CDMA cellular system, the receiving power of multiple access users is controlled so uniform as to solve the near-far problem. Errors in the power control results in severe degradation of the capacity [38]. However, if the power control is applied so as to make CCI cancellation more efficient, the power control errors can also be compensated by CCI cancellation.

6.3 TDL ARRAY ANTENNA BASED ON SPATIAL SPECTRAL ESTIMATION

A conventional updating algorithm of adaptive array antennas, such as LMS and Applebaum arrays, beamforms to track the desired signal and to suppress interference by nulls so as to maximize array output signal-to-noise ratio (SNR). Applebaum array is also useful when the arrival angle of the desired signal is known in advance. LMS array doesn't require any knowledge for the arrival angle of the desired signal. As long as reference signal correlated with the desired signal can be obtained, the array pattern will adaptively track the desired signal so as to maximize SNR. However, it is difficult to obtain a reliable reference signal in time-varying channels such as a CDMA mobile radio channel.

We have proposed and investigated an algorithm for controlling weights of antenna elements which are derived from spatial spectrum of spatially sampled signals by array of antenna elements. The arrival angles can be estimated from spatial frequency spectrum, which can be obtained by

Discrete Fourier Transform(DFT) or Maximum Entropy Method(MEM) for spatially sampled signals by antenna elements placed on a straight line. We have proposed and investigated an adaptive array antenna of which the weight coefficients are updated by Wiener solution derived from the estimated spatial spectrum. We name them DFT array[42] and MEM array[43].

(1) Updating Algorithm Based on Spatial Spectral Estimation

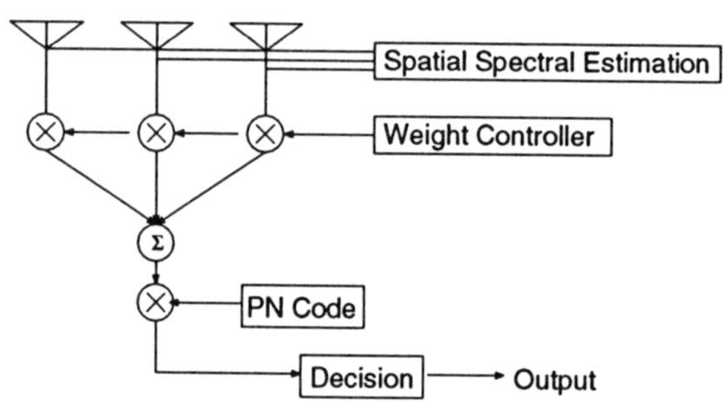

Figure 23: Adaptive Array Based on Spatial Spectrum Estimation

Fig.23 shows configuration of the adaptive array antenna based on spatial spectral estimation in receiver. An algorithm of updating antenna weights in receiver is described as follows.

(i) Each antenna element receives signals with different arrival angles, which are due to multipath, in the presence of additive white Gaussian noise(AWGN).

(ii) Power spectrum is calculated from spatially sampled signal at each time instant.

(iii) Arrival angles and SNR for the desired and its delayed signals are estimated by the power spectrum.

(iv) Optimal weight coefficients are derived by substituting the estimated arrival angles and SNR's into Wiener Solution.

(v) Antenna output is produced by weighted and sum of sampled signals.

This algorithm can repeatedly update the weight coefficients at each time instant. Therefore, the adaptive array antenna based on spatial spectral estimation can adapt quick variance in a channel comparing with LMS array. It has no problem on convergence of weight coefficients.

(2) Performance for Spread Spectrum Signals

When MEM array is applied to suppress CCI in CDMA, its spectral estimation performance corresponding to the bandwidth of spread spectrum signals or broadband signals should be investigated.

The spatial frequency f_s which is derived from spatial spectrum estimation of the spread spectrum signal is written by

$$f_s = (f_c + f_m)\frac{\sin\theta}{c} \tag{11}$$

where $\omega_c = 2\pi f_c$ is a carrier frequency, $\omega_c - \Delta W/2 < \omega_m = 2\pi f_m < \omega_c + \Delta W/2$. ΔW is the spread bandwidth and c is a velocity of an electromagnetic wave. It is noted that the estimated spatial frequency of a spread spectrum signal has also wide band corresponding to its temporal frequency bandwidth. Therefore, estimation of arrival angles and SNR in MEM array must has larger errors for a spread spectrum signal than that for a narrowband signal.

"MEM Broadband" in Fig.24 shows that estimation errors of arrival angle in MEM array increase as bandwidth of spread spectrum signals increases.

There is another algorithm based on spatial spectral estimation which is called MUSIC [41]. MUSIC estimates arrival angles(DOA) in noise subspace which is defined by eigen vectors of covariance matrix of spatially sampled signals, while MEM does it in signal subspace. MUSIC has better estimation performance for broadband signals than MEM if noise subspace is larger for uncorrelated signals than signal subspace. "MUSIC Broadband" in Fig.24 shows estimation errors in MUSIC corresponding to bandwidth or spreading ratio.

These algorithms based on spatial estimation are sensitive to correlation among signals. Spread spectrum signals in CDMA have cross-correlation between signature sequences assigned to CDMA users. In order to remove effect due to the correlation, we can estimate arrival angles by using spatial samples of despread signal instead. "MUSIC Correlation" and "MEM Correlation" in Fig.24 show that the estimation can be improved by this method in which the correlation can be reduced by the processing gain.

Another countermeasure to broadband signals is to use bandpass filters(BPFs) in estimation of arrival angles. If spatially sampled signals

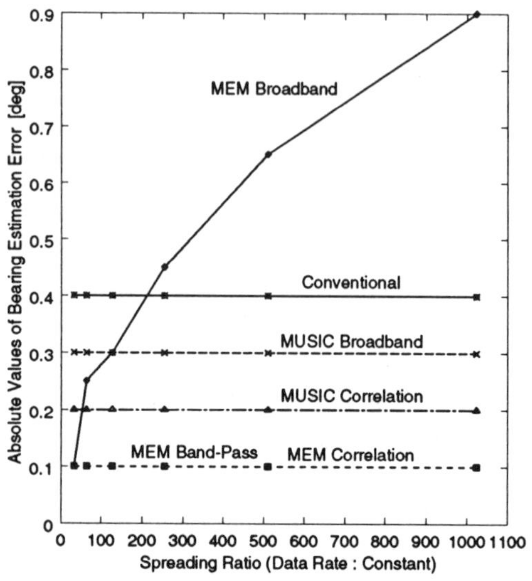

Figure 24: Estimation Errors of Arrival Angles vs. Spreading Ratio(Bandwidth Expansion)

pass through BPF before spatial estimation, effect of broadband signals can be reduced. "MEM Bandpass" in Fig.24 shows estimation performance of this method which achieves the same accuracy as "MEM Correlation". If BPF is used in MUSIC by the same manner, noise will be colored. MUSIC estimation performance must degrade because MUSIC operates in noise subspace.

7   CONCLUDING REMARKS

For CDMA, CCI cancellation schemes in temporal and spatial domains have been discussed.

For wider commercial applications of SS techniques, i.e. consumer and personal communications, improving transmission capacity of CDMA and combating with narrowband interference and fading are required compatibly. For this purpose, we should study total performance of capacity and complexity on practical mplementation, relationship between regulation and security, common use of frequency bands with other systems, development of ASIC for SS systems etc. It would be worthwhile to study these subjects for future consumer and personal communications.

REFERENCES

[1] Study Report on Standard for Wireless LAN, RCR July 1992.

[2] M. K. Simon, J. K. Omura, R. A. Scholtz and B. K. Levitt, Spread Spectrum Communications, Computer Science Press, 1985.

[3] R. Kohno,"A Trend of Research and Development Concerning Low Power Radio Systems for Consumer Communications," *IEICE Techn. Rep.*, SSTA89-1, pp.1-8, March 1989.

[4] G. Marubayashi, M. Nakagawa and R. Kohno,"Research and Development Activities on Spread Spectrum Communication Systems," *Journal of the IEICE*, 73, 2, pp.580-592, May 1989.

[5] R. Schneiderman, "Spread Spectrum Gains Wireless Applications," *Microwaves & RF*, May 1992.

[6] Federal Communications Commission (FCC) technical regulations 15.247

[7] K. S. Gilhousen *et al.* "On the Capacity of a Cellular CDMA System," *IEEE Trans. on Vehicular Technology*, vol. VT-40, no. 2, pp. 303-312, May 1991.

[8] R. Kohno,"Interference Cancellation and Information Security in Spread Spectrum Systems," *IEICE Techn. Rep.*, SST91-5, pp.29-36, April, 1991.

[9] R. Kohno,"Pseudo-Noise Sequences and Interference Cancellation Techniques for Spread Spectrum Systems," *IEICE Transactions on Communications, Electronics, Information and Systems*, vol.E74, No.5, May 1991.

[10] A. Salmasi and K. S. Gilhousen, "On the System Design Aspects of Code Division Multiple Access (CDMA) Applied to Digital Cellular and Personal Communications Networks." *Proc. of IEEE VTS Conference*, May 19-22, 1992.

[11] J. S. Lehnert and M. B. Pursley, "Multipath Diversity Reception of Spread - Spectrum Multiple - Access Communications," *IEEE Trans. Commun.*, vol. COM-35, pp. 1189-1198, Nov. 1987.

[12] H. V. Poor and S. Verdu,"Single-User Detectors for Multiuser Channels," *IEEE Trans. Commun.*, vol. COM-36, pp. 50-60, Jan. 1988.

[13] R. Kohno, H. Imai and M. Hatori,"Cancellation Techniques of Co-Channel Interference in Asynchronous Spread Spectrum Multiple Access," *IECE Tech. Rep.*, CS82-38, pp.29-35, June 1982.

[14] R. Kohno, H. Imai and M. Hatori,"Soft Decision Receiver Us-

ing Viterbi Algorithm for Asynchronous SSMA," *Proc. IECE National Conference*, 1322, p.5-193, March 1983.

[15] S. Verdu, "Minimum Probability of Error for Asynchronous Gaussian Multiple Access Channels," *IEEE Trans. Inform. Theory*, vol. IT-32, pp. 85-96, Jan. 1986.

[16] R. Kohno, H. Imai and M. Hatori, "Cancellation Techniques of Co-Channel Interference in Asynchronous Spread Spectrum Multiple Access Systems," *Trans. IECE, Japan*, vol. J66-A no. 5, pp. 416-423, May 1983 (in Japanese).
Electronics & Communications in Japan, Scripta Publishing Company, vol. 66, no. 5, pp. 20-29, May 1983, (English version).

[17] R. Kohno, H. Imai, M. Hatori and S. Pasupathy, "An Adaptive Canceller of Co-channel Interference for Spread Spectrum Multiple Access Communication Networks in a Power Line," *IEEE J. Select. Areas Commun.*, vol. SAC-8, pp. 691-699, May 1990.

[18] T. Masamura, "Spread Spectrum Multiple Access System with Intrasystem Interference Cancellation," *IEICE Trans., Japan*, vol. E71, no. 3, pp. 224-231, March 1988.

[19] Z. Xie, R.T. Short, C.T. Rushforth, "A Family of Suboptimum Detectors for Coherent Multiuser Communications," *IEEE J. Select. Areas Commun.*, vol. SAC-8, pp. 683-690, May 1990.

[20] M. K. Varanasi, B. Aazhang, "Multistage Detection in Asynchronous Code-Division Multiple-Access Communications," *IEEE Trans. Commun.*, vol. COM-38, pp. 509-519, April 1990.

[21] Y. C. Yoon, R. Kohno and H. Imai, "A Spread-Spectrum Multi-Access System with a Cascade of Co-Channel Interference Cancellers for Multipath Fading Channels," *Proc. IEEE Int. Symp. Spead Spectrum Tech. & Appl., (ISSSTA'92)*, pp.87-90, Nov. 1992.

[22] Y. C. Yoon, R. Kohno and H. Imai, "Cascaded Co-Channel Interference Cancelling and Diversity Combining for Spread-Spectrum Multi-Access over Multipath Fading Channels," *IEICE Trans. Commun., Japan*, vol. E76-B no. 2, Feb. 1993.

[23] Y. C. Yoon, R. Kohno and H. Imai, "A Spread-Spectrum Multi-Access System with Co-Channel Interference Cancellation over Multipath Fading Channels," *IEEE J. Select. Areas Commun.*, (to be published) Dec. 1993.

[24] P. M. Grant, S. Mowbray and R. D. Pringle, "Multipath and Co-Channel CDMA Interference Cancellation," *Proc. IEEE ISSSTA'92*, pp. 83-86, Nov. 1992.

[25] D. J. Goodman, P. S. Henry and V. K. Prabhu," Frequency-hopped Multilevel FSK for Mobile Radio," *Bell Syst. Tech. J.*, vol.59, No.7,

pp.1257-1275, Sept. 1980.

[26] T. Kawahara and T. Matsumoto,"Optimum Rate Reed Solomon Codes for FFH/CDMA Mobile Radio," *IEE Elec. Letts.*, vol.27, No.22, pp.2066-2067, Oct. 1991.

[27] T. Mabuchi, R. Kohno and H. Imai,"An Advanced Decoding Scheme of a Coded MFSK/FH-SSMA System," *IEICE Tech., Rep.*, SST92-25, June 1992.

[28] T. Mabuchi, R. Kohno and H. Imai,"A Scheme of Cancelling Co-Channel Interference in MFSK/FH-SSMA System," *Proc. 15th Symp. Information Theory and Its Applications (SITA'92)*, Sept. 1992.

[29] T. Kawahara and T. Matsumoto,"A DS/CDMA Mobile Radio System Featuring OR-Channel Characteristic," *Proc. IEICE National Conference*, A-131, p.1-133, Sept. 1992.

[30] D.V.Sarwate and M.B.Pursley,"Crosscorrelation Properties of Pseudorandom and Related Sequences," *Proc. IEEE*, vol.68, pp.593-619, 1980.

[31] H. Fukumasa, R. Kohno and H. Imai,"Design of Pseudo-Noise Sequences with Good Odd and Even Correlation Properties," *Proc. IEEE ISSSTA'92*, pp. 139-142, Nov. 1992.

[32] Y. C. Yoon, R. Kohno and H. Imai,"Performance of a Cascade of Co-Channel Interference Cancellers for DS/SSMA in the Presence of Amplitude and Phase Estimation Errors," *IEICE Tech. Rep.*, SST92-65, pp.31-36, Jan. 1993.

[33] R.T.Compton,Jr., Adaptive Antennas Concepts and Performance, Prentice-Hall, 1988.

[34] R. Kohno, H. Wang, and H. Imai,"Adaptive Array Antenna Combined with Tapped Delay Line Using Processing Gain for Spread-Spectrum CDMA Systems," *Proc. 3rd IEEE Int. Sym. Personal, Indoor and Mobile Radio Commu.*, pp.634-638, Oct. 1992.

[35] R. Kohno, H. Imai and M. Hatori "Design of an Automatic Equalizer Including a Decoder of Error-Correcting Code", *IEEE Trans. Commun.*, COM-33, pp. 1142-1146, (1985)

[36] R. Kohno, H. Imai, M. Hatori and S. Pasupathy,"Combination of an adaptive Array Antenna and a Canceller of Interference for Direct-Sequence Spread-Spectrum Multiple-Access System," *IEEE J. Selected Areas Comun.*, JSAC8, 4, pp.675-682, May 1990.

[37] A. J. Viterbi,"Very Low Rate Convolutional Codes for Maximum Theoretical Performance of Spread-Spectrum Multiple-Access Channels," *IEEE J. Select. Areas Commun.*, vol. JSAC-8, pp. 641-649, May 1990.

[38] E. Kudoh and T. Matsumoto,"Effects of Power Control Error on the System User Capacity of DS/SSMA Cellular Mobile Radios," *IEICE*

Trans. Commun., Japan, vol. E75-B, pp.524-529, June 1992.

[39] A. Saifuddin, R. Kohno and H. Imai,"Cascaded Combination of Cancelling Co-Channel Interference and Decoding of Error-Correcting Codes for CDMA," *Proc. 1994 IEEE International Symp. Spread Spectrum Tech. & Appli.* July 1994.

[40] R. Kohno, S. Pasupathy, H. Imai, and M. Hatori,"Combination of Cancelling Intersymbol Interference and Decoding of Error-Correcting Codes," *IEE Proceedings,* vol.133, No.3, pp.224-231, June 1986.

[41] T. Shan, M. Wax, and T. Kailath,"On Spatial Smoothing for Direction-of-Arrival Estimation of Coherent Signals," *IEEE Trans. Acoustics, Speech and Sig. Proc.,* vol.ASSP-33, No.4, pp.806-811, August 1985.

[42] C. Yim, R. Kohno and H. Imai,"An Adaptive Array Antenna Using DFT in Spatial Domain,"*IEICE Trans. Commun., Japan,* vol. J75-BII, No. 8, pp.556-565, August 1992.

[43] M. Nagatsuka, N. Ishii, R. Kohno, and H. Imai,"Adaptive Array Antenna Based on Spatial Spectrum Estimation Using Maximum Entropy Method", *IEICE Trans. Commun.,* vol. E77-B, No. 5, May 1994.

Interference Suppression for CDMA Overlays of Narrowband Waveforms

Laurence B. Milstein and Jiangzhou Wang

Abstract

An overview of the performance of CDMA networks which overlay narrowband signals in order to increase spectral efficiency is presented. Such systems employ interference suppression filters in the CDMA receivers to minimize the interference caused by the narrowband waveforms. The results presented in this paper include the effects of multipath fading, as well as both intracell and intercell interference for cellular systems.

Introduction

In this paper, an overview of the use of interference rejection techniques for code division multiple access (CDMA) systems which are designed to share common spectrum with narrowband waveforms is presented. It is assumed that these narrowband signals are being overlaid by a CDMA network in order to increase the overall spectral efficiency of the band, and that therefore they do not represent intentional jamming. Rather, the interference that they impose on the CDMA waveforms is the unavoidable result of attempting this type of spectral sharing, and it is the purpose of the interference suppression techniques to minimize the resulting degradation to system performance.

There have been both experimental results ([1]) and analytical results ([2-7]) obtained for this problem; most of the overview presented below is taken from either [2] or [3]. The idea is to make use of the key difference in the received waveforms, namely the amount of occupied bandwidth, to design a receiver to suppress most of the energy of the narrowband signal, while not causing excessive attenuation/distortion to the spread spectrum signals. The manner in which such suppression schemes operate in general interference environments has been documented in the literature (see, e.g., [8]), and thus will not be repeated here. Rather, the emphasis in this paper is on the specific scenario of CDMA overlays.

This work was partially supported by the Office of Naval Research under Grant N00014-91-J-1234, and by the National Science Foundation Industry/University Cooperative Research Center on Ultra-High Speed Integrated Circuits and Systems at UCSD.

S.G. Glisic and P.A. Leppänen (eds.), Code Division Multiple Access Communications, 147-160.
© 1995 *Kluwer Academic Publishers. Printed in the Netherlands.*

148

The CDMA system to be studied employs direct sequence (DS) spreading, binary phase shift keyed modulation, forward error correction coding with perfectly interleaved block codes, and ideal correlation detection. We consider both flat fading and frequency-selective fading channels, and we attempt to account for both intercell and intracell interference arising from a standard mobile cellular geometry.

Mathematical Model

The transmitted signal for the kth CDMA user is given by

$$S_k(t) = \text{Re}\left\{\sqrt{2P_k}\,b_k(t)a_k(t)\exp\left[j(2\pi f_0 t + \phi_k)\right]\right\} \tag{1}$$

where P_k is the average power of the signal, $b_k(t)$ is a random binary sequence representing the data, $a_k(t)$ is the spreading sequence, also modeled as a random binary sequence, f_0 is the carrier frequency, and $\varnothing_k$ is a random phase uniformly distributed in $[0,2\pi]$. Each data symbol has a duration of T seconds, each chip of the spreading sequence has duration T_c seconds, and the processing gain is defined as $N=T/T_c$.

The channel is modeled as a discrete multipath fading channel. The channel between the kth user and the receiver of interest (namely the receiver in the base station of what we refer to as the first cell) is modeled by the complex lowpass equivalent impulse response

$$h_k(t) = \frac{1}{\left(d_{1,k}\right)^{\gamma/2}} \sum_{l=1}^{L}\left[A_{kl}\exp\left(j\eta_{kl}\right) + \beta_{kl}\exp\left(j\eta_{kl}\right)\right]\delta(t-\tau_{kl}), \tag{2}$$

where $d_{1,k}\left(d_{1,k}\neq 0\right)$ is the distance between the kth user and the first cell base station, and γ is the propagation path loss exponent. In (2), A_{kl} and η_{kl} are the gain and the phase of the specular component of the lth path from the kth user, respectively; they are assumed to be deterministic and constant over the duration of at least one bit. The random gain β_{kl} and random phase μ_{kl} of the fading component of the lth path of the kth user have a Rayleigh distribution with $E\left[\beta_{kl}^2\right] = 2\rho_{kl}$, and a uniform distribution in $[0,2\pi]$, respectively. Consequently, the complex-valued gain $A_{kl}\exp\left(j\eta_{kl}\right) + \beta_{kl}\exp\left(j\mu_{kl}\right)$ of the lth path from the kth user is a Gaussian random variable, whose real and imaginary parts are statistically independent, with means $A_{kl}\cos\left(\eta_{kl}\right)$ and $A_{kl}\sin\left(\eta_{kl}\right)$, respectively, and common variance $E\left[\beta_{kl}^2\right]E\left[\cos^2(\mu_{kl})\right]=E\left[\beta_{kl}^2\right]E\left[\sin^2(\mu_{kl})\right]=\rho_{kl}$. The path delay, τ_{kl}, is uniformly distributed in $[0,T_b]$ We assume that there are L paths associated with each user. The gains, delays and phases of different paths and/or of different users are all statistically independent. We define the quantity H_{kl} as the ratio of the specular component power to the Rayleigh fading power. That is,

$$H_{kl} = A_{kl}^2 / E\left[\beta_{kl}^2\right] = A_{kl}^2 / (2\rho_{kl}). \tag{3}$$

The interference in the channel is assumed to be a non-fading narrowband BPSK signal, and is given by

$$J(t) = \text{Re}\left\{\sqrt{2J}d(t)\exp\left[j\left(2\pi(f_0+\Delta)t+\theta\right)\right]\right\}, \tag{4}$$

where Δ stands for the offset of the interference carrier frequency from the carrier frequency of the CDMA signals. The parameters J and Θ denote the received interference power and phase, respectively. The information sequence $d(t)$ has a bit rate of T_j^{-1}. Therefore, the interference bandwidth is approximately $B_{j=2}T_j^{-1}$ and we assume B_j is much less than B_s, the spread bandwidth of the CDMA waveform. The ratio of the interference bandwidth to the spread spectrum bandwidth, p, and the ratio of the offset of the interference carrier frequency to half of the spread spectrum bandwidth q, are defined as

$$p = B_j / B_s = T_c / T_j \tag{5}$$

and

$$q = \Delta / (B_s/2) = \Delta T_c, \tag{6}$$

respectively.

The received signal $r(t)$ can be represented as

$$r(t) = \text{Re}\left\{\sqrt{2P}\sum_{k=1}^{CK}\sqrt{\varepsilon(\gamma,c_k,k)}\cdot\sum_{l=1}^{L}\left[A_{kl}\exp(j\phi_{kl})+\beta_{kl}\exp(j\varphi_{kl})\right]\right.$$
$$\left. \cdot b_k(t-\tau_{kl})a_k(t-\tau_{kl})\exp(j2\pi f_0 t)\right\}+J(t)+n(t), \tag{7}$$

150

where $n(t)$ is an AWGN process with two-sided power spectral density $N_0/2$, $\phi_{kl} \equiv \theta_k + \eta_{kl} - 2\pi f_0 \tau_{kl}$, $\psi_{kl} \equiv \theta_k + \mu_{kl} - 2\pi f_0 \tau_{kl}$, C stands for the number of cells, each one containing K users, and c_k denotes the cell in which the kth user is located; the users are numbered such that C_k is the integer portion of $1+(k-1)/K$, $c_k = 1,...,C$. The first cell $(c_k = 1)$ is defined as the cell of interest, and P and $\varepsilon(\gamma,c_k,k)$ are defined as

$$P = P_k / \left(d_{c_k,k}\right)^{\gamma} \tag{8}$$

and

$$\varepsilon(\gamma,c_k,k) = \left(d_{c_k,k} / d_{l,k}\right)^{\gamma}, \tag{9}$$

respectively, where $d_{c_k,k}$ is the distance of the kth mobile user to its own base station (the c_kth cell), and $d_{c_k,k} \neq 0$. The fact that P is taken to be a constant implies that each base station provides adaptive power control to all K users of its own cell so that the received signals from its cell all arrive with the same power.

The receiver is shown in Fig. 1(a); it contains a narrowband interference suppression filter, a bandpass matched filter, a bank of U parallel PSK demodulators and diversity circuits, a hard decision device, a deinterleaver and a hard decision decoder. A detailed model of the PSK demodulators and diversity circuits is shown in Fig. 1(b).

Notice that, for simplicity, in the analysis, we have ignored any bandpass filtering at the front-end of the suppression filter. In most instances, we take the suppression filter to be a double-sided Wiener filter with M taps on each side. Its impulse response is $\sum_{m=-M}^{M} \alpha_m \delta(t-mTc)$, where $\alpha_0 = 1$ and $\alpha_m = \alpha_{-m}$. The resulting suppression filter output is given by

$$r_f(t) = \sum_{m=-M}^{M} \alpha_m r(t - mt_c) \tag{10}$$

where $f_0 T_c$ is taken to be an integer.

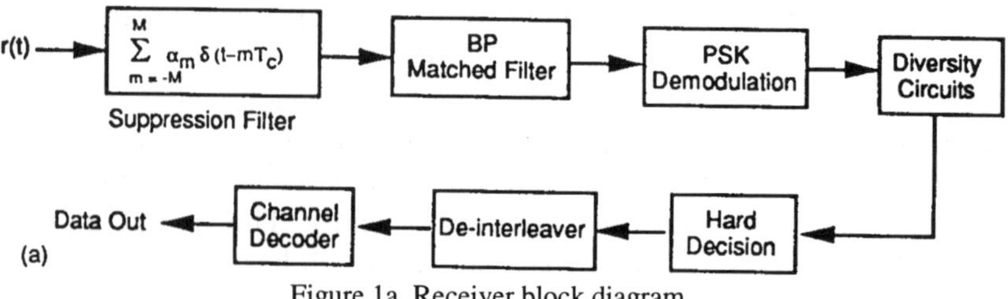

Figure 1a. Receiver block diagram.

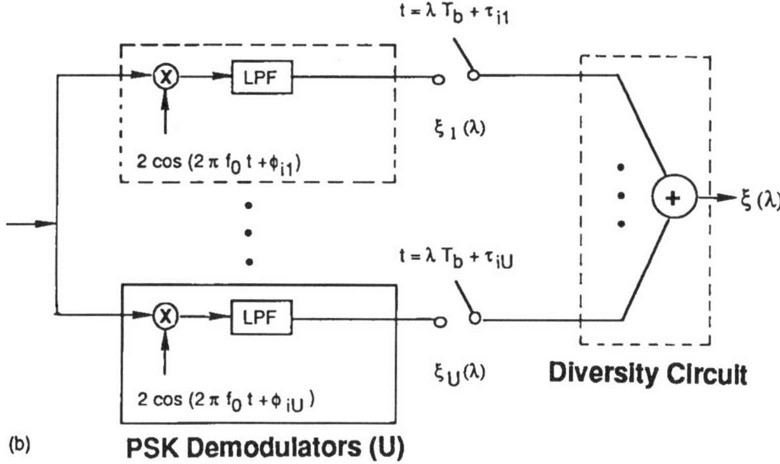

Figure 1b. Detailed model of the PSK demodulators and diversity circuits.

The signal $r_f(t)$ enters a bandpass matched filter, whose impulse response is variable and at any point in time is matched to a specific T_b-second segment of $2a_i(t)\cos(2\pi f_0 t)$. Also, it is assumed that the ith user of the first cell is the reference user. During each bit duration, the bandpass signal at the output of the appropriate matched filter contains $2M+1$ narrow pulses, each of width $2T_c$, centered at instants $\lambda T_b + \tau_{il} + mT_c$ for $m = -M,...,0,...,M$, which are caused by the $2M+1$ taps of the suppression filter and the lth $(l = 1, 2,..., L)$ path of the reference user. Of the $2M+1$ peaks, the middle peak $(m = 0)$ is the largest (or main) peak, due to the zero-th tap of the suppression filter, whereas sidelobe peaks are due to the taps other than the zero-th tap. We assume that

$$\left| \tau_{il} - \tau_{i\hat{l}} \right| \geq (2M+1)T_c \quad \text{for} \quad \hat{l} = 1,...,L, \hat{l} \neq l \quad \text{(see Fig.2), since, if}$$

$\left| \tau_{il} - \tau_{i\hat{l}} \right| < (2M+1)T_c$, the overlay of the outputs due to different paths might preclude the paths from being resolvable. We further assume that the receiver only uses the combination of the main peaks of the resolvable paths of the reference user to form the diversity combining.

In any given symbol interval, the output of the appropriate bandpass matched filter enters U conventional PSK demodulators, $1 \leq U \leq L$, (Fig. 1(b)), corresponding to U different resolvable paths of the reference user. The coherent receiver makes use of carrier reference signals of the form $2\cos(2\pi f_0 t + \phi_{il})$, which are in phase with the specular components of the different signals from the reference user. These carrier reference signals can be derived from the bandpass matched filter outputs by means of carrier synchronization circuits (one circuit per path). The output of the lth demodulator, $1 \leq l \leq U$, is a lowpass signal, which has $2M+1$ peaks at the instants $\lambda T_b + \tau_{il} + mT_c, m = -M,\ldots,0,\ldots,M$. The desired component of the lowpass output signal (i.e., the main peak, corresponding to $m=0$) of the lth demodulator is sampled at the instant $t = \lambda T_b + \tau_{il}$, and this gives rise to the random variable $\xi_l(\lambda)$. Note that the time difference between any two samples is at least equal to $(2M+1)T_c$. The U random variables, $\xi_l(\lambda), l = 1,\ldots,U$, are directly summed to form the decision variable $\xi(\lambda)$.

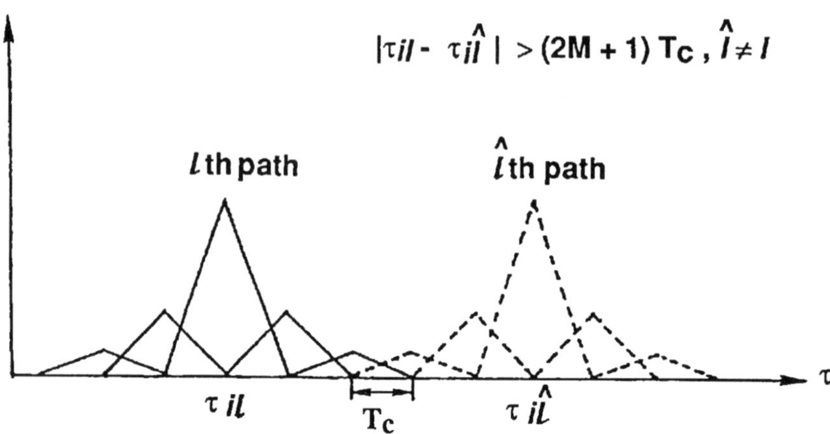

Figure 2. The BP matched filter output envelope.

Performance Results

To see how the use of the suppression filter enhances the system performance, consider the following results, taken from [2] and [3]. For simplicity, we initially consider just a single cell $(C=1)$, a nonselective Rayleigh fading channel $(L=1$ and $H=0$), and an interfering BPSK signal located at the carrier frequency of the CDMA waveforms. We later relax these assumptions.

Fig. 3 shows the BER performance of the system with a double-sided suppression filter as a function of the average signal-to-noise ratio, denoted by $\bar{E}_b / N_0$, and derived in [2]. Also, the BER of the system both with a single-sided suppression filter and without a suppression filter are plotted in order to allow for comparison. The systems use $K=10$ active users, $N=255$ chips per bit, and an interference power-to-signal power ratio $J/S=20dB$. It is assumed that the two-sided filter has three taps and is symmetric, and the single-sided filter has the same number of total taps. Finally, the narrowband interference occupies 10 percent $(p=0.1)$ of the spread spectrum bandwidth. It is seen from the figure that, as expected, the BER performance can be improved by using a suppression filter. In particular, tremendous improvement can be achieved by using a double-sided filter.

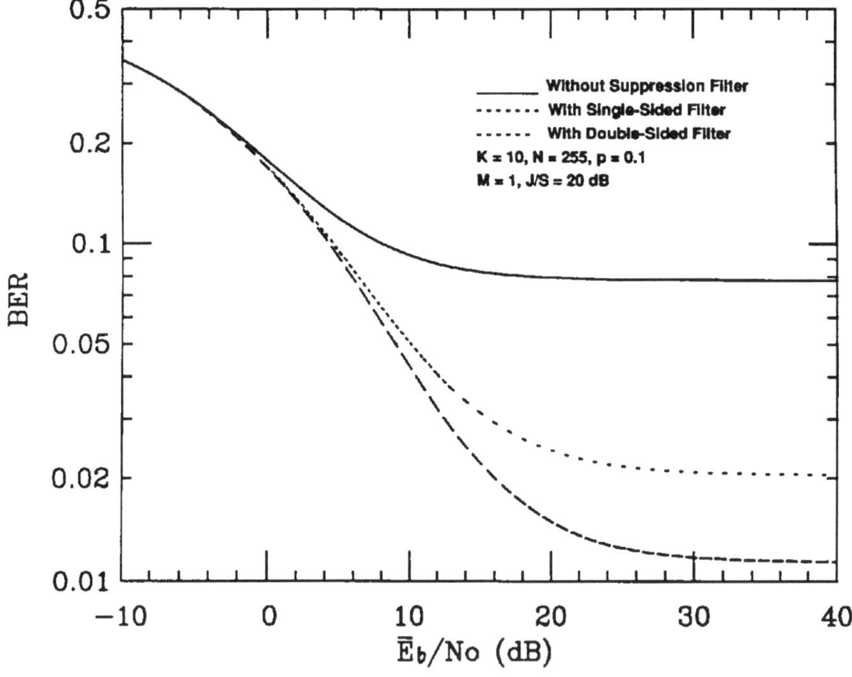

Figure 3. The BER performance of the system.

Fig. 4 illustrates the asymptotic $\left(\bar{E}_b / N_0 \to \infty\right)$ BER of the system with a double-sided filter, a single-sided filter and without a suppression filter, as a function of J/S. When J/S is large (i.e., $J/S \geq 10dB$), the system without a suppression filter degrades significantly, whereas the system is much more tolerant of the interference when the suppression filter is present.

In Fig. 5, the asymptotic BER of the DS-CDMA system with a double-sided filter is plotted as a function of the number of active users (K) for different values of J/S. The system parameters are $N = 255$, and $p = 0.1$, and a three-tap double-sided suppression filter is employed. Also, the BER of the system without the suppression filter is shown for comparison. It is seen that, for a given BER, the system with the double-sided filter can support many more users than can the system without the suppression filter, especially when the J/S is large. When the interference is not too large (i.e., $J/S \leq 10dB$) the system employing the suppression filter can support almost the same number of users as can the system in the absence of any interference to begin with.

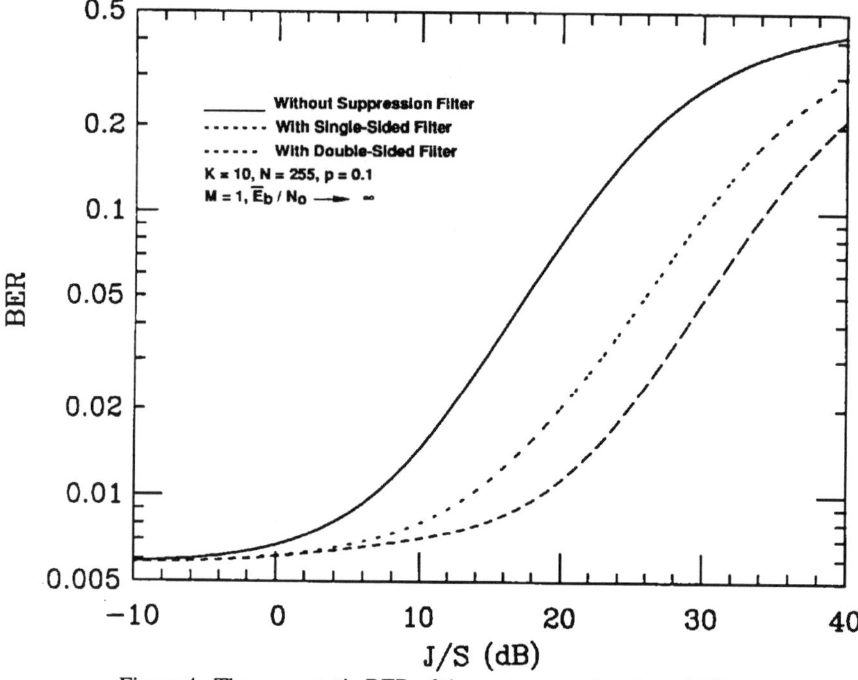

Figure 4. The asymptotic BER of the system as a function of J/S.

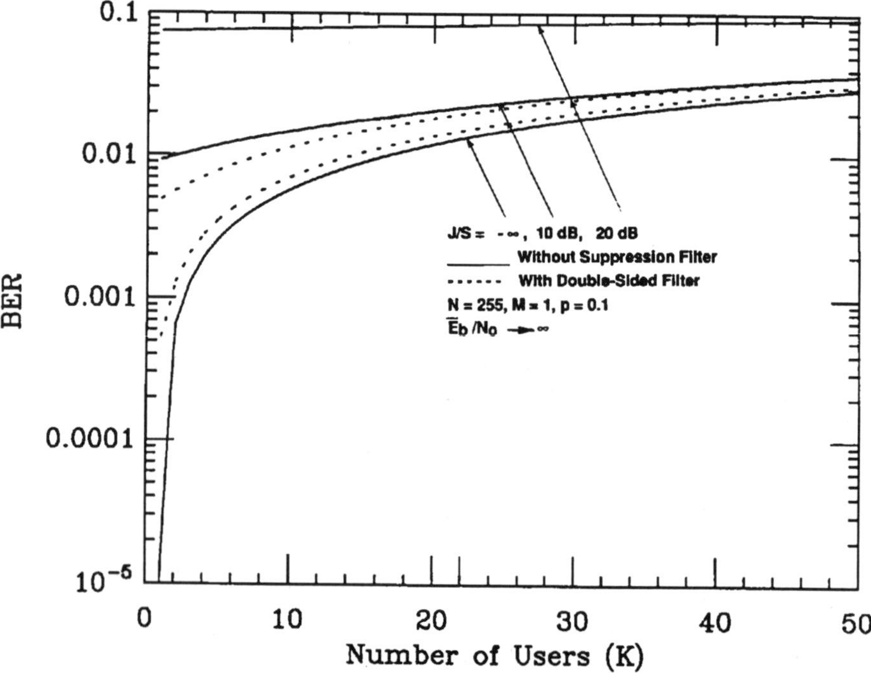

Figure 5. The asymptotic BER of the system as a function of the number of
active users (K).

Now consider the more general case, corresponding to multiple cells, a frequency selective channel, and Rician fading. Unless noted otherwise, it is assumed that the ratio of the interference bandwidth to the spread spectrum bandwidth is 10%, the ratio of the offset of the interference carrier frequency to half of the spread spectrum bandwidth is 20%, the ratio of the specular component power to the fading component power, H, is 7 dB, and the processing gain, N is 255. Note that where the BER is plotted as a function of the ratio $\bar{E}_i / N_0$, this corresponds to the energy-per-bit-to-noise spectral density of an uncoded system.

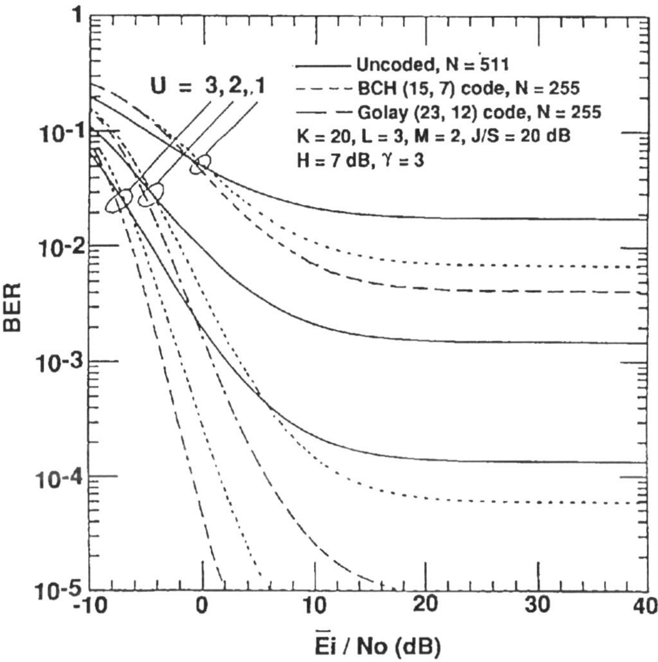

Figure 6. The asymptotic BER of the system as a function of the ratio $\bar{E}_i / N_0$

In Fig. 6, the decoded BER for $J/S = 20$dB is shown. Also, the uncoded BER is plotted in order to allow for comparison. Note that the uncoded and coded systems are compared for the same spread spectrum bandwidth. In our example, $N = 255$ for the coded systems, whereas $N = 511$ for the uncoded system. As expected, both the uncoded and coded BER decrease as the order, U, of diversity increases. The performance advantage obtained by using diversity becomes larger as the error correcting capability of the code increases for the same spread bandwidth.

In Fig. 7, the asymptotic BER of the system with and without the suppression filter is shown as a function of the ratio of the interference bandwidth to the spread spectrum bandwidth. It is seen from the figure that the CDMA system with a double-sided Wiener filter can suppress the narrowband interference very effectively (assuming, of course, that p does not become too large, in which case the interference is no longer narrowband).

156

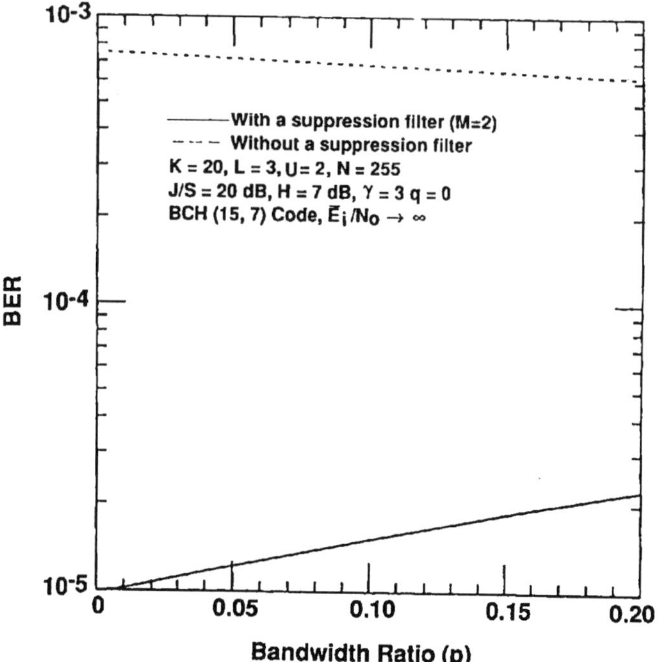

Figure 7. The asymptotic BER of the system as a function of the bandwidth ratio p.

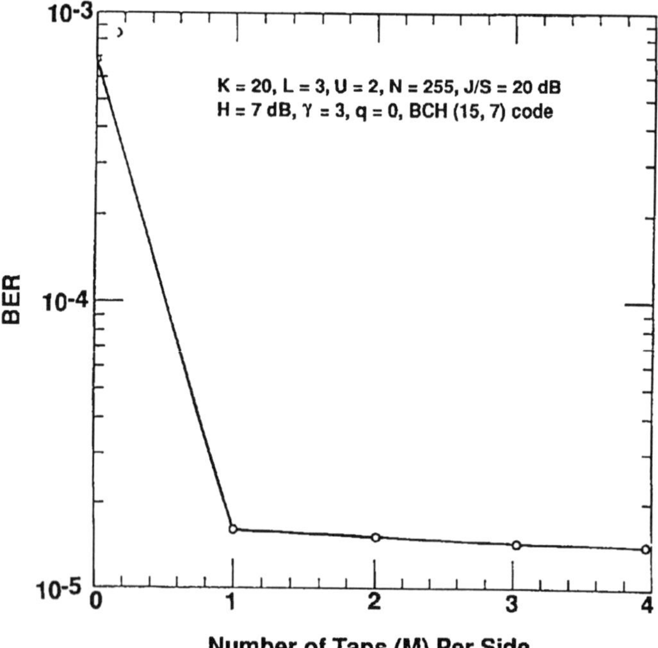

Figure 8. The asymptotic BER of the system as a function of the
number of taps (M) on each side.

Fig. 8 illustrates the asymptotic BER for the system employing the double sided suppression filter as a function of the number of taps on each side, M. It is seen from the figure that the BER decreases significantly when $M=1$, and then keeps a steady value with increasing M. Therefore, we conclude that when the carrier frequencies of the narrowband interference and CDMA signals are the same, the suppression filter with three total taps is sufficient to reduce the narrowband interference, and increasing the number of total taps beyond three is not necessary.

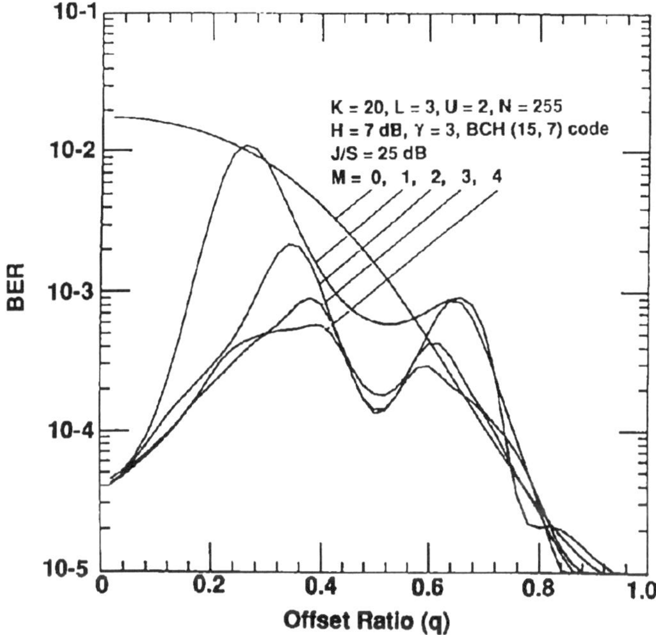

Figure 9. The BER performance of the system with the suppression filter as a function of the ratio (q).

Fig. 9 illustrates the BER performance of the system with the suppression filter as a function of the ratio (q) of the offset of the interference carrier frequency to the half spread spectrum bandwidth, for different number of the taps. It is seen that when q is very small, the BERs of the system with double-sided filters for $M \geq 1$ are almost identical, as was the case in Fig. 8. However, the BER performance improves as M increases for larger q, and the BER for large M is also more robust to a change in q.

In Fig. 10, the asymptotic BER of the DS-CDMA system with a suppression filter is plotted as a function of the number of active users, K, for various values of J/S. Also, the BER of the system without the suppression filter is shown for comparison. It is seen that for large J/S, for a given BER, the system with the suppression filter can support many more users than can the system without the suppression filter. When the interference is such that $J/S \leq 10 dB$, the multiple-access and adjacent-cell interference dominates, so that the suppression filter is not needed.

158

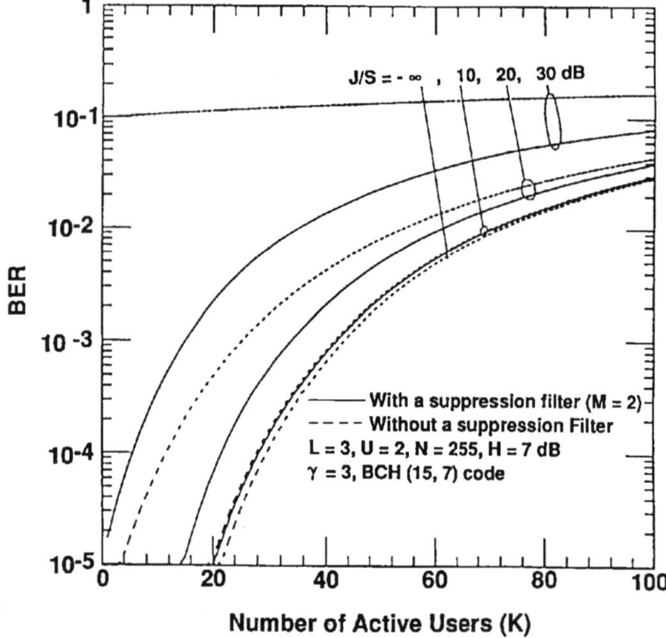

Figure 10. The asymptotic BER of the system as a function of the number of active users (K) for various values of J/S.

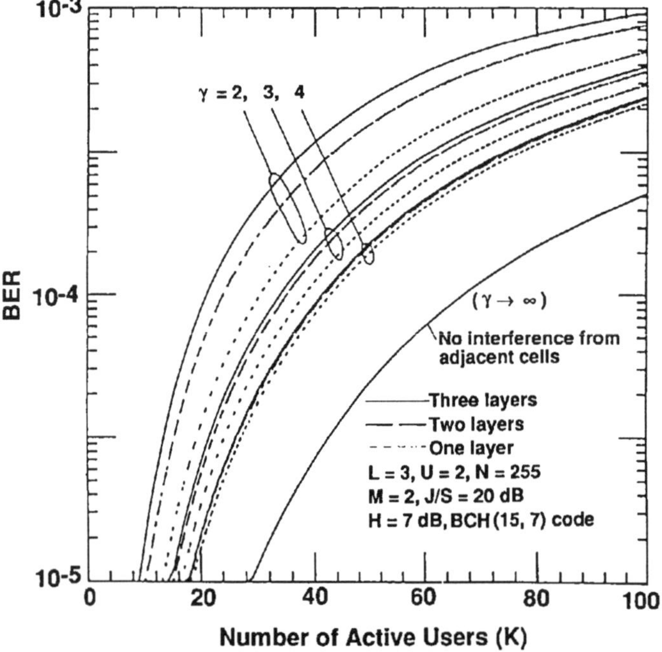

Figure 11. The asymptotic BER of the system versus the number of active users (K) for various values of the propagation loss exponent γ.

Fig. 11 illustrates the asymptotic BER of the CDMA system versus K, the number of active users, for various values of the propagation loss exponent, $\gamma, (\gamma = 2, 3, 4)$. Also, the BER in the absence of adjacent cell interference (i.e., $\gamma \to \infty$) is shown for comparison. In order to show the effect of the adjacent-cell interference on the performance of the system for different values of γ, a single layer of cells, two layers of cells and three layers of cells are considered. It is clear from this figure that when $\gamma = 2$, the BER computed by accounting for only a single layer of cells is too optimistic; alternately, if $\gamma \geq 3$, the difference in BER performance between accounting for only a single layer and accounting for more layers is insignificant. Further, for a given γ the gap in the BER between two and three layers is much less than that between one and two layers.

Conclusion

We have presented an overview on the use of interference suppression filters to enhance the performance of a CDMA network overlaying a narrowband conventional digital communications signal. Among our conclusions are the following:

1. When the narrowband interference power-to-signal power ratio J/S is small with respect to the multiple-access and adjacent-cellinterference, the suppression filter is unnecessary, whereas when J/S is on the order of (or greater than) the multiple-access and adjacent-cell interference, the suppression filter provides significant enhancement in performance.

2. When the ratio of the offset of the interference carrier frequency to half of the spread spectrum bandwidth is very small, that is, when the interference carrier frequency is very close to the spread spectrum carrier frequency, spread spectrum systems with double-sided suppression filters only need three total taps. However, a double-sided filter with a larger number of taps (i.e, $M > 1$) is preferable when the ratio is large.

3. For large narrowband interference, the system with a suppression filter can support many more users than can the system without a suppression filter.

REFERENCES

[1] Milstein, L. B., D. L. Schilling, R. L. Pickholtz, V. Erzeg, M. Kullback, E. Kanterakis, D. Fishman, W. H. Biederman, and D. Salerno, On the feasibility of a CDMA overlay for personal communications networks. *IEEE Journal on Selected Areas in Comm.,* Vol.10, pp. 655-668, May 1992.

160

[2] Wang, J. and L. B. Milstein, Applications of suppression filters for CDMA overlay situations. *1992 International Conference on Communications,* pp. 310.3.1-310.5, June 1992.

[3] Wang, J. and L. B. Milstein, CDMA overlay situations for microcellular mobile comunications. To appear in *IEEE Trans. on Comm.*

[4] Gottesman, L. D., and L. B. Milstein, The coarse acquisitionperformance of a CDMA overlay system. Submitted to *IEEE Journal on Selected Areas in Comm.*

[5] K. G. Filis and S. C. Gupta, Coexistence of cellular CDMA and FSM: Interferenced suppression using filtered PN sequences, *IEEE 1993 Global Telecomm. Conf.,* pp. 898-902.

[6] P. Wei, T. Soni, J. R. Zeidler, W. H. Ku and P. K. Das, Adaptive interference suppression for CDMA based PCN systems, *IEEE International Conference on Communications ,* May 1993, Geneva, Switzerland.

[7] L. A. Rusch and H. V. Poor, Narrowband interference suppression in CDMA spread spectrum communications, *IEEE Trans. Comm.,* pp. 1969-1979, April 1994.

[8] Milstein, L. B., Interference rejection techniques in spread spectrum communications. *Proceedings of the IEEE,* pp. 657-671, June 1988.

DS/CDMA SUCCESSIVE INTERFERENCE CANCELLATION

Jack M. Holtzman

ABSTRACT

Conventional DS/CDMA detectors operate by enhancing a desired user while suppressing other users, considered as interference (*multiple access interference, MAI*) or noise. A different viewpoint is to consider other users not as noise but to jointly detect all users' signals (multiuser detection). This has significant potential of increasing capacity and near/far resistance. Optimal multiuser detection is, however, too complex to implement, thus motivating the search for suboptimal algorithms. Our objective is to underline the need for simplicity and to discuss what is a relatively simple form of multiuser detection, successive interference cancellation. The cancellation scheme uses only components already present in a conventional detector.

I. INTRODUCTION

Conventional DS/CDMA detectors operate by enhancing a desired user while suppressing other users, considered as interference (*multiple access interference, MAI*) or noise. A different viewpoint is to consider other users not as noise but to jointly detect all users' signals (multiuser detection). This has significant potential of increasing capacity and near/far resistance. Since Ref. 1 discusses multiuser detection, we will only give enough background on multiuser detection here that is needed for our objective (also see [29] for more extensive discussion). Our objective is to underline the need for simplicity and to discuss what is a relatively simple form of multiuser detection, successive interference cancellation.

As mentioned, optimal multiuser detection has significant potential for performance improvement. It is, however, much too complex to implement. Much effort is currently in progress on developing suboptimal detectors which are not too complex to implement. The key question is: Is there a suboptimal version that is

(i)   Cost effective to build in a practical system and which
(ii)   Still retains enough advantage over the conventional detector?

Towards answering this question, the orientation of the paper is as follows. One can look at the optimal multiuser detector and successively make simplifications until implementability is attained. Instead of working our way downwards from the optimal, we discuss working upwards from the

S.G. Glisic and P.A. Leppänen (eds.), Code Division Multiple Access Communications, 161-180.
© 1995 *Kluwer Academic Publishers. Printed in the Netherlands.*

conventional detector. That is, we consider what is believed to be the simplest augmentation to the conventional detector, a form of successive interference cancellation. We will try to provide a perspective on such an augmentation.

The earliest paper on DS/CDMA multiuser detection is probably [2]. Another early paper, [3], specifically discusses multiuser detection in terms of cancellation. Key theoretical papers (with earlier references) are [4], [5]. There have been a number of other papers on interference cancellation, e.g. [6-17, 23] (not an exhaustive list).

The paper is organized as follows. Section II places the candidate for multiuser detection at the base station in a cellular system and Section III places bounds on the improvements to be expected. Section IV introduces optimal and suboptimal multiuser detection for a simplified system. Section V discusses issues associated with the successive interference cancellation scheme. Concluding remarks are given in Section VI.

II. MULTIUSER DETECTION IN CELLULAR SYSTEMS

In a cellular system, a number of mobiles communicate with one base station (BS). Each mobile is concerned only with its own signal while the BS must detect all the signals. Thus, the mobile has the information only about its own chip sequence while the base station has knowledge of all the chip sequences. For this reason, as well as less complexity being tolerated at the mobile (where size and weight are critical), multiuser detection is currently being envisioned mainly for the BS, or in the reverse link (mobile to BS). It is important to realize, however, that the BS maintains information only on those mobiles in its own cell and there are mobiles outside the cell causing interference. This plays a role in the limitations on improvements to be expected in a multiuser detection system, to be discussed next.

<u>Remark:</u>   Note that we assume the same information about chip sequences is available to the multiuser detector as is available for the conventional detector in current cellular systems. It is conceivable, and has indeed been proposed, that the chip sequences of other users be considered by explicitly communicating this information or by adaptive or blind methods (see [1]). Our assumption is based on making the minimum modification to current conventional detectors.

The next section on limitations on improvement from multiuser detection underlines the importance of maintaining simplicity.

III. LIMITATIONS TO IMPROVEMENTS

Before we discuss multiuser improvements to the conventional DS/CDMA detector, it is important to define factors that limit such improvement. One factor is intercell interference in a system that cancels only the intracell interference I. For intercell interference which is a fraction f of the intracell interference, the bound on capacity increase (with all of the intracell interference canceled) is $(1+f)/f$. For $f = 0.55$, this factor is 2.8 [18][1].

Another factor is the fraction, f_c, of energy captured by a Rake receiver. That is, a Rake receiver with L branches or "fingers" will try to capture the energy in the L strongest multipath rays but there will be additional received energy in additional rays. Ref. 19 gives examples of the fraction of captured energy. The fraction of captured energy is a function of chip duration times the number of Rake branches divided by the delay spread.

So, combining the two effects (measured by f and f_c), the total interference before cancellation is $(1+f)I$ (neglecting the smaller self-interference due to uncaptured multipath power of the desired user). Cancellation removes at most $f_c I$ so the bound on improvement is $(1+f)/(1-f_c+f)$. Depending on the system parameters, f_c can vary from near 1 to substantially less (see [19]). For $f_c \approx 1$, the above bound on capacity improvement of 2.8 remains. For $f_c = 0.5$, the bound is reduced to 1.5. Note that a small f_c also presents a problem to the conventional detector.

It should also be recognized that interference cancellation is used not only to increase capacity but also to alleviate the near/far problem, and the preceding bound does not account for that benefit. This actually translates into a capacity benefit which is, however, more difficult to quantify than by the above simple signal/interference argument. An interference cancellation scheme could recapture part of the reduction due to received power variability by reducing the variability (or it could be used to relax the requirements on power control).

To put these constraints on improvements into further perspective, we are assuming here that multiuser detection is a candidate primarily for the reverse link for reasons given in Section II. Since the reverse link is usually more limiting than the forward link [20], increasing the reverse link capacity will improve the overall system capacity. But increasing it beyond the forward link capacity will not further increase the overall system capacity. Thus,

(i)   The potential capacity improvements are not enormous (order of magnitude) but certainly non-trivial and worth pursuing.

(ii)   Enormous capacity improvements only on the reverse link (the candidate for multiuser detection) would only be partly used anyway in determining overall system capacity.

---

[1] With a sectorized antenna, it is conceivable to cancel users in another sector.

164

(iii)   Hence, the cost of doing multiuser detection must be as low as possible so that there is a performance/cost tradeoff advantage to multiuser detection.

The bottom line is that there are significant advantages to multiuser detection which are, however, bounded and a simple implementation is needed.

Remark: The bounds presented here are based on certain assumptions (see the Remark in Section II) and parameters and are, thus, not universal bounds. Rather, they attempt to indicate the *type of constraints* that need to be considered for the near-term simple adaptation of a conventional detector.

IV. OPTIMAL AND SUBOPTIMAL MULTIUSER DETECTION FOR A SIMPLIFIED DS/CDMA SYSTEM

We will explain the basic problem with a very simplified DS/CDMA system. There are a number of simplifications which will be relaxed in the rest of the article. In fact, each relaxation of simplification will represent another factor to consider for the multiuser detection system. Suppose there are K users sharing the same bandwidth and each user's baseband signal is

$$u_k(t) = A_k b_{ik} a_k(t - iT), \quad iT \le t < (i+1)T \tag{IV - 1}$$

where T is the bit interval and $A_k,\ b_{ik},\ a_k(t)$ are the amplitude, bit sequence, and spreading chip sequence (or signature waveform, or code), respectively, of user k. The b_{ik} are ± 1 and the objective is to detect those polarities, which contain the transmitted information. The received signal (at baseband) is the sum of all the users' signals plus noise:

$$r(t) = \sum_{k=1}^{K} u_k(t) + z(t) \tag{IV - 2}$$

where $z(t)$ is the additive noise. Note the simplifications which need to be relaxed:

(i)   Synchronous reception: all of the users' bits are aligned in time.
(ii)   Coherent reception: all of the relative phases are zero (coherent reception only requires knowledge of the phases, not that they are the same).
(iii)   The amplitudes are constant.
(iv)   No multipath.

The conventional DS/CDMA receiver detects the bit from user j by correlating the received signal with the chip sequence of user j. Since the system is synchronous, we can isolate each bit interval. Over the interval $[0, T]$, the decision statistic is:

$$y_{0j} = \int_0^T r(t)a_j(t)dt = \int_0^T a_j(t)[\sum_{k=1}^{K} A_k a_k(t)b_{0k} + z(t)]dt$$

$$= A_j b_{0j} + \sum_{k \neq j}^{K} A_k b_{0k} \int_0^T a_j(t)a_k(t)dt + \int_0^T a_j(t)z(t)dt$$

(IV - 3)

where it is assumed that the integral of $a_j^2(t)$ is unity.

The decision on the sign of b_{0j} is given by the sign of y_{0j}. Note that y_{0j} consists of three terms. The first is the desired information which gives the sign of the information bit b_{0j} which is exactly what is sought. The second term is the result of the multiple access interference, and the last is due to the noise. The second term typically dominates the noise so that one would like to remove its influence. Its influence is felt through the cross-correlations between the chip sequences. Using knowledge of the chip sequences (available at the BS), one could cancel the effect of one user upon another. This is, in fact, the intuitive motivation for interference cancellation schemes. It is important to also note that the factors $A_k b_{0k}$ are also needed. These can be obtained by either multiplying separate estimates of A_k and b_{0k} or by an estimate of the product itself. This choice will be further discussed below.

For the simplified problem discussed above, it may be shown that the maximum likelihood decision for the vector of bits, b, is given by

$$\hat{b}^* = \arg\left\{ \max_{b \in \{-1,+1\}^k} \left[2y^T b - b^T RAb \right] \right\}$$

(IV - 4)

where

$$y = RAb + \tilde{z}$$

(IV - 5)

where $\hat{z}$ is the noise term after correlation, A is a *diagonal* matrix of amplitudes and R is the matrix of crosscorrelations,

$$R_{jk} = \int_0^T a_j(t)a_k(t)dt$$

(IV - 6)

166

The exponential complexity motivates the need for suboptimal approaches. Multistage approaches are common. One, given in [21], uses (IV-7) instead of (IV-4).

$$\hat{b}_k(m+1) = \arg\left\{ \max_{\substack{b_k \in \{+1,-1\} \\ b_l = \hat{b}(m), \forall l \neq k}} \left[2y^T b - b^T R A b \right] \right\} \qquad \text{(IV - 7)}$$

In (IV-7), an estimate at stage $(m+1)$ is made using estimates at stage m. Comparing the two equations, one sees that while (IV-4) requires the simultaneous search over all components of the bit vector, (IV-7) searches over one component at a time. Decisions for the other bits are used from a previous stage of processing. This leads to two issues for multistage detectors:

(i)   How to choose the initial stage.
(ii)   How to choose the subsequent stages of processing.

Different alternatives will lead to different suboptimal algorithms discussed. One alternative is to use the conventional detector and another is the *decorrelator detector* which we now motivate. Inspection of (IV-5) immediately suggests a method to solve for b, whose components b_{0j} contain the bit information sought. If $\hat{z}$ was identically zero, we have a linear system of equations, $y = RAb$, the solution of which can be obtained by inverting R (assuming it is invertible) since A is diagonal. With a non-zero noise vector $\hat{z}$, inverting R is still an effective procedure (and actually optimal in certain restricted circumstances). This results in

$$R^{-1}y = Ab + R^{-1}\tilde{z} \qquad \text{(IV - 8)}$$

where it is seen that the information vector, b, is recovered in the first term but contaminated by a new noise term (which may be enhanced).

Consideration of alternative initial and subsequent stages of processing is given in [22].

V. SUCCESSIVE INTERFERENCE CANCELLATION

We have seen there can be a number of alternatives for suboptimum multiuser detection. They have different comparative performance and complexity. What they all have displayed is improved performance and increased complexity compared to the conventional detector. We shall discuss now what is probably the least complex of the schemes but which still retains superiority over the conventional detector.

As mentioned, most of the schemes could be explained in terms of an initial stage of processing followed by one or more stages of processing of a different type. The initial stage is typically a conventional detector or a decorrelating-type detector. The simpler initial stage is the conventional detector. A scheme for subsequent stages based on simplicity is to subtract off the contributions of the multiple access interference, with the order of the subtractions being given by the relative strength of the users. Contributing to its simplicity is the flexibility to limit the number of cancellations.

It is most important to cancel the strongest signal before detection of the other signals because it has the most negative effect. Also, the best estimate of signal strength is from the strongest signal for the same reason that the best bit decision is made on that signal: the strongest signal has the minimum MAI since the strongest signal is excluded from its own MAI. This is the twofold rationale for doing successive cancellation in order of signal strength:

(i)   Canceling the strongest signal has the most benefit.
(ii)   Canceling the strongest signal is the most reliable cancellation.

Ranking the Signal Strengths

Before the successive cancellation can proceed, it is necessary to rank the signal strengths. They could be obtained by separate channel estimates or they can be obtained directly from the outputs of the conventional detector. The sign of the output of the correlator for each user provides a bit decision for that user. The amplitude of that correlator output also provides an estimate of the received signal strength (from the square of the amplitude) of that user. There is noise associated with this estimate but we shall find its accuracy sufficient. The channel estimate method corresponds to using separate estimates of A_k and b_{0k} while using the correlator output corresponds to using an estimate directly of the product $A_k b_{0k}$ as discussed in Section III. For specificity in the rest of the article, we shall assume that the rankings are based on the correlator outputs.[2] This is also consistent with trying to maintain simplicity -- no components are needed beyond what is already provided for the conventional detector.

---

[2] Refs.[14] - [17] provide analytic results for using the correlator outputs for amplitude estimates, which form the basis for much of the present paper. Ref. 7 uses the despread signal to select the strongest and Ref. 12 also uses the despread signals for amplitude estimates. Channel estimates are suggested in [10], [11] which uses a novel method of cancellation in the spectral domain. Successive cancellation in conjunction with the decorrelator detector is discussed in Ref. 24.

Successive Cancellation with a Coherent BPSK System

We shall first describe the method for a coherent BPSK system to fix ideas. These results will be unrealistic for the following reasons:

(i)   A non-coherent system would be used on the reverse link because of the lack of a pilot signal used for a coherent signal [20].

(ii)   Fading and multipath are not included.

Each of these factors can be introduced after the basic idea is illustrated. One important point can be made at this point regarding non-coherence. To do the cancellation, phase information as well as amplitude information (to detect relative strengths) is needed. In a coherent system, the phase information is directly available. In a non-coherent system, it is not, but the information needed for the cancellation can be obtained from I and Q channels.

The successive cancellations are carried out as follows:

(i)   Recognize the strongest user (one with maximum correlation value).

(ii)   Decode the strongest user.

(iii)   Estimate the amplitude of the decoded user from the output of the correlator.

(iv)   Regenerate strongest user's signal using its chip sequence and the estimate of its amplitude.

(v)   Cancel the strongest user.

(vi)   Repeat (until all users are decoded, or until a permissible number of cancellations are achieved).

The received signal is:

$$r(t) = \sum_{k=1}^{K} A_k b_k(t - \tau_k) a_k(t - \tau_k) \cos(w_c t + \phi_k) + z(t)$$

(V - 1)

where

$r(t)$ - received signal

K - Total number of active users

A_k - Amplitude of kth user

$b_k(t)$ - bit sequence of kth user

$a_k(t)$ - spreading chip sequence of kth user

$n(t)$ - Additive White Gaussian Noise (two sided power spectral density $= N_0 / 2$)

$N = T / T_c$ where, T - bit period and T_c - chip period

τ_k and ϕ_k are the time delay and phase of the kth user, which are assumed to be known, i.e., tracked accurately. (Both these assumptions will be relaxed.)

The bits and chips are rectangular. Their values are all i.i.d. random values with probability 0.5 of ±1. The τ_k and ϕ_k are i.i.d. uniform random variables in [0,T] and [0,2π], respectively, for the asynchronous case.

It will be assumed that the users are labeled from 1 to K in accordance with their estimated signal strengths.

At the output of the low pass filter (LPF) of the I-Channel, we get:

$$d^I(t) = LPF\{r(t)\cos(\omega_c t)\}$$

$$= \sum_{k=1}^{K} A_k a_k(t - \tau_k) b_k(t - \tau_k) \frac{\cos(\phi_k)}{2} + \frac{n_c(t)}{2} \qquad \text{(V - 2)}$$

where $n_c(t)$ is the in-phase component of the low pass filtered Gaussian noise $n(t)$. Similarly $d^Q(t)$ is obtained. The first decision variable (correlation value) in the I-channel is given by:

$$\hat{Z}_1^I = \frac{1}{T} \int_T d^I(t) a_1(t - \tau_1) \cos(\phi_1) dt \qquad \text{(V - 3)}$$

Similarly, $\hat{Z}_1^Q$ is obtained.
The decision on the bit is then made by using the decision variable

$$\hat{Z}_1 = \hat{Z}_1^I + \hat{Z}_1^Q \qquad \text{(V - 4)}$$

The correlation value is then used for cancellation:

$$d_1^I(t) = d^I(t) - \hat{Z}_1 a_1(t - \tau_1) \cos(\phi_1) \qquad \text{(V - 5)}$$

Generalizing for the jth cancellation, we get:

$$d_j^I(t) = d_{j-1}^I(t) - \hat{Z}_j a_j(t - \tau_j) \cos(\phi_j) \qquad \text{(V - 6)}$$

where $\hat{Z}_j$ is the correlation (decision variable) after the $(j-1)^{st}$ cancellation.

As the users are canceled, the multiple access interference decreases but the noise due to imperfect cancellation in the previous stages increases. After j

170

cancellations, the signal to noise ratio conditioned on the ordered amplitudes A_k (for the asynchronous case) is found in [14] to be given by:

$$\gamma_{j+1} = \frac{A_{j+1}^2}{\dfrac{1}{3N}\sum_{k=j+2}^{K} A_k^2 + \dfrac{N_0}{T} + \dfrac{1}{3N}\sum_{i=1}^{j}\eta_i} \qquad (V - 7)$$

where η_i represents noise accumulated from previous cancellation stages.

The bit error probability at the $j+1^{st}$ cancellation stage, conditioned on the ordered set of amplitudes A_k, is given by:

$$\begin{aligned} P_e^{j+1} &= P\left\{\hat{Z}_{j+1} < 0 | b_{j+1} = +1\right\} \\ &= Q\left(\sqrt{\gamma_{j+1}}\right) \end{aligned} \qquad (V - 8)$$

The Gaussian approximation used in the above was found to be sufficiently accurate by comparison with simulations for the simple case considered when the received powers are not too dissimilar. Additional analysis is needed for unequal powers as mentioned in [27]. Further analysis is also needed for other cases such as Rayleigh fading and multipath.

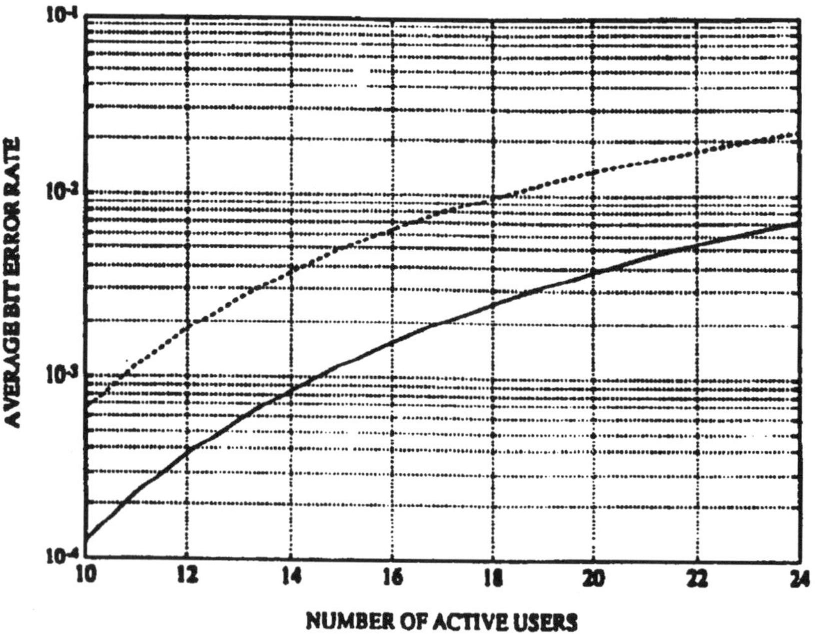

Figure 1. BER Performance of Interference Canceller and Conventional Detection

An illustrative result is given in Figure 1. The example is for equal received powers (ideal power control) which we shall see is a pessimistic case for successive cancellation. Figure 1 shows the BER averaged over all the users calculated using the Gaussian approximation and comparison with the conventional detector. Substantial improvement is shown even with equal powers which does not take full advantage of a scheme based on ranked power.

Comparison with Using Channel Estimates

As mentioned, in a conventional detector, the output of the correlator is used for bit detection -- the sign of the correlator output is taken as the bit decision. We are also using the correlator output to obtain an estimate of the signal amplitude. and are thus making extra use of a quantity already available. It is of interest to compare the accuracy of this method of estimation with using separate channel estimates. In the above, we assumed that the amplitude estimate was obtained from only one bit. We shall here assume that n successive bits are used for estimation using the correlator outputs. For the case studied in [14], it was found that there is a rough equivalence in estimation accuracy if the independent power estimate's accuracy in estimating *each* bit energy is around 1 dB if $n = 10$. An example of averaging is given in Figure 2. The averaging improves the amplitude estimation by a factor of about $1 / \sqrt{n}$.

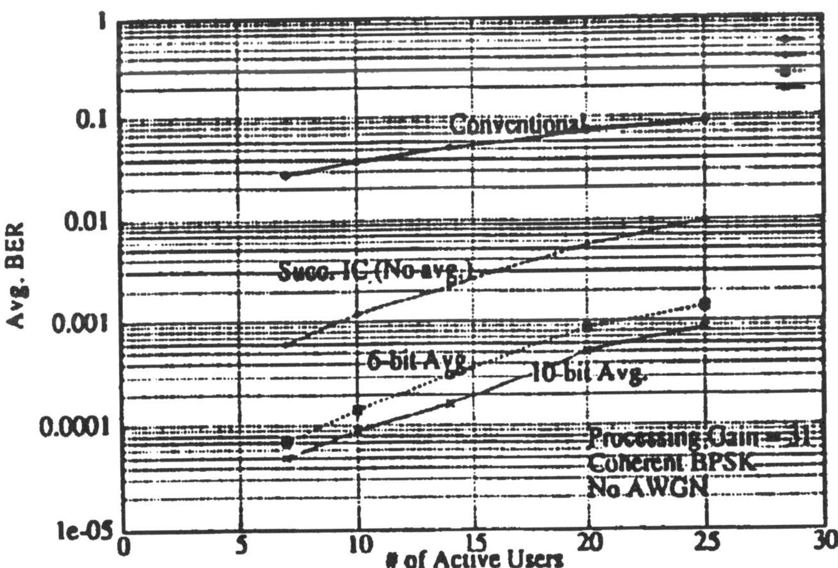

Figure 2. Average BER vs. # of Active Users with no Averaging,
6-Bit Averaging and 10-Bit Averaging

Alternate Ways of Implementing

There are two ways of implementing this type of interference cancellation. One, which has been assumed in explaining the method, involves identifying the strongest user and feeding back that user's chip sequence to the (baseband) received signal and then the conventional detector is used on all the users who have not been selected as strongest. An alternate method is to feed back the cross-correlations between the strongest user and each remaining user and subtracting that off the correlation output of each of those users. This is discussed in [23] in terms of precorrelation reconstruction vs postcorrelation reconstruction (p. 517). It is easily verified that both are equivalent as far as the equation leading to BER calculation. The latter method is more in line with the other methods which use these cross-correlations. The implementation implications are being compared.

Synchronous $\rightarrow$ Asynchronous

In synchronous reception, all the bits are aligned when received. In the reverse link, in which multiuser detection is a candidate, the different users' bit sequences would be asynchronous.[3] This presents a problem, the solution of which provides an added benefit. For synchronous systems, it is non-ambiguous as to what is meant by "strongest signal" as all of the users are aligned. For asynchronous systems, bits from one user overlap partially with bits of other users. To take into account all of the interactions between all of the bits actually would require an infinite memory, or a truncation to reasonably capture the interactions. A simple heuristic scheme is to group n bits of each user into a cancellation frame, where the maximum time between the first bit start and last bit end is $(n+1)$ bit durations. After the entire frame is received, the correlations of the n bits of each user are averaged and the ranking of the users is obtained from these averaged correlations. Then the n chip sequences of the strongest user are canceled. This introduces a delay and some small end effects of parts of a chip sequence not being canceled. The benefit is from the averaging of the correlations which, as pointed out before, improves the accuracy of the ranking and of the cancellation.

Coherent $\rightarrow$ Noncoherent

As previously mentioned, multiuser detection is appropriate on the reverse link (mobile to base station). On the reverse link, there is no pilot signal for coherent detection [20]. So non-coherent detection is used, e.g, M-ary orthogonal modulation [20].[4] The basic issue is that in order to cancel chip sequences in the coherent system, the phase as well as amplitude is used.

---

[3] If all of the users are sufficiently near and the delays small enough, synchronous operation may be possible.

[4] For consideration of a coherent uplink in future mobile communication systems, see [25].

Coherent detection assumes knowledge of the phase but non-coherent detection is based on the lack of this information. The cancellation can still be accomplished by canceling the inphase and quadrature components. The M-ary modulation case has been studied for single path Rayleigh fading [17] and for multipath resolution and combining [15]. In both those cases, the Gaussian approximation is not adequate and additional analysis was required.

Rayleigh Fading

When all of the users' signals are received with independent Rayleigh distributions, this method of successive cancellation can be readily analyzed by considering the order statistics of the received signals. The amplitudes are assumed to be Rayleigh distributed with unit mean square value, i.e. its pdf is given by,

$$f_A(x) = 2xe^{-x^2} \tag{V - 9}$$

The pdf's of the ordered A_k (where A_1 is the strongest and A_k is the weakest) is denoted by $f_{A_k}(x)$ and is obtained as follows:

$$f_{A_k}(x) = \frac{K!}{(K-k)!(k-1)!} F^{K-k}(x)[1 - F(x)]^{k-1} f(x) \tag{V - 10}$$

The error probability expression after the jth cancellation is then unconditioned using the pdf of the $j+1^{th}$ strongest amplitude as follows:

$$P_e^{j+1} = \int_0^\infty Q\left(\frac{A_{j+1}}{\sqrt{E_{A_k}[\eta_{j+1}]}}\right) f_{A_{j+1}}(x) dx \tag{V - 11}$$

where $E_{A_K}\left[n_{j+1}\right]$ represents the equivalent noise.

The average probability of error is then obtained as the average of the BER resulting from all stages of cancellation.

An interesting question is how the method performs compared to ideal power control where all the signals are received with equal energies. There are really two questions:

(i)   How does the cancellation scheme's performance compare under Rayleigh fading or ideal power control?

(ii)   How does the *improvement* over the conventional detector compare under Rayleigh fading or ideal power control?

The answers are displayed in Figure 3. The answer to (ii) is that the improvement is better for the Rayleigh fading case because the ranking takes advantage of the unequal powers. Nevertheless, in answer to (i), the scheme still works better under ideal power control because that is an easier environment. Thus, the scheme does not have complete near/far resistance but does allow for relaxation of power control requirements.

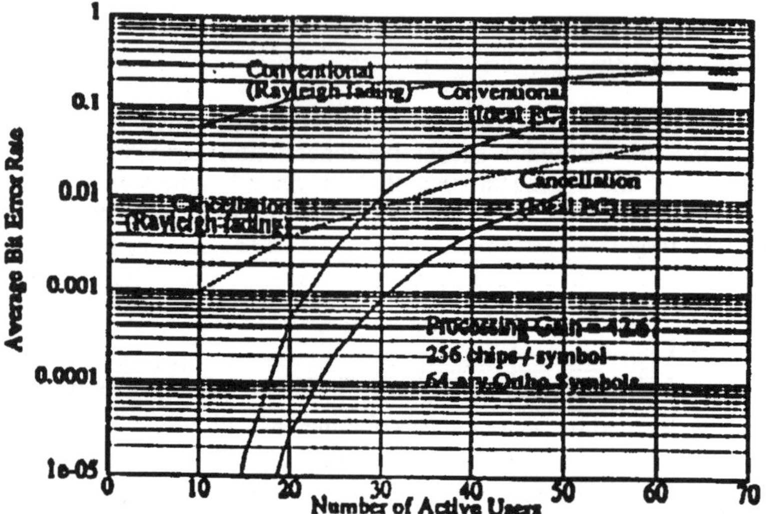

Figure 3 Average Ber (Bit Error Rate) Under Ideal Power
Control and Under Rayleigh Fading

Multipath

Each user's signal is typically received via several multipaths. If the chip duration is smaller than the time between the multipaths, the individual multipaths can be resolved and diversity reception can improve performance. To do cancellation with multipath reception is conceptually simple but increases the complexity. Cancellation is done with all of the multipaths being tracked. Multipath cancellation is done in [26]. Multipath cancellation with M-ary modulation is analyzed in [15].

Limiting the Number of Cancellations

The successive cancellation must operate fast enough to keep up with the bit rate and not introduce intolerable delay. For this reason, it will presumably be necessary to limit the number of cancellations. The ability to limit the number of cancellations is consistent with the objective of controlling complexity by choosing an appropriate performance/complexity tradeoff.

Parallel vs. Successive Cancellation [16]

Successive cancellation works by successively subtracting off the strongest remaining signal. An alternative (the parallel method) is to simultaneously subtract off all of the users' signals from all of the others. It is found that when all of the users are received with equal strength, the parallel method outperforms the successive scheme (Figure 4). When the received signals are of distinctly different strengths (the more important case), the successive method is superior in performance (Figure 5). The important thing to note is that in both cases, both outperform the conventional detector.

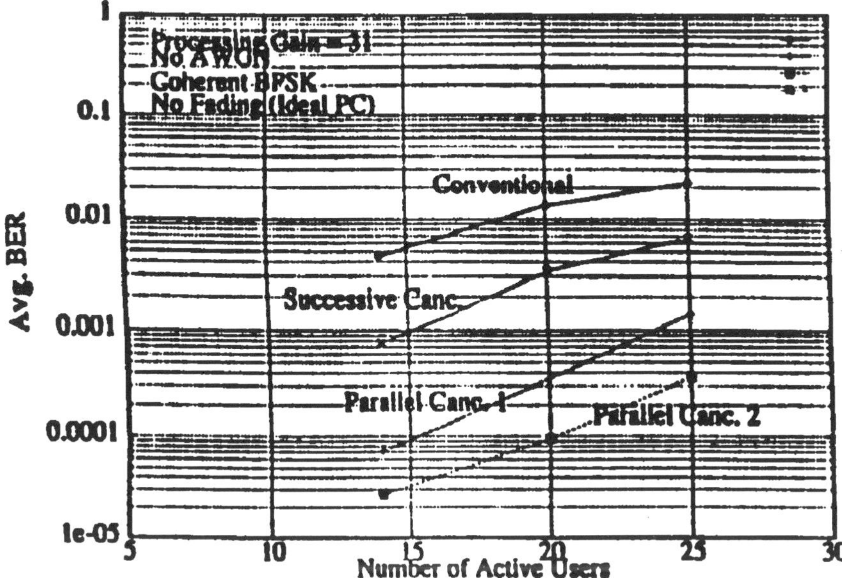

Figure 4 Parallel vs. Successive Cancellations (Ideal Power Control)

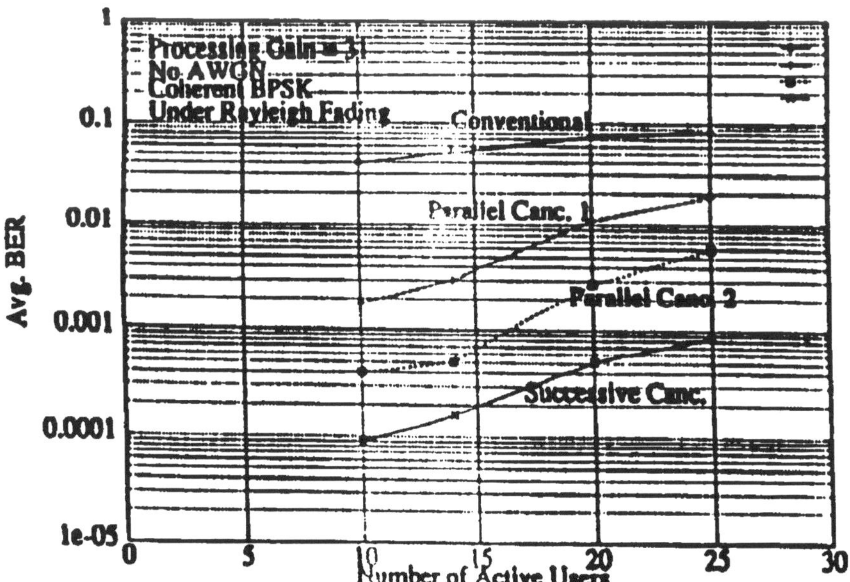

Figure 5 Parallel vs. Successive Cancellations (Rayleigh Fading)

Sensitivities and Robustness

Most of the discussion and analyses of multiuser detection have assumed a number of idealizations. For example, in canceling out a chip sequence, it has been assumed that there is perfect synchronization of each user. Clearly, if there are tracking errors, the chip sequence being canceled will be offset and doing an imperfect cancellation. The pertinent question is whether the tracking error tolerable for the conventional detector is tolerable for the cancellation, or how much tighter it must be for the interference cancellation. For the conventional detector, the synchronization must be within a fraction of a chip duration. This problem is analyzed in [27]. For numerical results presented here, the interference cancellation scheme was subjected to pessimistic conditions for the interference canceller:

(i)   Did not use averaging of the correlator outputs for amplitude estimates which significantly improves cancellation performance.

(ii)   Assumed equal received powers (perfect power control). The improvement over the conventional detector is greater in the more realistic case of unequal received powers.

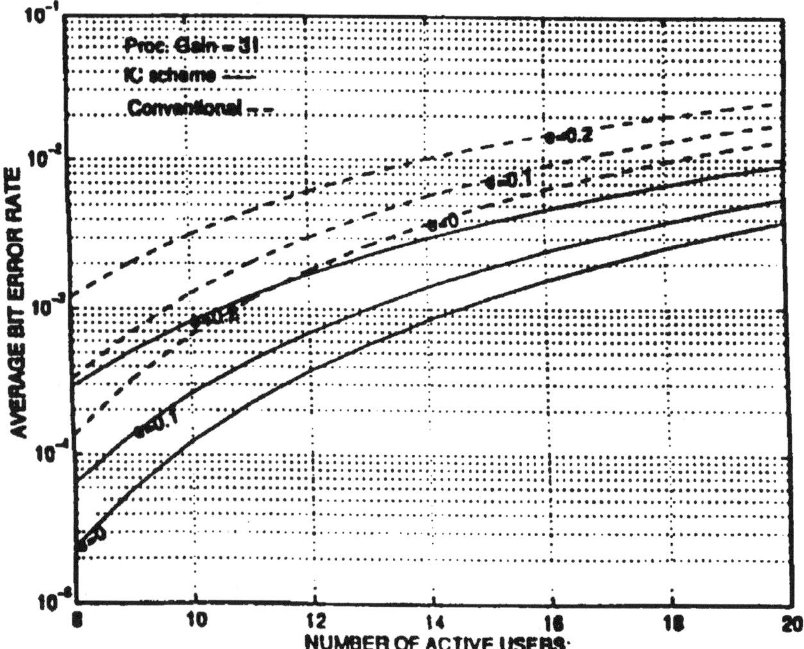

Figure 6 Analysis Results of Conventional Detector vs. IC
Scheme for Asynchronous, Random Phase and $N_0 = 0$.

The processing gain of spread spectrum is set to N=31 and total numbers of users are ranged from 10 to 24. Figure 6 shows a result for no AWGN. There are 3 curves each for the interference cancellation scheme and the conventional detector. Each curve represents different standard deviation e's of tracking error normalized with respect to ratio of chip duration (e = 0 is zero tracking error). The interference cancellation scheme retains superiority over the conventional detector. Similar types of results were found in [28] from simulation of the scheme of [10] and [11]. While it is premature to draw any general conclusions about robustness at this point, these results are promising.

VI. CONCLUDING REMARKS

The theoretical foundations of multiuser detections are fairly well understood now. There is enough potential advantage to investigate the feasibility of practical implementations.

While the potential advantages are significant, they are bounded as discussed in Section III so that a relatively simple implementation is dictated. We have discussed what may be the simplest version, which still retains significant performance advantage over the conventional detector. More work is needed on questions of robustness and actual implementability in a real system including questions of complexity and delay.

REFERENCES

[1]   S. Verdu, "Adaptive Multiuser Detection," *Proceedings of ISSSTA '94*, Oulu, Finland, July 1994, pp. 43-50.

[2]   K. S. Schneider, "Optimum Detection of Code Division Signals," *IEEE Trans. on Aerospace and Electronic Systems,* Vol. AES-15, No. 1, p. 181-185, Jan. 1979.

[3]   R. Kohno, M. Hatori, and H. Imai, "Cancellation Techniques of Co-Channel Interference in Asynchronous Spread Spectrum Multiple Access Systems," *Electronics and Communications,* Vol. 66-A, No. 5, pp. 20-29, 1983.

[4]   S. Verdu, "Minimum Probability of Error for Asynchronous Gaussian Multiple Access Channels," *IEEE Trans. on Info. Theory,* Vol. IT-32, No. 1, Jan. 1986.

[5]   S. Verdu, "Optimum multiuser asymptotic efficiency," *IEEE Trans. on Communications,* vol. 34, no.9, pp.890-897, Sept. '86.

[6]   A.J. Viterbi, "Very Low Rate Convolutional Codes for Maximum Theoretical Performance of Spread-Spectrum Multiple Access Channels," *IEEE Journal on Selected Areas in Communications*, Vol. 8, no. 4, pp. 641-649, May 1990.

[7] S. Kubota, S. Kato, and K. Feher, "Inter-channel interference cancellation technique for CDMA mobile/personal communications base stations," in *Proceedings. International Symposium on Spread Spectrum Techniques and Applications (ISSSTA), Yokohama, Japan,* pp. 91-94, Dec '92.

[8] Y.C. Yoon, R. Kohno, and H. Imai, "Combination of an adaptive array antenna and a canceller of interference for direct-sequence spread-spectrum multiple access system," *IEEE Journal on Selected Areas in Comm.,* vol.8, no. 4, May '90.

[9] A. Kajiwara and M. Nakagawa, " Spread spectrum block demodulator with high capacity crosscorrelation canceller," in *Proceedings, Globecom,* 1991.

[10] P. Dent, B. Gudmundson, and M. Ewerbring, "CDMA-IC: a novel code division multiple access scheme based on interference cancellation," in *Proceedings, PIMRC (Boston, MA),* pp. 4.1.1-4.1.5, Oct. '92.

[11] P. Teder, G. Larsson, B. Gudmundson, and M. Ewerbring, " CDMA with Interference Cancellation: a technique for high capacity wireless systems:, in *Proceedings, IEEE International Conf. on Comm. (ICC, Geneva),* 1993.

[12] M. Kawabe, T. Kato, T. Sato, A. Kawahashi, and A. Kukasawa, "Advanced CDMA scheme for PCS based on interference cancellation," in *record, ICUPC,* pp.1000-1003, 1993.

[13] Y.C. Yoon, R. Kohno, and H. Imai, "Cascaded co-channel interference canceling and diversity combining for spread spectrum multi-access over multipath fading channels," *IEICE Transactions on Comm.,* no.2, pp. 163-168, Feb '93.

[14] P. Patel and J. Holtzman, "Analysis of a simple successive interference cancellation scheme in DS/CDMA system using correlations," in *Proceedings, Globecom '93,* pp. 76-80, Houston, TX, December 1993.

[15] P. Patel and J. Holtzman, "Analysis of a simple successive interference cancellation scheme in DS/CDMA system," *IEEE Journal on Selected Areas in Communications,* Vol. 12, No. 5, pp. 796-807, June 1994.

[16] P. Patel and J. Holtzman, "Performance comparison of a DS/CDMA system using a successive interference cancellation (IC) scheme and a parallel IC scheme under fading," *ICC '94,* pp. 510-515, New Orleans, LA, May 1994.

[17] P.R. Patel and J.M. Holtzman, "Analysis of Successive Interference Cancellation in M-ary Orthogonal DS/CDMA System with Single Path Rayleigh Fading," *Proceedings of 1994 International Zurich Seminar on Digital Communications,* pp. 150-161, March 1994, Zurich Switzerland.

[18] A.J. Viterbi, "The Orthogonal-Random Waveform Dichotomy for Digital Mobile Personal Communications," *IEEE Personal Communications,* pp. 18-24, First Quarter 1994.

[19] L.F. Chang,"Dispersive Fading Effects in CDMA Radio Systems," *Proc. of ICUPC '92,* Dallas, pp.185-189, 9/29-10/2/92.

[20] K.S. Gilhousen, I.M. Jacobs, R. Padovani, A.J. Viterbi, L.A. Weaver, and C.W. III, "On the capacity of a cellular CDMA system," *IEEE Transactions on Vehicular Technology,* vol. 40, no. 2, pp. 303-311, May '91.

[21] M. Varanasi and B. Aazhang, "Multistage detection in asynchronous code-division multiple access communications," *IEEE Transactions on Communications*, vol.38, pp.509-519, April '90.

[22] T.R. Giallorenzi and S.G. Wilson, "Decision Feedback Multiuser Receivers for Asynchronous CDMA Systems," *Proceedings Globecom '93,* pp. 1677-1682.

[23] R.S. Mowbray, R.D. Pringle, P.M. Grant, "Increased CDMA System Capacity Through Adaptive Cochannel Interference Regeneration and Cancellation," *IEE Proceedings-I,* Vol. 139, No.5, pp. 515-524, October 1992.

[24] A. Duel-Hallen, "Decorrelating Decision-Feedback Multiuser Detector for Synchronous Code-Division Multiple-Access Channel," *IEEE Transactions on Communications*, vol.41, No.2, pp.285-290, February 1993.

[25] P. Teder and T. Sandin, "A Performance Study of the Uplink in the CODIT Concept" *Proceedings of Race Mobile Telecommunications Workshop,* Amsterdam, Holland, May 1994, pp. 176-179.

[26] P.M. Grant, S. Mowbray and R.D. Pringle, "Multipath and Co-Channel CDMA Interference Cancellation," *Proceedings of ISSSTA '94,* Oulu, Finland, July 1994, pp. 83 - 86.

[27] F.C. Cheng and J.M. Holtzman, "Effect of Tracking Error on DS/CDMA Successive Interference Cancellation," *Proceedings of Globecom '94, 3rd Comm. Theory Mini-Conference* San Francisco, CA, November 1994.

180

[28] L. Levi, F. Muratore and G. Romano,"Simulation Results for a CDMA Interference Cancellation Technique in a Rayleigh Fading Channel," *Proceedings of 1994 International Zurich Seminar on Digital Communications*, pp. 162-171, March 1994, Zurich Switzerland.

[29] S. Verdu, "Multiuser Detection," in V. Poor, ed., Advanced Signal Processing, Vol. 2, pp. 369-409, JAI Press Inc., 1993.

Chapter 4

Performance Analysis

Rake Reception for a CDMA Mobile Communication System with Multipath Fading

Daniel L. Noneaker and Michael B. Pursley

Abstract

The performance of a rake receiver for direct-sequence (DS) spread spectrum communications on multipath fading channels depends on the number and type of multipath components present at the output of the channel. The focus in this paper is on noncoherent combining of the outputs of the taps of the rake receiver in a code-division multiple-access (CDMA) system that employs DS spread spectrum with binary DPSK data modulation. It is shown that the advantages and disadvantages of a rake receiver, as compared with a standard correlation receiver, depend greatly on the fading characteristics of the individual multipath components. In general, rake receivers are more appropriate for DS systems with a smaller number of chips per data symbol and for channels that have only diffuse, Rayleigh fading multipath components. On the other hand, a correlation receiver is preferred for DS systems with a large number of chips per data symbol if the channel has a strong specular multipath component in addition to one or more diffuse components. In this paper the probability of error on the channel is determined as a function of the chip rate for doubly selective fading channels, and the pairwise error probability is evaluated for a CDMA system with binary convolutional encoding and Viterbi decoding. The effects of the Doppler spread and the interleaving depth on the performance of the system and the choice of chip rate are discussed.

This research was supported by the Holcombe endowment at Clemson University.

S.G. Glisic and P.A. Leppänen (eds.), Code Division Multiple Access Communications, 183-201.
© 1995 *Kluwer Academic Publishers. Printed in the Netherlands.*

I Introduction

One of the most significant sources of potential degradation to a mobile communication system is fading due to multiple propagation paths in the communication channel [1] . Multipath propagation is common in urban areas and in many indoor environments. Direct-sequence (DS) spread-spectrum modulation has been proposed for use with code-division multiple-access (CDMA) systems for cellular and personal communications, and this choice is motivated in part by the ability that is provided by DS spread spectrum to mitigate the effects of multipath fading. It provides the ability to resolve the individual components of a multipath signal [2], it enables a standard correlation receiver to discriminate against unwanted multipath components, and it permits the combination of these components in a rake receiver [3].

The ability of the receiver to resolve multipath components is determined largely by the channel characteristics, the chip rate of the DS spread-spectrum system, and the number of taps in the rake receiver. In this paper the error probability is examined for different multipath channel models in a CDMA system that uses DS spread-spectrum modulation, binary differential phase-shift-key (DPSK) modulation, noncoherent demodulation, and rake reception with equal-gain square-law combining of the outputs of the taps of the rake receiver. The use of noncoherent demodulation is dictated by the characteristics of the reverse link (i.e., the link from the mobile to the base-station), which is generally conceded to be the limiting link for the CDMA system. The effects of the chip rate and the number of taps on system performance are investigated.

For our purposes, the *chip rate R* is defined as the number of chips per channel symbol. Chip rates in the range of tens of chips per bit to hundreds of chips per bit are considered in this paper. A CDMA system with a chip rate of 50 is taken as a representative low-chip-rate system, and a chip rate of 400 is used as an example of a high-chip-rate system. The results of this paper demonstrate that a low chip rate is superior to a high chip rate only if the taps in the rake receiver for the low-chip-rate signal capture approximately equal amounts of energy from the output of the multipath channel. Even then, the high chip rate is better if more than half of the available energy is contained in a small number of resolvable components. If one of the taps captures much more energy than each of the others, the high-chip-rate system is superior to the low chip-rate system. In some cases this is true even if as little as one-tenth of the total received energy is contained in the strongest component of the channel output. The results of this paper show that the selection of the key parameters of the system (e.g., chip rate and number of taps) and the performance that results depend critically on the characteristics of the multipath channel.

II Doubly Selective Fading Channels

Narrowband measurements of the mobile communication channel have led to the use of a channel model in many analyses that consists of a continuous delay spectrum and Rayleigh fading. A Rayleigh-fading model of the channel is often adequate to reflect the effect of the channel on a narrowband signal. However, wideband measurements of the channel [4]-[6] indicate that the channel often consists of a few strong paths (sometimes only one) together with a number of weaker paths. Other measurements [7]-[9] show that indoor channels may exhibit similar characteristics.

For a DS spread-spectrum system, the relative delays for several components of the received signal may fall within one chip interval. Then the composite signal formed by those components may be modeled as a single fading signal that has a Rician fading amplitude. The parameters of the Rician fading depend on the *chip rate* of the DS signal. The chip rate is measured in chips per channel symbol, and it is given by

$$R = T/T_c$$

where T_c is the chip interval of the DS signal and T is the duration of a channel symbol. A DS signal with a low chip rate will result in a strong Rayleigh component, and a signal with a high chip rate will result in a weak Rayleigh component. Therefore the use of a purely Rayleigh-fading model of the channel can mask the ability of a high-chip-rate CDMA system to resolve stable signal components.

We model the channel in a manner which reflects the phenomenon described above and allows for tractable analysis of system performance. The channel is viewed as consisting of several clusters of paths, with each cluster consisting of a single specular path plus a continuum of Rayleigh-fading paths with path delays that are centered at the delay of the specular path. The channel is the sum of several such non-overlapping clusters.

The channel considered is a special case of the Gaussian wide-sense-stationary uncorrelated-scattering (WSSUS) channel [10]. It is represented as a Gaussian random process $h(t, \xi)$, which is characterized by its specular part

$$E[h(t, \xi)] = \sum_{i=1}^{M} \rho_i \delta(\xi - \xi_i) \tag{1}$$

and the autocovariance function of its diffuse part

$$Cov[h(t, \xi), h(x, \alpha)] = \sum_{i=1}^{M} 2\sigma_i^2 d(t - x) g_i(\xi) \delta(\alpha - \xi). \tag{2}$$

where t and x are time variables and ξ and α are path delay variables. The number of path delay clusters in the channel is M. The parameter ρ_i is the amplitude of the ith specular component, d is the *time-correlation function* of the channel, and g_i is the *delay spectrum* of the ith cluster of paths. We assume that the g_i are non-overlapping functions. The Rayleigh-fading channel is equivalent to the one given by equations (1) and (2) for the special case in which the specular component of each cluster has an amplitude of zero.

We consider rectangular delay spectra [11] for each cluster, and we denote by μ_i the normalized *total delay spread* of the ith cluster. The total delay spread of a cluster is defined as the difference between the maximum delay for which the delay spectrum of the cluster is nonzero and the minimum delay for which it is nonzero. For each time-correlation function d, the key parameter is the half power bandwidth B_d of the Fourier transform of d. The normalized half-power bandwidth $D_T = B_d T$ is referred to as the *Doppler spread* of the channel. In our examples we consider exponential time-correlation functions. The *coherence time* of the channel is denoted t_c , and it is given by

$$t_c = (2\pi \, B_d)^{-1}.$$

Another parameter of interest is the *energy ratio*, defined by $\zeta_i = \rho_i^2/(2\sigma_i^2)$, which is the ratio of the received specular energy to the average received diffuse energy in the ith cluster of the channel. The multiple-access interference, multipath interference, and thermal noise are modeled as additive white Gaussian noise (AWGN) with single-sided power spectral density N_0. If $\mathcal{E}$ denotes the transmitted energy per channel symbol, the signal-to-noise ratio *for the specular component* of the received signal *due to the first cluster* is defined as

$$SNR = \rho_1^2 \mathcal{E}/N_0. \tag{3}$$

The term "energy" is used in later sections to refer to the energy in the component of the received signal that is of interest at that point in the discussion. The component to which the term refers is clear from the context of the discussion.

III   Error Probabilities for the Rake Receiver

We consider a DS system with binary DPSK modulation and rake reception. The ith tap of the rake receiver is synchronized to the received signal component arriving with path delay ξ_i. The differentially coherent rake receiver employs equal-gain square-law diversity combining. Suppose that (U_k, V_k) are the sampled matched-filter outputs at the kth receiver tap during the ith and $(i-1)$th channel symbol intervals, respectively. Then the rake receiver makes a binary decision based on the sign of the real-valued decision statistic

$$Z_i = \sum_{k=1}^{L} \mid U_k + V_k \mid^2 - \mid U_k - V_k \mid^2, \tag{4}$$

where L is the number of receiver taps.

Analysis of the performance of the receiver can be simplified by modeling the multipath interference as Gaussian noise and by assuming that the aperiodic autocorrelation function of the spreading sequence is ideal [11]. For good spreading sequences and a small number of taps, the influence of the autocorrelation function on the bit error probability is minimal [12]. The probability of error is determined in [11] and [13] for a rake receiver and the channel given by equations (1) and (2). The result is applicable to a system with uncoded transmission, and it also gives the probability of error at the input to the decoder of a system with hard-decision decoding of a binary error-correcting code.

We also consider a DS system that employs binary convolutional encoding with soft-decision Viterbi decoding. The output of the data source is convolutionally-encoded, the binary code symbols are interleaved, and the output of the interleaver is transmitted by binary DPSK modulation. The output of the square-law combiner in the rake receiver is given by equation (4), and it is passed to a deinterleaver without a hard decision on the code symbol. The output of the deinterleaver is employed by the Viterbi decoder to determine a branch metric that is referred to as the *square-law metric*. If $\mathbf{c} = (c_{i+1}, \cdots, c_{i+n})$ are the binary code symbols on a branch of the code trellis, the branch metric is given by

$$M_{\mathbf{c}} = \sum_{j=1}^{n} (-1)^{c_{i+j}} Z_{i+j}.$$

Let $\mathbf{e}$ denote an error event in the convolutional code. The pairwise error probability [14] associated with this error event is given by

$$Pr(\mathbf{0} \rightarrow \mathbf{e}) = Pr(Z < 0)$$

where the random variable Z is a Hermitian quadratic form in jointly Gaussian random variables [15]. If $\mathbf{e}$ has a Hamming weight of d, Z can be represented as

$$Z = \sum_{i=1}^{dL} |Y_i|^2 - \sum_{i=dL+1}^{2dL} |Y_i|^2,$$

where the complex-valued random variables $\{Y_1, \cdots, Y_{2dL}\}$ are independent and each has a Gaussian distribution. The expected value and variance of Y_i are denoted by m_i and $2\beta_i^2$, respectively, for each i. For codes that employ binary DPSK modulation and the square-law metric, the mean of Y_i is zero for $i \geq dL + 1$.

An expression for the probability that Z is less than zero is obtained in [16], and the form of the expression depends on the number of repeated values in the set $\{\beta_1, \cdots, \beta_{2dL}\}$. If, for example, each β_i is distinct for $i \geq dL + 1$, then

$$P(Z < 0) = 1 - \sum_{j=dL+1}^{2dL} A_j \prod_{i=1}^{dL} \frac{\beta_j^2}{\beta_j^2 + \beta_i^2} \exp\left(\frac{-m_i}{2(\beta_j^2 + \beta_i^2)}\right) \quad (5)$$

where

$$A_i = \prod_{j \geq dL+1, j \neq i} \left(\frac{\beta_i^2}{\beta_i^2 - \beta_j^2}\right).$$

Note that the expression given here can also be used to determine the bit error probability for an uncoded DS system that employs time diversity of order d in addition to rake reception.

The effect of the Doppler spread and the interleaving depth on the pairwise error probability for the minimum distance error event of a convolutional code gives some insight into the effect of these parameters on the performance of the code. However, the bit error probability for convolutional encoding and Viterbi decoding is determined by a large number of dependent error events. In some instances it is possible to express in closed form a tight upper bound on the bit error probability by employing an improved transfer-function bound that uses equation (5) for the pairwise error probability of low-weight error events [16]. However, the transfer-function bound does not converge for some values of Doppler spread and interleaving depth that are are interest, and it is sometimes necessary to examine the performance of the system by means of simulation. Thus, in this paper we use both simulations and the analytical result of equation (5) to examine the performance of systems with convolutional coding.

IV   Performance of the Rake Receiver

A number of previous analyses of CDMA for mobile communications have been based on the assumption that the channel exhibits purely Rayleigh fading. However, as noted in Section II, a Rician-fading model of the channel may more accurately reflect the effect of the channel on wideband CDMA signal. In this section we consider the performance of rake receivers, and examine the manner in which the channel model affects conclusions about the performance. In particular, we consider the effect of the chip rate on the performance of a system for both Rayleigh-fading channels and Rician-fading channels.

The effect of the chip rate on the bit error probability depends on the strength of the specular component, the energy ratio, and the total delay spread of each cluster, the number of receiver taps, and the signal-to-noise ratio. Three channel models — two Rician-fading models and one purely Rayleigh-fading model — are employed to examine the influence of the chip rate. Each channel consists of three clusters, and each cluster has a total delay spread of 0.05. The first cluster of Channel A has a specular component of amplitude ρ and an energy ratio of ζ. Neither the second nor the third cluster has a specular component ($\rho_2 = \rho_3 = 0$), but each has a diffuse component with the same average power as the diffuse component of the first cluster ($\sigma_2 = \sigma_3 = \sigma$). The second and third clusters of channel B have no specular components, but each has a diffuse component with an average power that is 6 dB less than that of the diffuse component of the first cluster ($\sigma_2 = \sigma_3 = \sigma/2$). Some measurement data suggest that channel B may be a more realistic model than channel A. None of the three clusters of Channel C contains a specular component, so that the channel exhibits purely Rayleigh fading. The diffuse component of the first cluster of channel C has an average power of $2\sigma^2 + \rho^2$, and the power of either of the other two clusters is $\sigma^2/2$. Thus, each signal component of channel C has the same average energy as the corresponding signal component of channel B, but the first component of channel C is subject to more severe fading than the first component of channel B. Channels A, B, and C are illustrated in Figure 1.

In each of these channels, each cluster has a total delay spread of 0.05. If the chip rate is less than 20, the total delay spread is less than T_c, so most of the diffuse energy in a cluster is captured by a single tap. The envelope of the received signal at the output of the tap has a fading distribution which is approximately Rayleigh. For a chip rate greater than 20, the total delay spread is greater than T_c, and less of the diffuse energy is utilized by the receiver. In the limit as the chip duration decreases to zero (i.e., the chip rate increases) the amount of diffuse energy that is captured at each tap goes to zero, but the amount of specular energy that is captured, if any, remains constant. In the limit there is no fading. For channel C, of course, there is also no energy collected from the signal in the limit as $T_c \to 0$.

Our approach is to compare the performance of different systems by determining the signal-to-noise ratio that is required for each system to achieve a specified probability of error. Forward error-correction codes often provide adequate performance if the probability of a code symbol error is 10^{-2} or less, so we select 10^{-2} as the specified error probability for the first two examples. The effect of the multipath diversity provided by rake reception is illustrated in Figure 2 for channel A with $\zeta = $ -3 dB and for receivers with one and three taps. With a correlation receiver (a one-tap rake receiver), the performance of the high-chip-rate system is 5.2 dB better than the low-chip-rate system, but if a three-tap rake

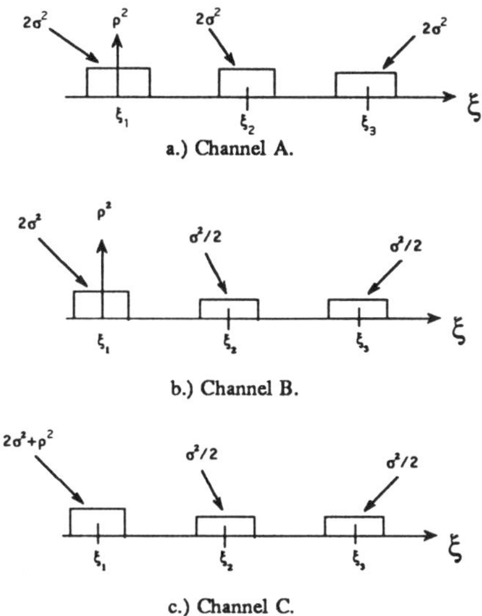

Figure 1. Delay spectra of channels A, B, and C. ($\mu_1 = \mu_2 = \mu_3 = 0.05$)

receiver is employed the low chip rate is 1.0 dB better than the high chip rate. Thus, if there are several strong diffuse components and only weak specular components, the rake receiver is preferred over the correlation receiver for the low-chip-rate system. For such a channel, the low-chip-rate system with the rake receiver gives better performance than the high-chip-rate system.

If the total energy in the second and third clusters is small and there is a strong specular component, the relative performance of the two systems changes in favor of the high chip rate. Contrast Figure 2 with Figure 3, which shows the performance for channel B and an energy ratio of -3 dB for the first cluster. The system with the chip rate of 50 obtains a significant diversity gain for this channel. Yet the correlation receiver with the chip rate of 400 is superior to either receiver with the low chip rate.

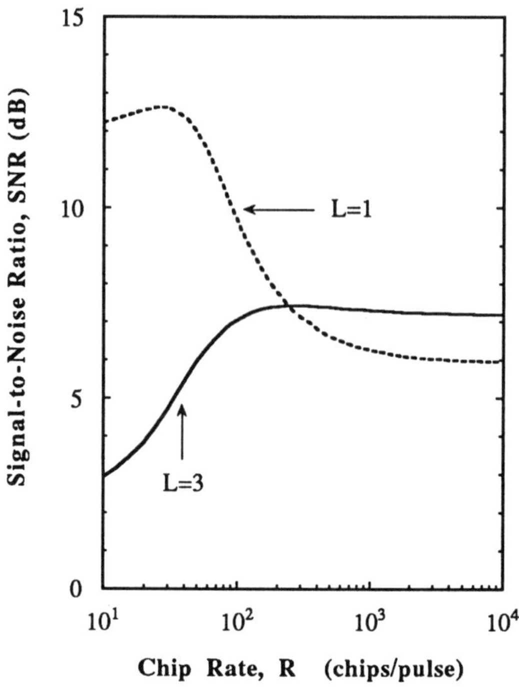

Figure 2. SNR required to achieve $P_e = 10^{-2}$ for channel A with $\zeta =$ -3 dB.

If the channel is modeled as one that exhibits frequency-selective Rayleigh fading with no specular component, the conclusions that are drawn concerning the performance of rake receivers may differ considerably from the conclusions obtained from the two previous examples. Performance results for rake receivers with one and three taps are shown in Figure 4 for channel C. Unlike the results for channel B, the performance of the system degrades with increasing chip rate for either receiver. For channel C, the low chip rate is superior to the high chip rate for a given number of taps.

As the specified error probability is decreased, the performance of the high-chip-rate system improves relative to that of the low-chip-rate system for either channel A or channel B. Suppose the channel and the number of taps are the same as in Figure 2, but the target error probability is 10^{-3}. With a correlation receiver, an improvement of more than 10 dB is obtained by using a chip rate of 400 in place of a chip rate of 50. Even if a three-tap rake receiver is employed, the high-chip-rate

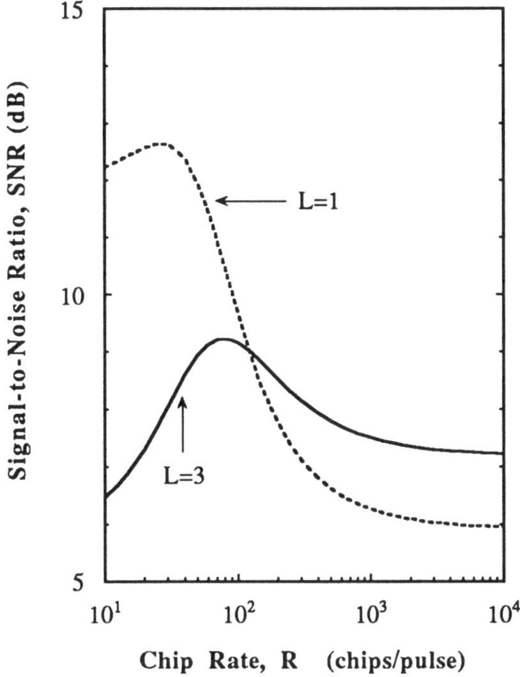

Figure 3. SNR required to achieve $P_e = 10^{-2}$ for channel B with $\zeta = $ -3 dB.

system is 1.1 dB better than the low-chip-rate system. In contrast, for channel C the relative merits of the two chip rates are not affected by the target error probability. Similarly, if the first cluster of channel A or B contains a very strong specular component, the relative performance of the two systems changes in favor of a chip rate of 400. If channel B is modified to have an energy ratio of 3 dB in the first cluster, a chip rate of 400 yields significantly better performance than a chip rate of 50 for either a correlation receiver or a three-tap rake receiver. However, if the strong specular component is modeled as additional Rayleigh-fading energy in channel C, the low chip rate is superior to the high chip rate for either a correlation receiver or a three-tap rake receiver.

V   Receiver Complexity and Performance

If the signal at the receiver can be resolved into several components,

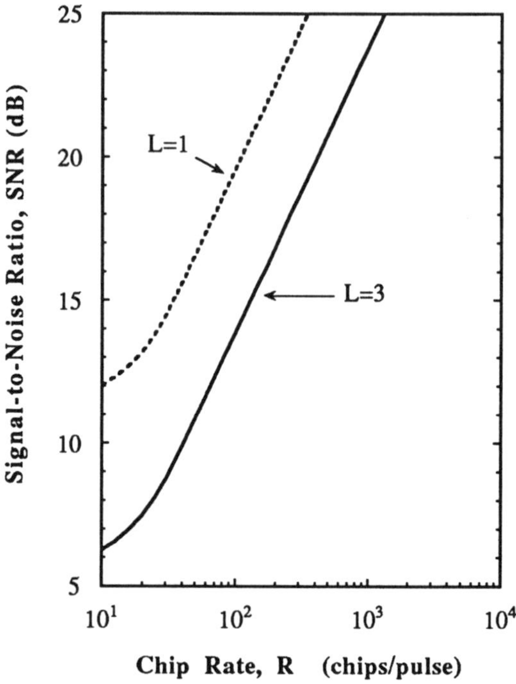

Figure 4. SNR required to achieve $P_e = 10^{-2}$ for channel C with ζ = -3 dB.

a decision must be made as to which components, and how many, the rake receiver should attempt to capture. One approach is to collect the greatest possible amount of energy by using the same number of taps as the number of signal components that can be resolved, up to the limit of the number of taps available. If square-law combining is employed, it cannot be stated *a priori* that this approach gives the best performance. The optimal selection of multipath components to be combined depends on the characteristics of the multipath channel; in particular, the inclusion of additional components does not necessarily inprove performance [11, 13]. In addition, even if the use of extra taps improves performance, the performance gains may not justify the required increase in receiver complexity.

Consider a CDMA system with a chip rate that is sufficiently high to resolve a specular signal component at each tap of the rake receiver. If a specular path exists in the channel, the signal envelope at the output of the matched filter at each tap has a low variance. So the reduction in the frequency of deep fades obtained by diversity combining does not have a significant impact on the performance. The primary benefit of diversity combining in this situation is the increase in the average energy that is collected, balanced against the increase in thermal noise that is collected. If a chip rate of 400 is used in conjunction with channels A or B, the additional energy collected by the second and third taps is insignificant, and thus the additional thermal noise introduced by the extra taps degrades the performance. For these channels, a single tap gives better performance than multiple taps for systems with a high chip rate. This is seen in Figures 2 and 3.

A CDMA system with a lower chip rate suffers from more frequent deep fades of the matched filter output than a system with a higher chip rate, so there is more motivation to employ diversity combining if the chip rate is low. The benefit of a three-tap rake receiver for a low-chip-rate system is illustrated in Figures 2 and 3 for channels A and B.

If there is no specular component in the received signal, a stable signal envelope cannot be achieved even with a high chip rate. The consequences of this are illustrated in Figure 4. If a correlation receiver is used in conjunction with channel C, adequate performance cannot be obtained with a chip rate of 400. It is also indicated in Figure 4 that the use of a three-tap rake receiver significantly improves the performance of a high-chip-rate system operating on channel C. This is in marked contrast to the conclusions that are drawn for channels A and B.

These examples demonstrate that the benefit of employing multiple taps in a rake receiver depends upon the characteristics of the channels that will be encountered by the system. For a low chip rate the use of multiple taps is beneficial with many channel containing strong specular components. In contrast, the correlation receiver may be the best choice for use with a high chip rate if the channel has a significant specular component, even if less than one-half of the energy is in that component.

VI   Error-control Coding

The comparisons made thus far are applicable to a system with error-correction coding, if the receiver employs binary hard-decision demodulation and the interleaving is deep enough that consecutive symbols at the decoder input are subject to independent fading. For a given code, the target error probability for a code symbol is the value necessary to achieve the desired error probability at the output of the decoder. However, many receivers that employ Viterbi decoding make use of soft decisions to achieve better performance than can be obtained with hard-decision decoding. In addition, the interleaving depth is often limited by the allowable decoding delay, and for some channels the depth is insufficient to achieve independence of fading for consecutive decoder inputs. Consequently, we examine the effect of the interleaving depth on the performance of a CMDA system that employs rake reception and soft-decision decoding.

Consider a CDMA system with convolutional encoding and an interleaving depth of r channel symbols. A rake receiver is used in conjunction with soft-decision Viterbi decoding and a square-law metric. The effect of finite interleaving depth is examined by determining the signal-to-noise ratio required to achieve a pairwise error probability of 10^{-4} for an error event with Hamming weight equal to the free distance of the code. As an example, consider an error event with a Hamming weight of 10. The CDMA system is used in conjunction with channel B. If a correlation receiver is used, a chip rate of 50 is 0.4 dB better than a chip rate of 400 in the case of infinite interleaving depth. However if the interleaving depth is one (that is to say, there is no interleaving), the low-chip-rate system is 19 dB poorer than the high-chip-rate system. If a three-tap rake receiver is employed, the low-chip-rate system provides superior performance with infinite interleaving depth but 7 dB poorer performance with no interleaving.

The most signifcant change in the performance due to a change in the interleaving depth occurs if the depth is on the order of the coherence time of the channel. Consider an error event with a Hamming weight of five. The pairwise error probability is shown in Figure 5 for channel B with rake receivers of one and two taps. The Doppler spread is 0.004, so that the coherence time of the channel is equal to the duration of approximately 40 channel symbols. The performance is shown for interleaving depths of 25 and 150. A low-chip-rate system that uses a correlation receiver experiences severe fading, and it therefore requires the time diversity provided by coding and interleaving in order to provide an acceptable pairwise error probability. For a given Doppler spread the interleaving depth determines the effective order of time diversity

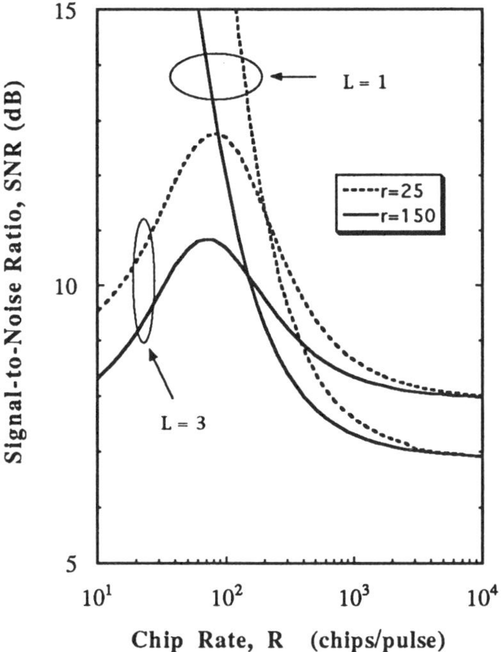

Figure 5. SNR required to achieve code performance target for channel B with ζ = -3 dB.

provided by the code, so the performance of the low-chip-rate system is highly sensitive to the interleaving depth. For this example the performance is 3.3 dB poorer for an interleaving depth of 25 than for a depth of 150. In contrast, the high-chip-rate signal is not subject to severe fading, and the performance differs by less than 0.7 dB for the two interleaving depths. The use of multiple taps in the rake receiver reduces the sensitivity of the low-chip-rate system to the interleaving depth. It is shown in Figure 5 that changing the interleaving depth from 150 to 25 degrades performance by only 1.7 dB for a system with the low chip rate and a two-tap rake receiver. However, the interleaving depth affects this system more than it does a system with the high chip rate and a correlation receiver.

The effects of the chip rate and the Doppler spread on the bit error probability are illustrated by the simulation results shown in Figure 6 for channel B with ζ = -3 dB. The DS spread-spectrum system employs a 32 by 18 block interleaver and the NASA-standard convolutional code of rate one-half and constraint length seven [17]. The performance of

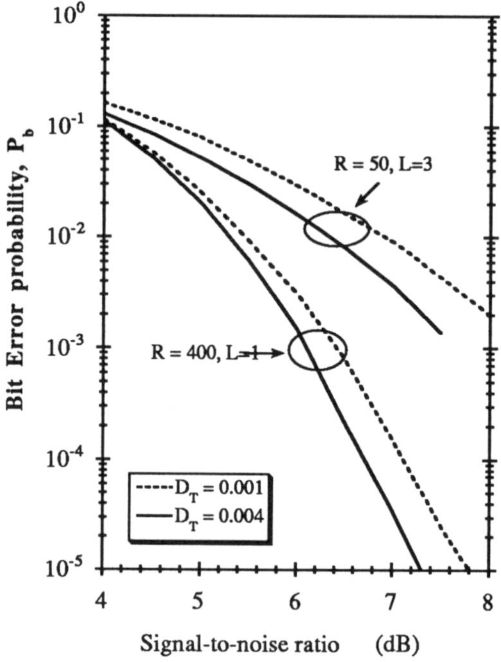

Figure 6. Performance of two systems with convolutional code and block interleaver.

the low-chip-rate system and a three-tap rake receiver is compared with the performance of the high-chip-rate system and a correlation receiver. If the target probability of error is 10^{-2} and the Doppler spread is 0.004, the high-rate system gives 1.2 dB better performance than the low-rate system. The difference in performance increases to 1.5 dB if the Doppler spread is 0.001. If a lower probability of error is required, the superiority of the high chip rate over the low chip rate with this channel is even more pronounced.

VII.   Conclusions

We have examined the manner in which the channel model influences the conclusions that are drawn from an analysis of the performance of rake receivers. CDMA systems with chip rates of 50 and 400 are compared for both Rician-fading channels and purely Rayleigh-fading channels. The comparison shows that the high chip rate provides better performance if most of the received energy is contained in a few strong specular components. However, if the channel exhibits purely Rayleigh fading, then the low-chip-rate system obtains better performance than with the high chip rate for a given number of taps.

If the channel contains a strong specular component, the low-chip-rate system benefits from the use of multiple taps in the rake receiver for a wider range of channel conditions than the high-chip-rate system. With a chip rate of 400, a rake receiver is of value only if there are two or more strong specular components that are of nearly equal strength. However, if the multipath components exhibit purely Rayleigh fading, a rake receiver is necessary for adequate performance with either chip rate.

The contrast in results for the two models demonstrates the need to select channel models that accurately reflect the circumstances that are likely to be encountered in mobile communications. Since wideband channel measurements indicate the presence of strong specular components in many urban environments, it appears that the Rayleigh-fading channel model is not sufficiently general for the analysis of high-chip-rate CDMA systems in those environments.

We have also examined the effect of coding and interleaving on the performance of CDMA systems with chip rates of 50 and 400. The performance of the low-chip-rate system is shown to be highly sensitive to the effect of interleaving depth. The degree of sensitivity is less with a rake receiver than with a correlation receiver. In contrast, it is shown that for a channel with a strong specular component, the high-chip-rate system is much less sensitive to the effect of interleaving depth than is the system with the high chip rate.

References

[1] G. L. Turin, "A statistical model of urban multipath propagation," *IEEE Trans. Veh. Technol.*, vol. VT-21, pp. 1–9, Feb. 1972.

[2] G. L. Turin, "Introduction to spread-spectrum antimultipath techniques and their application to urban digital radio," *Proc. IEEE*, vol. 68, pp. 328–353, March 1980.

[3] R. Price and P. E. Green, "A communication technique for multipath channels," *Proc. IRE*, vol. 46, pp. 555–570, March 1958.

200

[4] D. C. Cox and R. P. Leck, "Correlation bandwidth and delay spread multipath propagation statistics for 910-MHz urban mobile radio channels," *IEEE Trans. Commun.*, vol. COM-23, pp. 1271–1280, Nov. 1975.

[5] D. M. J. Devasirvatham, "Radio propagation studies in a small city for universal portable communications," *Proc. 1988 Vehicular Technology Society Conference*, pp. 100–104, June 1988.

[6] D. Schilling, "Broadband spread spectrum multiple access for personal and cellular communications," *Proc. of the 43rd IEEE Vehicular Technology Conf.*, pp. 819–822, May 1993.

[7] D. M. J. Devasirvatham, "A comparison of time delay spread and signal level measurements within two dissimilar office buildings," *IEEE Trans. Ant. and Prop.*, vol. AP-35, pp. 319–324, Mar. 1987.

[8] A. A. M. Saleh and R. A. Valenzuela, "A statistical model for indoor multipath propagation," *IEEE Journal on Selected Areas in Communications*, pp. 128–137, Feb. 1987.

[9] T. S. Rappaport, "Delay spread and time delay jitter for the UHF factory multipath channel," *Proc. 1988 Vehicular Technology Society Conference*, pp. 186–189, June 1988.

[10] P. A. Bello, "Characterization of randomly time variant linear channels," *IEEE Trans. Commun. Syst.*, vol. CS-11, pp. 360–393, Dec. 1963.

[11] D. L. Noneaker and M. B. Pursley, "The effect of chip rate on the performance of a CDMA system," *Proc. 1993 Global Telecommunications Conference*, vol. 4, pp. 45–49, Dec. 1993.

[12] D. L. Noneaker and M. B. Pursley, "The effects of spreading sequence selection on DS spread-spectrum with selective fading and two forms of rake reception," *Proc. 1992 Global Telecommunications Conference*, vol. 4, pp. 66–70, Dec. 1992.

[13] D. L. Noneaker and M. B. Pursley, "On the chip rate of CDMA systems with doubly selective fading and rake reception," *IEEE Journal on Selected Areas in Communications*, vol. 12, pp. 853–861, June 1994.

[14] A. J. Viterbi and J. K. Omura, *Principles of Digital Communications and Coding.* New York: McGraw-Hill, 1979.

[15] G. L. Turin, "The characteristic function of Hermitian quadratic forms in complex normal variables," *Biometrika*, vol. 47, pp. 199–201, June 1960.

[16] D. L. Noneaker and C. Frank, "The effect of finite interleaving depth on the performance of convolutional codes in Rician-fading channels," *Proc. of the 1994 IEEE International Symposium on Information Theory*, p. 29, June 1994.

[17] I. M. Onyszchuk, "On the performance of convolutional codes," PhD. dissertation, Dept. Elec. Eng., California Institute of Technology, May 1990.

Frequency Hopped Systems for PCS

David E. Borth, Phillip D. Rasky, Greg M. Chiasson and
James F. Kepler

Abstract -   In this paper, the application of frequency hopping methods to
cellular and personal communication systems (PCS) is studied. The
performance enhancements due to the use of frequency hopping (FH) is
initially reviewed followed by a brief summary of previously designed
cellular/PCS systems that have incorporated FH methods. A recently
designed slow frequency hop (SFH) prototype which was developed for
PCS applications is then described. Performance results using the SFH
prototype in laboratory and field trials are provided. These results
demonstrate the excellent performance that may be obtained in PCS environments using a SFH architecture which has a complexity level that is
suitable for PCS implementation.

I.   Introduction

In a series of rulemaking proceedings beginning in October of 1993, the U.
S. Federal Communications Commission (FCC) released 140 MHz of
spectrum in the 1.85-1.99 GHz region for use in personal communication
systems (PCS) applications [1]. Anticipating the release of this spectrum,
over the last several years, the Communication Systems Research
Laboratory designed, constructed and field tested a prototype PCS system
which incorporated a hybrid form of multiple access - in particular slow
frequency hopping CDMA combined with time division multiple access
(TDMA). This system was designed to permit true portable/vehicular
mobility with large radius cells (1-4 miles) while also allowing for coexistence with lower-cost PCS systems that operate in pedestrian-only
environments. In contrast to several other systems proposals for the PCS
bands, it was deemed desirable to support the full range of PCS services,
including medium-to-high data rate voice and data services, by incorporating these services into the initial system design. This paper describes this
system and indicates the level of performance that may be obtained from
such a system in a PCS environment.
The paper is organized into 5 sections. Following the introduction, the
attributes of slow frequency hopped systems, as applied to cellular and PCS
applications, are described. Next, related work in the area of FH systems for
PCS and cellular applications is reviewed. A description of the slow-
frequency hop prototype system is then given followed by the results of field
tests conducted with this system.

S.G. Glisic and P.A. Leppänen (eds.), Code Division Multiple Access Communications, 203-223.
© 1995 *Kluwer Academic Publishers. Printed in the Netherlands.*

II.   Performance Attributes of SFH Systems

Slow frequency hopped (SFH) CDMA digital cellular systems can benefit from two means of performance improvements over non-frequency hopped systems. First, at slow vehicle speeds, slow frequency hopping provides a form of frequency diversity through exploitation of frequency selectivity over the system bandwidth. From hop-to-hop the fading process is decorrelated either over the span of an interleaver, or over the duration of long fades [2]. This frequency diversity results in improved performance which is independent of vehicle speed, provided the hopping dwell period is sufficiently small.

Second, by careful assignment of small correlation frequency-hopping patterns, a given source of interference can affect the interference only a small portion of time. This interference reduction effect has been termed interferer diversity in [3]. Interference diversity causes the interference experienced by any user to be reduced as the interference experienced by the user comes not from a single dominant interferer, as is the case in analog cellular systems, but rather from the aggregate of all users, each sampled one at a time. Provided that coding with adequate interleaving is employed in the SFH system, individual samples of extreme interference may be corrected in the decoding process.

Several systems that take advantage of both of these attributes of SFH systems are discussed in the next section.

III.   Related Work in FH Systems

A number of different groups of researchers have previously investigated the use of frequency hopping in cellular and PCS applications. In this section, this earlier work is briefly summarized.

Cooper and Nettleton DPSK FHMA System

In a series of papers beginning in 1977 [4], Cooper and Nettleton described what has become recognized as the first serious approach to applying spread-spectrum code division multiple access methods to the design of a digital cellular system. The Cooper and Nettleton proposal was a frequency-hopped multiple access cellular system employing differential PSK (DPSK) as the modulation method. Hence, this system will subsequently be called a DPSK FHMA system.

The DPSK FHMA system employs fast frequency hopping (FFH) over a 200 kHz bandwidth. FFH was employed as a diversity mechanism to overcome the time-selective fading effects present on the land mobile radio channel. The hopping rate of 200 khops/s provided 6.4 hops per information bit. Cooper and Nettleton proposed the use of the Yates-Cooper sequences, a set of sequences which has the one-coincidence property, i.e., when a frequency hop sequence is shifted in time by one dwell period with respect to any other frequency hop sequence in the set, there will be at most one dwell period for which two sequences will occupy the same frequency. No means for overcoming frequency selective fading effects was included in the system design. Block orthogonal coding was used as a means of correcting random errors due to both thermal noise and collisions between users. The system was designed to support single cell reuse. A simplified, but accurate analysis of this system is provided in [5].

Bell Labs Multilevel FSK Frequency Hop System

The work of Cooper and Nettleton prompted a number of researchers at Bell Labs during 1979-1982 [6] to investigate other forms of spread spectrum multiple access methods that were suitable for digital cellular systems. In particular, a team of researchers designed a fast frequency-hopped spread spectrum multiple access system that employed multilevel (M-ary) frequency-shift keying (FSK) as the modulation method. The capacity of this system was shown to be roughly three times the capacity of the Cooper and Nettleton proposal.

In contrast to the DPSK FHMA system, the Bell Labs FH-FSK system employed repetition coding as the means of correcting errors due to noise and frequency collisions. Goodman et al. provided an analysis of the FH-FSK system in [6]. Additional enhancements to the FH-FSK system are described in [7-8].

S900D System

Prior to the development of the GSM system in Europe, several proposals were put forth as trial systems to be used in selecting the access and channelization approaches for GSM. ANT and Bosch developed a "narrow band" TDMA proposal as one of these trial systems in which slow frequency hopping was offered as a means of overcoming the slow-moving vehicle problem. Further details of the S900D system are given in [9].

SFH900 System

Another trial system that was proposed prior to the acceptance of the current GSM specification was the SFH900 system, which was developed by Laboratoire Central de Te,'le,'communications (LCT) in France [10]. This system is interesting because it combined slow frequency hopping, concatenated coding and channel equalization to combat the effects of both time selective and frequency selective fading. An experimental prototype of this system was constructed and was demonstrated in the field tests in Paris during late 1986 which were used to test various component parts of the GSM system.

The SFH900 system employs TDMA within a cell and SFH-CDMA between cells as the access methods. The TDMA frame format consists of a transmit slot, a receive slot and a frequency switching slot, denoted by T, R, and S slots, respectively. All slots and frames are synchronized within the system to a common clock. The use of the T-R-S slot structure within a cell permits easy implementation of a SFH system without requiring a fast switching synthesizer. Between cells, SFH is employed as the multiple access method. The SFH900 system hops at a rate of 250 hops/s. Both Reed-Solomon codes and cyclotomicaly shortened Reed-Solomon codes are employed to combat errors in the system. To overcome the effects of frequency selective fading, a maximum likelihood sequence estimator (MLSE) equalizer is used.

The GSM SFH Digital Cellular System

The GSM SFH digital cellular system represents another slow frequency hopping approach to digital cellular systems. It has been commercially deployed at 800/900 MHz in many of the western European nations including the U.K., France, Germany, Denmark, Sweden, plus others. In addition, an extension of the GSM standard, defined by ETSI as DCS1800 for the 1800 MHz band, is currently being deployed in several European countries. A modification of the latter system has been proposed for the PCS frequency bands in the U.S. The GSM system is described in a formal set of documents that is available directly from ETSI [11].

The GSM system is an 8 slot per frame TDMA system operating at a data rate of 270.8333 kbps over the air. The modulation format employed is $B_b T=0.3$ GMSK. Speech information is convolutionally coded using a rate 1/2, constraint-length 5 convolutional code. At the channel data rates employed, the transmitted signal is expected to undergo moderate to severe frequency-selective fading of up to 4 symbol times (≈ 15 μsec). Hence the GSM specifications implicitly require that some form of equalizer be incorporated into both the mobile and base receivers. To combat slow time-selective fading and to provide interference diversity, slow frequency hopping at the frame rate of 217 hops/sec is specified in the GSM specifications.

Summary of Related Work

Table I summarizes both the related work on FH cellular and PCS systems as well as many of the parameters of the SFH PCS prototype system that is described below.

IV. Prototype SFH PCS System Description

The prototype SFH PCS system will now be described. In designing the SFH PCS prototype system, it was deemed desirable to develop a system that would both work in the large cell mobile environment and would exhibit minimum complexity. Accordingly, a study was conducted in which possible tradeoffs between coding/equalization/diversity were determined. In particular, the study looked at constraint-length 6-9 convolutional codes, maximum likelihood sequence estimation (MLSE) equalization, frequency hopping and antenna diversity. The study determined that, for a wide range of the type of channels that were expected to be encountered by the system, the combination of a constraint-length 6 or 7 convolutional code plus slow frequency hopping plus antenna diversity yielded the best performance overall at relatively low complexity. Reference [12] gives further details of this study. Surprisingly enough, channel equalization via MLSE yielded little to no performance improvement over even strongly frequency selective fading channels (1-2 symbol delay, equal gain paths), when the other diversity/coding elements were also present.

Table I - Summary of parameters of FH cellular systems

Parameter	FH-DPSK	FH-FSK	S900D	SFH900	GSM	SFH Prototype
FH type	fast	fast	slow	slow	slow	slow
Access method:						
- Intercell	CDMA	CDMA	TDMA	CDMA	TDMA	CDMA
- Intracell	CDMA	CDMA	TDMA	TDMA	TDMA	TDMA
Modulation	DPSK	FSK	4-CPFSK	GMSK	GMSK	4-QAM
FEC method	Block orthogonal	Repetition	None	Reed-Solomon	Convolutional	Convolutional
BW/channel	200 kHz		250 kHz	200 kHz	200 KHz	400 kHz
Cell reuse pattern	1 cell	1 cell	7 cell	1 cell, 3 sector /cell	3 cell, 3 sector/cell	1 cell, 3 sector/cell
Slots/frame	1	1	10	3	8	10
Hop rate	200 kh/s		31.25 h/s	250 h/s	217 h/s	500 h/s
FSF mitigation	none	none	none	MLSE equalizer	equalizer	diversity
Diversity method	frequency diversity	frequency diversity	none	frequency diversity	none	antenna diversity
Speech coder rate	31.25 kbps	32.89 kbps	9.6 kbps	16 kbps	13 kbps	16 kbps
Overall channel rate	31.25 kbps	32.89 kbps	128 kbps	201 kbps	270.83 kbps	500 kbps

Although not obvious, the hybrid multiple access approach does offer several advantages over pure TDMA or SFH-CDMA schemes. SFH CDMA is employed as the intercell multiple access method in the hybrid system. Within a cell, orthogonality is strictly maintained by requiring that no two users occupy the same frequency at the same time. This requirement results in absolutely *no* intracell interference under *all* channel conditions. The SFH CDMA approach allows the system to incorporate frequency diversity, which minimizes the performance degradation due to slow fading channels, as well as interference diversity, which ensures that the system is not subject to a worst case interference problem.

TDMA is employed as the intracell multiple access method in this system. TDMA readily permits the introduction of such system concepts as "bandwidth on demand" (via concatenation of time slots to support higher data rates) and asymmetric uplink and downlink data rates (via assignment of different numbers of slots to both paths). From a base site implementation viewpoint, TDMA allows a reduction in the total number of RF transceivers required. Finally, from a portable/mobile transceiver implementation viewpoint, TDMA inherently permits "reasonable" synthesizer switching times and handoff measurement times during the periods when the transceiver is neither transmitting or receiving.

Table II - Hybrid SFH TDMA/CDMA PCS system specifications

Frequency band:	1850-1970 MHz
Slots/frame	10
Channel bit rate	500 kbps
Frequency hopping rate	500 Hz
Duplex method	FDD
Intracell multiplexing	TDM/FDM (orthogonal SFH)
Intercell multiplexing	SFH/CDMA
Modulation	QPSK
Pulse Shaping	raised cosine, α=0.5
Time slot duration	0.2 msec
Frame duration	2.0 msec
Gross data rate/slot	34 kbps
Channel coding:	
Type	Rate 1/2, K=6 Convolutional, soft decision
Interleaving span	40 msec
Maximum differential multipath delay	16 μsec
Speech coder	32 kbps ADPCM

The parameters of the hybrid SFH TDMA/CDMA system (as implemented in the prototype) are shown in Table II. A block diagram of the system appears in Figure 1. In the following, a concise description of the operation of the transmitter and receiver is given. Further details of the transmitter/receiver operation and actual implementation are given in [13].

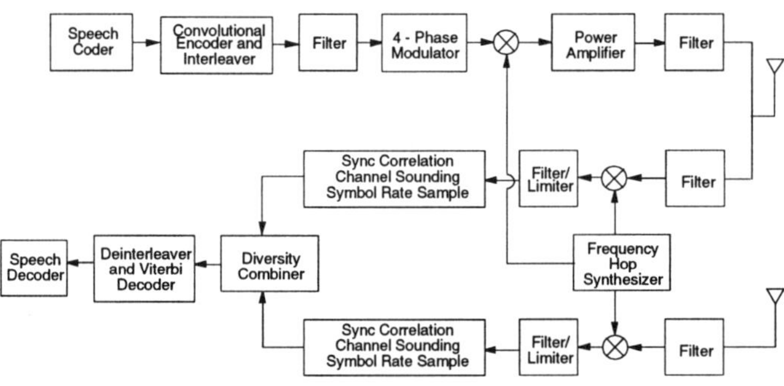

Figure 1. Prototype SFH PCS system block diagram.

Transmitter Description

As Table II indicates, a 32 kbps ADPCM speech coder utilizing the G.721 algorithm is employed as the speech coder in the prototype system. This speech coding algorithm permits "toll-quality" speech at moderate complexity. Other sub-multiple speech coding rates are also supported in the system and yield correspondingly higher spectral efficiency. The output of the speech coder is supplied to a rate 1/2, constraint length 6 convolutional coder with generator polynomials:

$$g_1(D) = 1 + D + D^2 + D^3 + D^5$$
$$g_2(D) = 1 + D^2 + D^4 + D^5 \ .$$

The output of the convolutional coder is supplied to a convolutional interleaver with parameters (I,j)=(18,9) which has an end-to-end delay corresponding to 40 msec of 32 kbps speech. The interleaved symbols are then fed to a filtered QPSK modulator; the output of which is upconverted with a frequency hop synthesizer, amplified, and transmitted.

Receiver Description

The receiver employs two diversity antennas and two independent receiver branches which are subsequently combined. In each of the branches, the signal is de-hopped, using a synthesizer time-locked to the transmit synthesizer, and filtered/hard-limited prior to undergoing sync acquisition, channel sounding, and symbol rate down-conversion. Hard-limiting is used in the receiver to simplify the receiver signal processing in several ways. First, hard-limiting inherently reduces the dynamic range of the arithmetic required in subsequent signal processing stages. Second, hard-limiting removes the requirement for an automatic gain control (AGC) subsystem in each of the diversity branches. Finally, because AGC is not required in the receiver, initial acquisition of the frequency-hopped transmitted signal is easier, since dynamic range searching is eliminated as a parameter in the initial time-frequency search process which is necessary to acquire the signal. Although hard-limiting any signal (in particular, multiple access signals) generally results in a degradation in performance, the performance degradation due to hard limiting is less than 1 dB for this system.

Following hardlimiting, sync acquisition, and downsampling, the two diversity branches are combined in a "near" maximal ratio combining fashion. The diversity combining process is no longer strictly maximal ratio combining due to the presence of the hard limiters in each of the branches. In [13] it is shown that an estimate of the optimal channel gain g to interference power σ_n^2 diversity weighting coefficient is given by

$$\frac{g}{\sigma_n^2} \sim \frac{R_{rx}}{\sigma_r^2 - R_{rx}^2}$$

where R_{rx} is the synchronization word cross-correlation and σ_r^2 is the variance of the received signal. The combiner also tracks timing and frequency offsets in the receiver. Following combining, the received signal is sent to a convolutional deinterleaver and a soft decision Viterbi decoder. The decoded data is then transferred to the ADPCM speech decoder.

210

In accordance with this system definition, transceiver hardware suitable for vehicular field testing has been constructed. The SFH prototype mobile and base station hardware are assembled in 19" racks with modules that are 10" high and at most 12" deep. The base and mobile units can transmit a peak power of 800 mW. The system is fully capable of establishing both mobile originated and mobile terminated telephone calls with an interface to the PSTN through an echo-canceled conventional analog line.

V. Experimental Results

Introduction

Considerable special purpose testing software has been integrated into the SFH prototype units to allow both subjective and objective quality assessments of the full-duplex link. The slow frequency-hop prototype is capable of logging statistics on the signal received by each of the diversity branches and bit and frame error rate statistics. In this way the transceiver can be simultaneously used to validate the performance of the system and provide channel sounding capabilities.

The following statistics on the received signal envelope from each antenna branch are measured: the received signal power, the standard deviation, the auto-covariance function (for lags up to the span of the interleaver, i.e., 20 hops), and the RMS delay spread. In addition, the cross-covariance function between the branches is estimated.

The auto-covariance function of the envelope aids in determining the extent of the frequency selectivity across the band or equivalently, the extent to which frequency hopping has decorrelated the slow fading channel. The cross-covariance between branches aids the determination of the diversity gain over fading that may be attainable. In order to isolate the fading envelope as the variable of interest, thereby removing the dependency of signal strength from the statistics, the correlation coefficient is calculated. This is computed as

$$\rho_{xy}(\tau) = \frac{R_{xy}(\tau) - \mu_x \mu_y}{\sigma_x \sigma_y},$$

where $R_{xy}(\tau)$ is the cross-correlation between variables x and y at a lag of τ frames, μ is the mean, and σ is the standard deviation. For the cross-covariance function x and y are the envelopes of each diversity branch. For the auto-covariance function x and y are substituted by the same envelope offset by the appropriate number of dwells. The lag-1 auto-correlation coefficient is typically analyzed as it is least susceptible to time decorrelation. The coefficient of variation,

$$C_x = \frac{\sigma_x}{\mu_x},$$

is calculated as a normalized standard deviation. The coefficient of variation is useful in assessing the severity of the channel in terms of the depth of fading.

As the symbol duration for transmission is on the order of the differential delay of multipath components over many propagation paths, the performance of the link is dependent on delay spread. The digital baseband processor contains an FIR filter matched to the 10-symbol synchronization pattern which is the preamble for each TDMA slot. In addition to its function of sounding the amplitude, phase, and timing epoch of the received signal, the matched filter output facilitates the estimation of the RMS delay spread. This is calculated from a power-delay profile accumulated over the measurement sample interval. The measured RMS delay spread is subject to bias from noise and, due to the symbol rate and filtering, the breadth of the autocorrelation function of the preamble. In order to remove this bias, an estimator which is a function of the measured delay spread and signal-to-noise ratio was developed to more accurately reflect the true RMS delay spread. The estimator was trained in the laboratory utilizing a multipath channel simulator. The resulting estimate of RMS delay spread is accurate to within 0.5 μsec for channels having an RMS delay spread value between 0 and 6 μsec.

During the field testing, all the aforementioned statistics are logged to a file on a notebook computer together with information from a Global Positioning System (GPS) receiver integrated into the mobile station apparatus. The GPS information consists of the time, the velocity of the mobile, and the location (in latitude and longitude) of the mobile. These statistics have typically been averaged over a 4 second interval which corresponds to 2000 frequency hops. The base and mobile stations utilize dual omnidirectional antennas. The outbound system hopping bandwidth used for field testing is 1960-1970 MHz. The base antenna site is deployed at a height of approximately 100 feet at the Motorola Center in Schaumburg, Illinois (suburban Chicago). The terrain in this area could be classified as a dense suburban environment. The area surrounding the base site over a radius of several miles includes residential neighborhoods (not more than 2 stories high), business and industrial parks, shopping malls, and office buildings. There are approximately 10 buildings of 10 to 20 stories in height to the south and east of the base site and at a distance greater than one mile away.

The SFH PCS field testing effort has two primary objectives: the assessment and validation of the technologies incorporated into the system and the accurate characterization of the PCS propagation environment. To accomplish these goals, it is not only necessary to measure a number of signal and channel parameters but also to interpret the results via comparisons to expected theoretical and simulation performance. Within this section, a preliminary review of current field testing results will be presented. The performance benefits attainable from the use of technologies such as diversity and frequency hopping will be investigated, and the characteristics of the PCS channel examined. Within this framework, theoretical and simulation results will be provided as necessary to aid in the analysis of the experimental data. Results presented are for the outbound link (base to mobile).

Field Data and Analysis

As discussed earlier, statistics on the coefficient of variation of the envelope of the received signal are accrued. This statistic is useful in assessing the degree to which the fading process matches ideal Rayleigh or Rician fading. Figure 2 plots the coefficient of variation over a range of Rician parameters, K, defined as the ratio of the power in the direct ray to that in the diffuse rays. When the Rician parameter equals zero, the fading is purely Rayleigh, and as indicated in the figure, the coefficient of variation equals 0.52. If a direct, line-of-sight path exists, the fading process becomes Rician, and as the power in the direct ray increases, the value of the coefficient decreases. Figure 3 depicts the probability density function for the coefficient of variation measured in a predominantly shadowed environment. In this area, very few line-of-sight paths to the transmitter exist; therefore, the fading is expected to be Rayleigh in nature. Figure 3 confirms this hypothesis with the large majority of the coefficients falling in the 0.4 to 0.55 range.

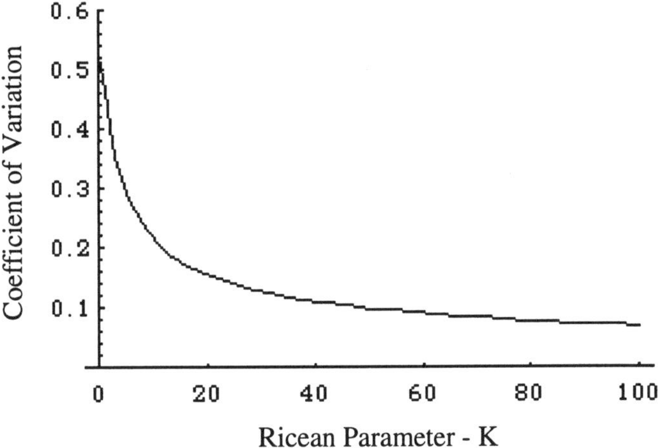

Figure 2. Coefficient of variation versus the Rician parameter, K.

Statistics are also accrued on the correlation coefficient of the envelope of the received signal. Of primary interest is the lag one coefficient. This statistic is of benefit in determining the degree of correlation in the fading process between consecutive frames. Since the frame duration is 2 ms, there will always be a vehicle-speed-dependent time decorrelation in the fading process. In addition, if frequency hopping is employed, the frequency selectivity of the channel can result in additional decorrelation. (Recall that there is one frame per frequency hop.)

To establish a baseline for interpreting the correlation coefficient results, Figure 4 illustrates the relationship between the correlation coefficient and vehicle speed for Rayleigh and Rician faded channels. In producing this figure, the Rayleigh fading curve was generated from the approximation [14]: $\rho(\tau) \doteq J_0^2(\omega_m \tau)$

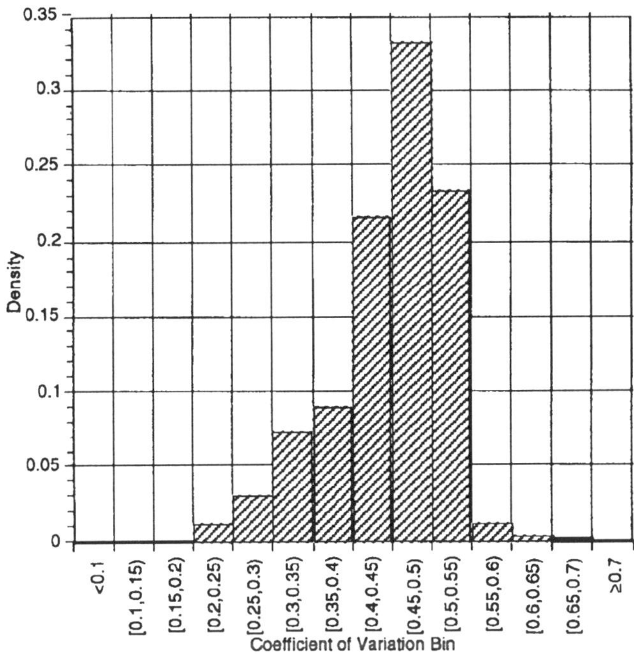

Figure 3. Probability density function for the coefficient of variation in a predominantly shadowed environment.

where τ is the frame duration and ω_m is the maximum Doppler frequency, calculated at a carrier frequency of 1.9 GHz. Rician fading, however, proved to be more intractable; therefore, a Monte Carlo simulation was employed to generate the Rician results. In the figure, the curve denoted by Corr. Coef - Theory' depicts the theoretical result for Rayleigh fading, and the curve denoted by 'Corr. Coef. Rician - K = 5' shows the expected results for Rician fading with a Rician parameter of 5. It should be noted that the general shape and level of the results was relatively constant over a wide range of Rician parameter values. In addition, the correlation coefficient for Rician fading exceeds that of Rayleigh fading at all but the highest speeds. Also shown in the figure is a scatter plot of correlation coefficients for Rayleigh fading that was produced by the Rician simulation with the Rician parameter set equal to zero. Since the simulation results are in good agreement with the theoretical results, the simulation model is believed to function correctly.

The benchmark results depicted in Figure 4 used a flat fading (nonhopping) channel in which the fading process is decorrelated only in time. Additional decorrelation may be obtained by employing frequency hopping. The amount of decorrelation will depend on the frequency separation between hops as well as on the coherence bandwidth of the channel, which is a function of delay spread. For Rayleigh fading, a generalized approximation to the envelope correlation [14] can be written as:

$$\rho(s,\tau) \doteq \frac{J_0^2(\omega_m \tau)}{1 + s^2 \sigma^2}$$

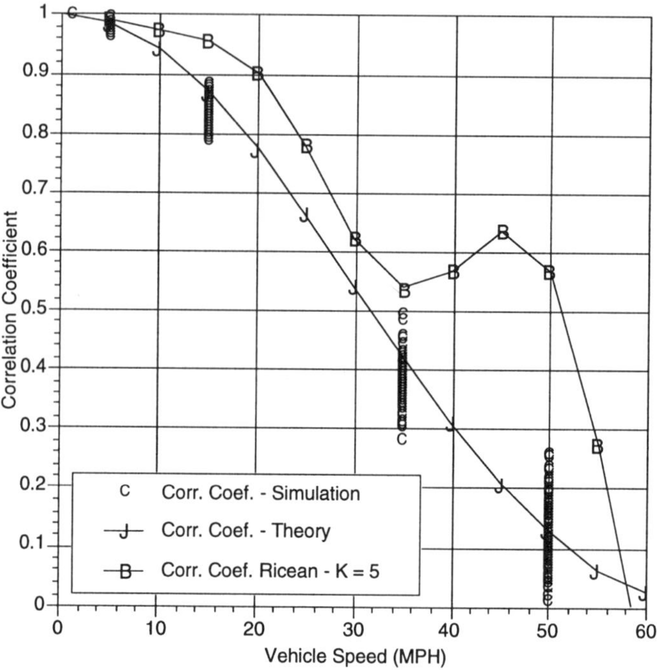

Figure 4. Correlation coefficient on flat fading (nonhopped) channels.
The results include only the effects of time decorrelation.

where s is the frequency separation between hops and σ is the RMS delay spread of the channel. Figure 5 plots the correlation coefficient as a function of vehicle speed and frequency hop distance. In this figure, an RMS delay spread of 1 μs has been assumed. As can be seen, the envelope correlation coefficient falls off quite rapidly as the frequency separation increases. Even hops separated by as little as 500 kHz provide nearly complete decorrelation across the entire range of vehicle speeds. In the field testing of the SFH prototype, an average frequency hop separation of 2.8 MHz was used. Therefore, correlation coefficients near zero are expected.

Using these expected correlation coefficient values, the experimental results can be interpreted. Figures 6 and 7 depict a scatter plot of the envelope correlation coefficient versus vehicle speed for hopped and nonhopped measurements on a variety of roads surrounding the basesite. Each scatter point was calculated from data corresponding to received signal envelopes accrued over a four second interval. In examining the figures, it is apparent that the coefficients for the nonhopped measurements are at least as large as those predicted by theory and that in both cases, a sizable number of correlation coefficient values are greater than that predicted by theory. In particular, the values for the frequency hopped measurement span a wide range.

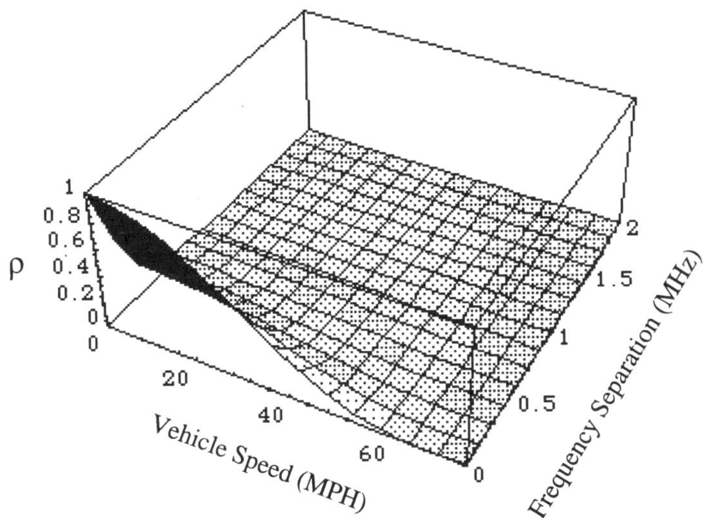

Figure 5. Correlation coefficient versus vehicle speed and frequency separation. The results in this plot use a 2 ms time separation and an RMS delay spread of 1 μs.

Therefore, it would seem that the predicted correlation coefficients form a lower bound on the true coefficient values but do not adequately explain the larger values. Several possible explanations exist for this discrepancy. First, the lowest predicted correlation values were based on the assumption of pure Rayleigh fading. The predictions under Rician fading were somewhat greater. In addition, a strong direct path diminishes the effective frequency selectivity in the hopped environment by dominating the signal level over all frequency hops. Therefore, if the environment were characterized by significant instances of Rician fading, larger correlation coefficients would be expected. Figure 8 confirms this hypothesis by depicting the correlation coefficient (taken over frequency hopped measurements) on a map of local roads. Note that the base site is located at the center of the innermost concentric circle and that the radii of each circle differ by 1 mile. Furthermore, the surrounding terrain is relatively flat with few other large buildings; therefore, nearby locations are likely to experience a strong, line-of-site path. As can be seen, these locations do in fact experience greater correlations than locations in which there is no direct ray. (For example, consider the values along the road which runs just to the left of the innermost circle. Immediately west of the basesite, a line-of-sight path exists, and the correlation values fall in the upper bin. However, as the road continues in the north or south directions, obstructions block the direct path, and the correlation values decline.) Thus, the Rician fading may be responsible for a portion of the larger than expected correlation values. Additionally, the measurement interval over which the correlation coefficient is averaged is fairly long (4 seconds). Variations in the log-normal shadowing process (a non-frequency selective phenomenon) may also be corrupting the correlation coefficient resulting in values which are greater than that which would occur due to fading alone.

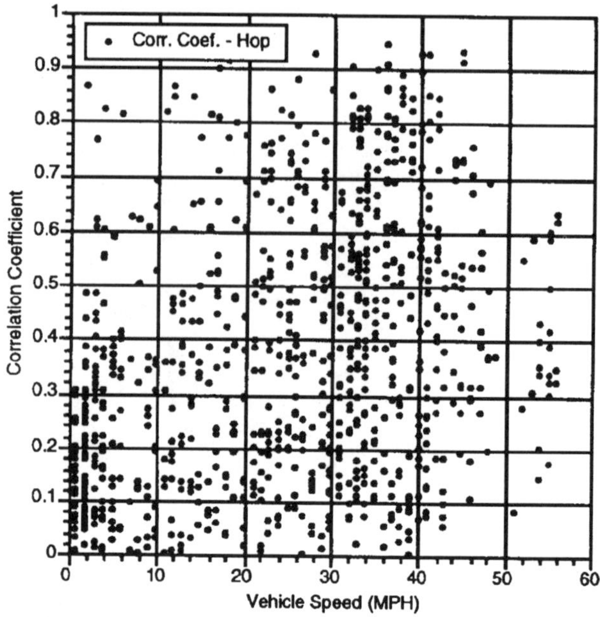

Figure 6. Envelope correlation coefficients taken with hopping measurements.

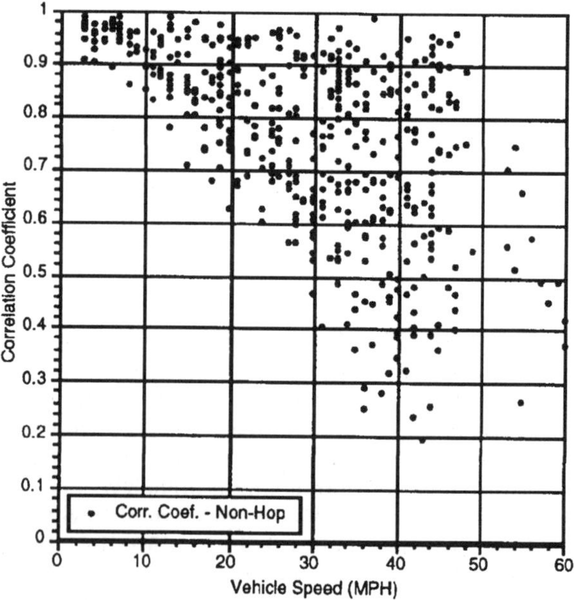

Figure 7. Envelope correlation coefficients taken with nonhopping measurements.

Figure 9 shows decoded BER data from several field tests overlaid on a map of the area surrounding the basesite. The figure shows data collected with both diversity and frequency hopping enabled. The BER data has been grouped into three bins, each of which represent a certain level of speech quality. The first bin, 0 to 0.1% BER, represents those areas where the speech quality is very high. The second bin, 0.1% to 1% BER, represents areas where the speech quality for 32 kbps ADPCM ranges from good to marginal. The third bin, greater than 1% BER, represents areas where the speech quality is unacceptable. Given these bins, Figure 9 can be interpreted as a coverage map in terms of speech quality. Except for the area to the northeast, which is severely shadowed, the SFH prototype system has a coverage area of about two miles around the base. If the received signal strength data were overlaid on a map, and the appropriate bins were chosen, the resulting figure would look very similar to Figure 9. This similarity indicates that the BER performance is, for the most part, a function of received signal power. No outages were observed at strong signal due to severe multipath or Doppler spread. Furthermore, no dependence on vehicle speed was noted in the frequency-hopped BER performance.

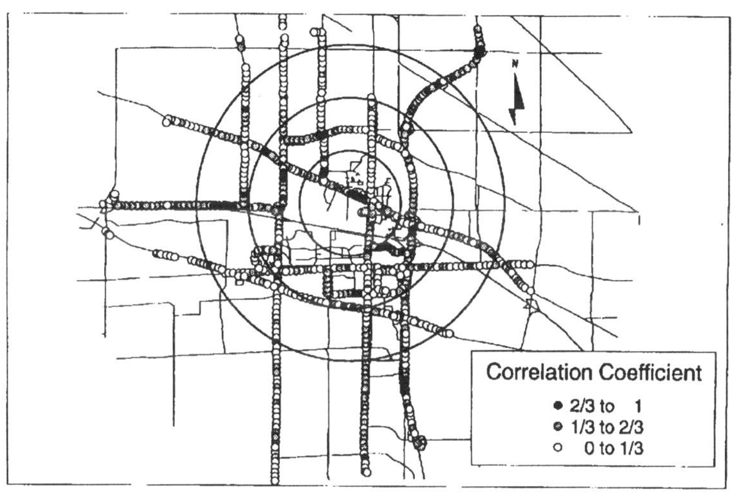

Figure 8. Correlation coefficient values on local roads surrounding the Schaumburg basesite. Note that locations with a line-of-site path have larger correlation coefficient values.

Although the format of Figure 9 is good for displaying the coverage, it is not suitable for displaying the diversity gain. A more appropriate format is the typical BER curve, as shown in Figure 10. The noise floor for each receiver is approximately -109 dBm, assuming an 8.5 dB noise figure and a 400 kHz noise bandwidth. Figure 10 shows a scatter plot of the frequency hopped decoded BER performance for both a single branch receiver and a diversity receiver. The test route for this data is roughly circular, with the base station in the center and a radius that ranges from 2 to 3 miles. The curves shown on the figure are generated by processing the data with a "median filter." This filter partitions the data into 1 dB bins and outputs only the median value in each bin. The median is preferred to the average because it is less

218

sensitive to outliers. At a 10^{-3} BER, the median curves show the antenna diversity gain to be ~ 7 dB. The diversity gain is larger when samples with more severe fading are isolated.

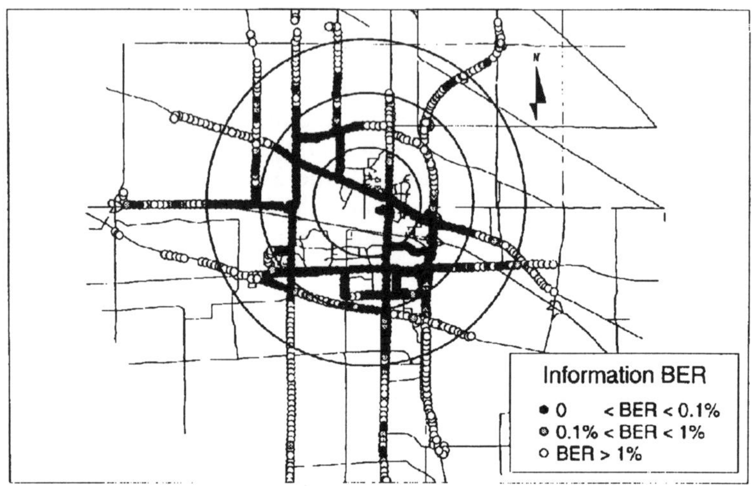

Figure 9. Field test results from the area surrounding the base station. The concentric circles have radii of 1, 2, and 3 miles and area centered at Motorola's Schaumburg campus. Conditions: frequency hopping and diversity enabled.

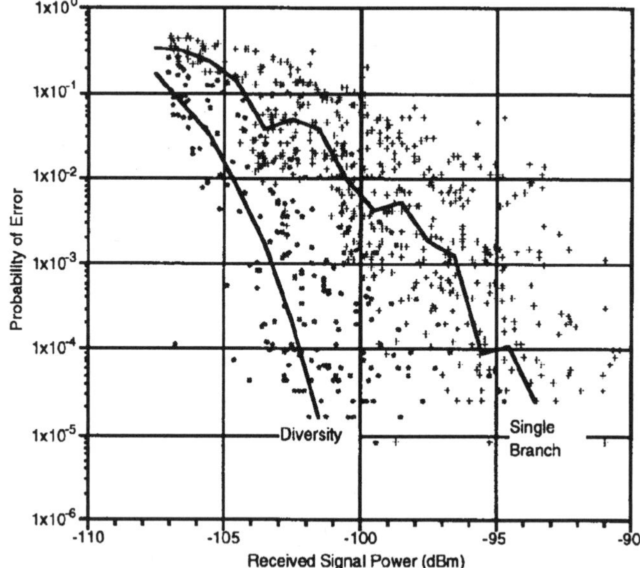

Figure 10. Field test results from a circular route surrounding the base station. Curve fit generated by the median filter. This plot shows the diversity gain to be about 7 dB. Conditions: frequency hopping enabled.

As mentioned earlier, the performance of the link is a function of the delay spread. In an attempt to quantify the impact that delay spread has on the coded BER performance, the data from the area surrounding the base was partitioned into three delay spread bins: 0 to 1 µs, 1 µs to 2 µs and greater than 2 µs. The data was taken with both frequency hopping and diversity enabled. Figure 11 shows the cumulative distribution of the delay spread for this area. Note that most of the data (77%) falls within the first bin, 0 to 1 µs. Figure 12 shows the channel and the decoded BER data after the delay spread partitioning. For clarity, only the median curves are shown. As expected, performance degrades with increasing delay spread. The curve for the low delay spread decoded data is within 1 dB of the simulation results using the GSM Typical Urban channel profile. This channel profile is characterized by an RMS delay spread of 1 µs. Note that the coding gain for delay spreads less than 1 µs is 14 dB at a BER of 10^{-3} (minus 3 dB to account for the added energy in a coded bit). As in the case of the diversity gain, the coding gain is larger for samples with more severe fading. In the case of the two higher delay spread bins, the channel BER performance never reaches 10^{-3}, that is, coding gain for high delay spread can not be measured as it is essentially infinite at this level of BER performance.

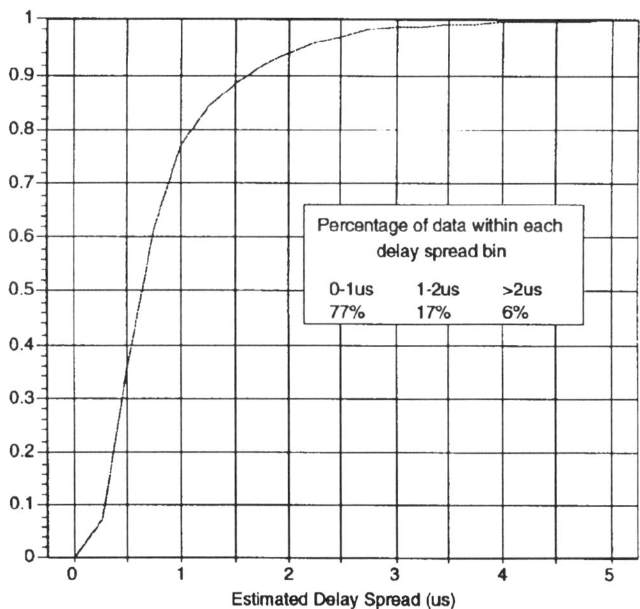

Figure 11. Cumulative distribution of the estimated delay spread for the Schaumburg, Illinois area.

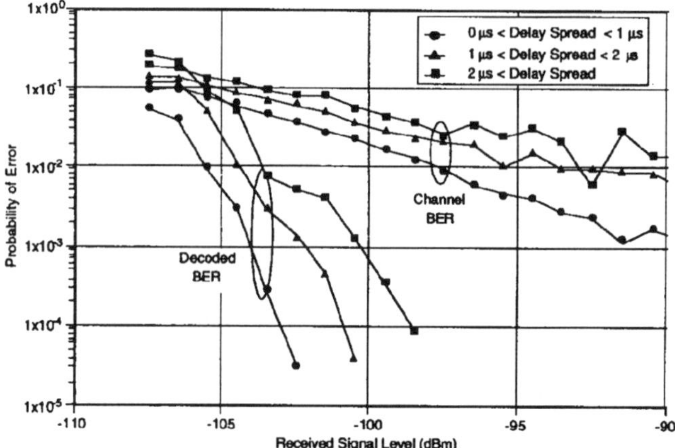

Figure 12. Field test results from the area surrounding the base station. Curve fit generated by the median filter. The data is partitioned into delay spread bins. Conditions: frequency hopping enabled, diversity enabled.

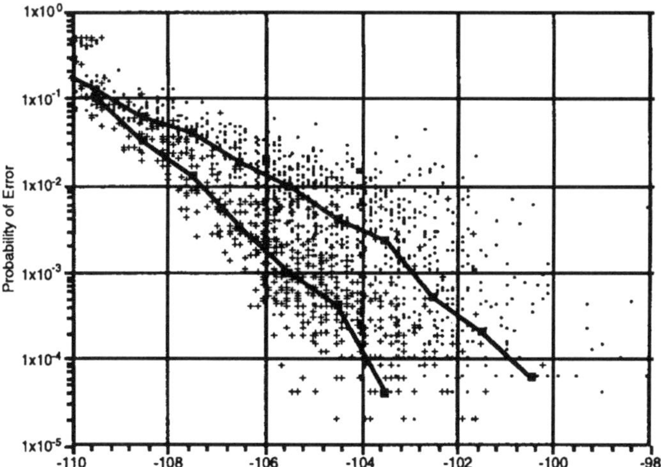

Figure 13. Field test results. Decoded BER performance. Curve fit generated by the median filter. This plot shows the frequency hopping gain to be about 2.4 dB. Conditions: diversity enabled, slow vehicle speed.

Since the SFH prototype described herein employs frequency hopping, one objective of the field tests is to measure the performance gain due to the hopping. There are two necessary conditions in order to show a performance improvement due to hopping. The first condition is that the hopping system bandwidth is greater than the coherence bandwidth of the channel, resulting in independent fading from hop-to-hop. With the hopping disabled as the baseline to measure the gain, the second condition is that the fading process is slow enough such that it is correlated from frame-to-frame. In other words, the second condition states that the gain due to hopping will be largest for low vehicle speeds. Simulation results employing an ideal frequency selective fading model, i.e., one in which the Rayleigh fading process is independent between hops, show that for a low delay spread (~ 1 μs), 5 mph faded channel, the frequency hopping gain for a diversity receiver at a 10^{-3} decoded BER is about 5.6 dB. Without diversity, the hopping gain increases to around 10 dB to 12 dB. Figure 13 shows the decoded BER performance, with diversity enabled, for a route located in a residential subdivision. The route was chosen because it has little delay spread, and it can be traveled at low vehicle speeds. Again, the curves show the median fit to the data. At a 10^{-3} BER, the hopping gain is only ~2.4 dB. Figure 14 shows the decoded BER performance with diversity disabled for the same route. In this case, the hopping gain at a 10^{-3} BER is ~5 dB. The hopping gain in both cases is much lower than that of the simulation results. One possible explanation for this result is that the fading process in the field is not as slow as the vehicle speed would indicate due to the changing environment such as motion in the scatterers. Other possible causes may be attributable to insufficient frequency selectivity across the hopping bandwidth due to the absence of delay spread or the dominance of line-of-sight paths.

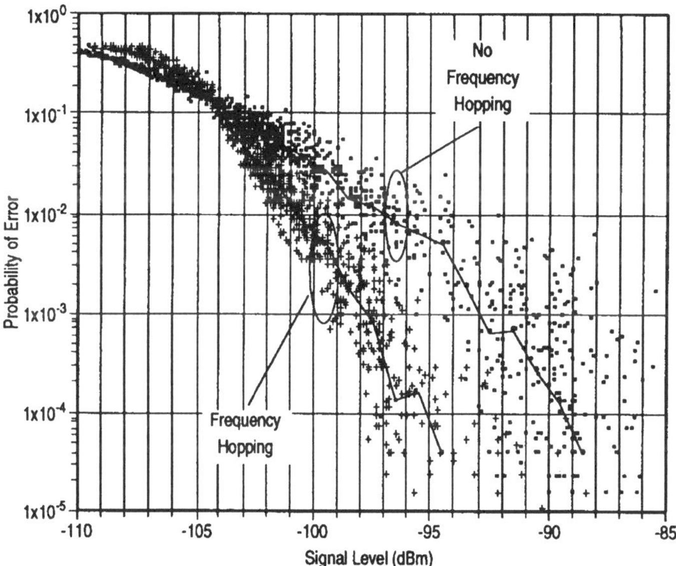

Figure 14. Field test results. Decoded BER performance. Curve fit generated by the median filter. This plot shows the frequency hopping gain to be about 5 dB. Conditions: <u>diversity disabled</u>, slow vehicle speed.

Acknowledgments

The authors gratefully acknowledge the efforts of Mark Cudak, Dave Gurney, Steve Kuffner, and Nick Tolli, all of the Communication Systems

Research Laboratory at Motorola, for the design and development of the research prototype described in this article.

References

[1] Federal Communications Commission, "In the matter of amendment of the Commission's rules to establish new Personal Communications Services," *Second Report and Order*, GEN Docket 90-314, FCC 93-451, October 22, 1993; subsequently revised in "Amendment of the Commission's rules to establish new personal communications services," *Memorandum Opinion and Order*, GEN Docket 90-314, FCC 94-144, June 13, 1994.

[2] M. Mizuno, "Randomization effects of errors by frequency-hopping techniques in a fading channel," *IEEE Transactions on Communications*, vol. COM-30, pp. 1052-1056, May 1982.

[3] D. Verhulst, M. Mouly, and J. Szpirglas, "Slow Frequency Hopping Multiple Access for Digital Cellular Radiotelephone," *IEEE Transactions on Vehicular Technology*, vol. VT-33, pp. 179-190, August 1984.

[4] G. R. Cooper and R. W. Nettleton, "A spread spectrum technique for high capacity mobile communications," *IEEE Transactions on Vehicular Technology*, vol. VT-27, pp. 264-275, November 1978.

[5] P. S. Henry, "Spectrum efficiency of a frequency-hopped-DPSK spread-spectrum mobile radio system," *IEEE Transactions on Vehicular Technology*, vol. VT-28, pp. 327-332, November 1979.

[6] D. J. Goodman, P. S. Henry, and V. K. Prabhu, "Frequency-hopped multilevel FSK for mobile radio," *Bell System Technical Journal*, vol. 59, pp. 1257-1275, September 1980.

[7] G. Einarsson, "Address assignment for a time-frequency-coded, spread-spectrum system," *Bell System Technical Journal*, vol. 59, pp. 1241-1255, September 1980.

[8] U. Timor, "Improved decoding of frequency-hopped multilevel FSK system," *Bell System Technical Journal*, vol. 59, pp. 1839-1855, December 1980.

[9] H.-P. Ketterling, "The Digital Mobile Radio Telephone System Proposal S 900 D," *Proceedings of the Second Nordic Seminar on Digital Land Mobile Radio Communication*, pp. 126-132, Stockholm, Sweden, October 14-16, 1986.

[10] J. L. Dornstetter and D. Verhulst, "The Digital Cellular SFH 900 System," *Proceedings of the IEEE International Conference on Communications,* pp. 36.3.1-36.3.5, June 1986.

[11] European Telecommunications Standards Institute, GSM Specifications, ETSI TC-SMG, Sophia Antipolis, France.

[12] P. D. Rasky, G. M. Chiasson and D. E. Borth, "Hybrid slow frequency-hop/CDMA-TDMA as a solution for high-mobility, wide-area personal communications," *Proceedings of the Fourth Winlab Workshop on Third Generation Wireless Information Networks,* pp. 199-215, East Brunswick, New Jersey, October 19-20, 1993.

[13] P. D. Rasky, G. M. Chiasson, and D. E. Borth, "An experimental slow frequency-hopped personal communication system for the proposed U.S. 1850-1990 MHz band," *Proceedings of the Second International Conference on Universal Personal Communications,* pp. 931-935, Ottawa, Canada, October 12-15, 1993.

[14] W. C. Jakes (ed.), *Microwave Mobile Communications*, New York: John Wiley & Sons, 1974.

CODE SYNCHRONIZATION: A REVIEW OF PRINCIPLES AND TECHNIQUES

Andreas Polydoros and Savo Glisic

Abstract: We review the main concepts underlying the derivation of appropriate receivers for code synchronization in spread spectrum systems, emphasizing a likelihood theory viewpoint. Practical structures for the separate tasks of code acquisition and tracking are discussed and categorized. A brief discussion of analysis tools and performance measures is presented, along with design considerations for threshold settings, and impact of fading. An extensive bibliography is also included.

1. Fundamentals of the Code Synchronization

In its most general form, the code synchronization problem can be posed as follows: Let us first assume a biphase modulated Direct-Sequence (DS) spread-spectrum receiver, whose input waveform r(t)is given by

$$r(t) = \sqrt{2S}d(t)c(t + \zeta T_c)\cos(\omega_c t + \omega_D t + \theta_c) + n(t) \tag{1}$$

In eq. (1), the parameters S, ω_c and θ_c stand for the waveform's average power, carrier radian frequency and phase, respectively, ω_D is a (possibly present, in mobile environments) radian frequency shift (Doppler offset) between transmitter and receiver, c(t) is a ± 1-valued spreading code with chip rate $R_c = T_c^{-1}$ chips/sec, delayed by ζT_c sec with respect to an arbitrary time reference, d(t) is a binary data sequence (which might or might not be present during the acquisition mode), and n(t) is additive noise or interference or both. The synchronizer's task consists of providing the receiver with reliable estimates $\hat{\zeta}$, $\hat{\omega}_D$ and $\hat{\theta}_c$ of the corresponding unknown quantities so that despreading and demodulation can follow. For the code synchronization part, in particular, the equivalent problem is to align the unknown phase ζT_c of the incoming code as closely as possible with the known (and controllable) phase τT_c of a locally generated (at the receiver) identical despreading code $c(t + \tau T_c)$; this is the topic of this article.

As mentioned, eq.(1) refers to direct-sequence spreading. The corresponding received waveform in the $i^{\underline{th}}$ hopping interval $((i-1)T_h + \zeta T_h, iT_h + \zeta T_h)$ for a typical Frequency Hopping (FH) system with, say, MFSK modulation would be similar to eq. (1):

S.G. Glisic and P.A. Leppänen (eds.), Code Division Multiple Access Communications, 225-266.
© 1995 *Kluwer Academic Publishers. Printed in the Netherlands.*

226

$$r(t) = \sqrt{2S}\cos(\omega_{c;i}t + \omega_{d;i}t + \omega_D' t + \theta_{c;i}) + n(t) \tag{2}$$

where $\omega_{c;i}$ is the hopping radian frequency generated by a PN code sequence during the $i\underline{^{th}}$ hop, $\omega_{d;i}$ is the data symbol radian frequency (we assume for simplicity an one-hop-per-symbol system), ω_D is the Doppler offset, assumed independent of the hop index, and ζT_h is the unknown timing offset of the received waveform with respect to the receiver time reference. Again, the job of the synchronizer would be to align the incoming and local hopping patterns (sequences) and provide an estimate $\hat{\zeta}$ for the timing offset, along with the other parameters as above. Thus, from a conceptual standpoint, the FH synchronization problem is identical to the previously stated DS one. Furthermore, it is typically true that FH code acquisition is faster and easier to perform than DS in benign noise environments due to higher SNR and smaller code uncertainty regions encountered (this might not be true in the presence of jamming or other strong interference).

The uncertainty about the values of the synchronizable parameters and the resulting need for estimating them can be attributed to a multitude of factors, whose importance varies with the application. For example, in asynchronous communication, there is no timing information provided initially to the transmitter/receiver pair. Even in the synchronous case, however, i.e., when the users are connected to some sort of universal timing system, uncertainties arise due to the relative range and dynamic movement (velocity, acceleration) of the transmitter with respect to the receiver, parameters which may only be vaguely known. This gives rise to the delay tracking problem, typically handled by one of many types of code tracking loops (see below). Moreover, it may be necessary to account for time differences arising from clock instability and the relative drift. Even for very accurate clocks, relative phase shifts are inevitable in the long run.

The uncertainty in range, denoted by ΔR, implies an unknown propagation delay for the spread signal which, in turn, translates into a temporal uncertainty region. The corresponding number N_{unc} of code chips in that region, for an one-way propagation link, will be given by

$$N_{unc} = \Delta R \frac{R_c}{V_{EM}} \tag{3}$$

where R_c is the code rate and V_{EM} is the velocity of the electromagnetic wave. The value of N_{unc} in eq. (3) could vary from a few chips to a whole code-length or many thousands of chips, depending on the application.

The other important aspect is the relative frequency translation (Doppler) ω_D, whose cause is the relative radial velocity between transmitter and receiver. If ΔR denotes the difference in range-rate of the pair, then the corresponding relative Doppler shift is approximately given by

$$\omega_D = \frac{\left(\Delta \dot{R}\right)\omega_c}{V_{EM}} \tag{4}$$

For predictably moving objects, such as satellites [Spil77], ω_D can be pre-computed and thus compensated for, whereas for fast and unpredictably moving objects (terrestrial high-speed vehicles, fighter aircraft, etc.), Doppler offset might be a significant consideration. In either case, bounds on ΔR are usually specified, from which bounds on ω_D can be derived. From now on, we let Δt and Δf denote the total time and frequency uncertainties, respectively.

Reasons similar to the above explain the ignorance of the receiver about the carrier phase θ_c. Typically, an estimate $\hat{\theta}_c$ is provided to the receiver (if needed for coherent demodulation) via one of the familiar carrier tracking loops. In most applications, code synchronization and despreading are performed prior to carrier tracking due to the fact that the spread signal does not provide enough SNR for the carrier tracking loop to operate successfully [AlHuHoUd78]. Code synchronization in the absence of carrier phase information is designated as *noncoherent* code synchronization. In a few circumstances, enough SNR may exist before despreading for the carrier loop to track; in that case, code synchronization could be performed with θ_c (and ω_D) assumed known (*coherent* code synchronization), resulting in a slightly different detector structure.

To motivate the detector structure, let us first assume for simplicity a coherently received waveform with no data modulation (i.e., let d(t) = 1 and assume ω_D, θ_c known in eq. (1)). The question then is: how should the receiver go about estimating ζ? Given a T-sec observation $r(t); 0 \leq t \leq T$, with T assumed fixed, then a desirable answer might be the conditional mean $\hat{\zeta} = E\{\zeta | r(t); 0 \leq t \leq T\}$ which is the well-known solution to the *minimum mean squared error* criterion and, under mild symmetry conditions, to the *maximum a-posteriori probability* (MAP) criterion, as well. Both of these criteria, however, require knowledge of the prior Probability Density Function (pdf) of ζ, which is often unavailable; furthermore, they tend to be computationally intensive. An attractive and popular alternative is the *maximum-likelihood* (ML) criterion, which is known to possess a number of attractive properties and is also equivalent to the MAP one for a uniform prior [VanT68]. The corresponding estimate, $\hat{\zeta}_{ML}$, is the argument that maximizes the *Likelihood Functional* $\Lambda\{r(t)|c(t;\zeta); 0 \leq t \leq T\}$, where the notation $c(t;\zeta)$ is used to denote

228

$c(t + \zeta T_c)$. For the case where the additive waveform n(t) in eq. (1) represents only white Gaussian noise with one-sided Power Spectral Density (PSD) N_0 W/Hz, and assuming that the energy term $\int_0^T c^2(t; \zeta) dt$ conveys no information on ζ (as is typically the case), then the coherent LF is proportional to $exp\left\{ \dfrac{2\sqrt{S}}{N_0} Y(\zeta) \right\}$ where S is the signal power and

$$Y(\zeta) = \int_0^T r(t)c(t; \zeta) dt$$

$$(5a)$$

is the time-domain (coherent) correlation operation between the observed waveform and the local code, positioned at the candidate offset ζT_c. This correlation result is thus a sufficient statistic for estimating ζ. If the environment is the noncoherent one mentioned previously, then a similar development shows that the corresponding sufficient statistic will be the familiar envelope

$$Y(\zeta) = \sqrt{Y_c^2(\zeta) + Y_s^2(\zeta)};$$

$$\left\{ \begin{array}{l} Y_c(\zeta) = \sqrt{2} \int_o^T r(t)c(t; \zeta) cos(\omega_c t + \omega_D t) dt \\[3mm] Y_s(\zeta) = \sqrt{2} \int_o^T r(t)c(t; \zeta) sin(\omega_c t + \omega_D t) dt \end{array} \right\}$$

$$(5b)$$

We note in eq. (5b) that the Doppler offset is assumed known, since it is used in the carrier-mixing operations in the in-phase and quadrature correlations. If it is not, then it becomes one additional parameter to be estimated (or averaged out), meaning that the LF is now a function of two parameters: ζ and ω_D.

The above formulation regarded DS systems, whose modeling tends to be a bit easier. For FH systems, the designer of the code synchronization circuits is faced with a multitude of modeling details regarding the presence of Fast FH (FFH) or Slow (FFH), a distinction that depends on the number of data symbols per hop, the exact type of data modulation in the SFH case, the presence or not of a known data preamble[1], etc. As is well known, in a FH system with a total of G carrier hopping frequencies, the hopping pattern is dictated by a PN-code derived, G-valued indicator function $I_h(t)$ whose purpose is to determine the carrier hopping frequency at each hop; thus, a timing uncertainty manifests itself in the shifted argument $I_h(t + \zeta T_c) \equiv I_h(t; \zeta)$. As an example of the likelihood formulation in the

SFH case, assume that a total of N_h hops is observed, each hop carrying $N_{s/h}$ symbols, where each symbol is drawn randomly from an MFSK alphabet and is unknown to the synchronizing unit (no preamble). Furthermore, the carrier phase is assumed to shift randomly from symbol to symbol through the channel (taking into account any deliberate phase continuity is possible but would just escalate the complexity of the formulation). Then, the LF conditioned on an assumed ζ (with Doppler ambiguity neglected) would average over all possible data-filter envelopes in each symbol, and do that for all symbols in a hop as well as for all hops:

$$\Lambda\left\{r(t)|c_h(t;\zeta); 0 \leq t \leq T = N_h T_h = N_h N_{s/h} T_s\right\} =$$

$$= \prod_{i=1}^{N_h} \prod_{l=1}^{N_{s/h}} \left\{ \frac{1}{M} \sum_{m=1}^{M} I_0\left(\frac{2\sqrt{S}}{N_0} Y^{(i,l,m)}(\zeta) \right) \right\} \tag{6a}$$

where $c_h(t;\zeta)$ is the PN-derived hopping code dictating the values of the indicator function, $I_0(.)$ is the familiar zeroth order modified Bessel function, and the envelope $Y^{(i,l,m)}(\zeta)$ is defined via the following two quadrature components, similar to (5b):

$$\left\{ \begin{array}{l} Y_c^{(i,l,m)}(\zeta) = \sqrt{2} \int\limits_{(i-1)T_h+(l-1)T_s}^{(i-1)T_h+lT_s} r(t) \cos(\omega_{I_h(t;\zeta)}t + \omega_m t)dt \\[2em] Y_s^{(i,l,m)}(\zeta) = \sqrt{2} \int\limits_{(i-1)T_h+(l-1)T_s}^{(i-1)T_h+lT_s} r(t) \sin(\omega_{I_h(t;\zeta)}t + \omega_m t)dt \end{array} \right\} \tag{6b}$$

Since eq.(6) appears complicated, simpler implementations would just energy-detect over the total M/T_s-Hz data band around each candidate carrier hopping frequency (assuming an orthogonal T_s^{-1} symbol-frequency spacing), and then create quasi-likelihood variants such as, for instance, choosing the maximum over each hop as the ML estimate of the hopping frequency and then accumulate such maxima over the observed hops. Still, this should be done, in principle, for all candidate offsets ζ.

The ML synchronizing receiver implied by eqs. (5) and (6) should, in principle, create all possible time-offset versions of the known code waveform, correlate all of them with the received chunk of data and choose the ζ corresponding to the largest correlation as its estimate, $\hat{\zeta}_{ML}$. In view of the continuous range of values of ζ (and ω_D, if also required), this is an impossibility and some type of range quantization is necessary, which is indeed what is done in practice. The resulting candidate values are called "bins" or "cells", and the initial parameter estimation problem is translated into a multiple-hypothesis problem: locate the cell most likely to contain the unknown offset, given this piece of data. This is exactly the _coarse-code synchronization_ or _code acquisition_ problem, the result of which is to resolve the code phase (or "epoch") ambiguity within the size of the cell. Since this remaining error is typically larger than desired, further operations are required in order to reduce it to acceptable levels. This remaining part of the synchronization task, namely, that of _fine synchronization_ or _code tracking_, is performed by one of the available code-tracking loops, which we discuss later.

For the two-dimensional problem mentioned previously (time plus frequency uncertainty), the corresponding cells will also be two-dimensional. Their total number will equal $q = (\Delta t / \delta t)(\Delta f / \delta f)$, where δt and δf stand for the cell dimensions in the respective directions. For accurate estimation, their sizes should be as small as possible; however, they cannot be too small, as that would either overburden the receiver complexity or extend the search time, since the total search space is obviously proportional to q. Typical choices for δt are of the order of the code chip time T_c or a large fraction thereof (such as 1/2 or 1/4), and of the order of the data bandwidth for δf. Similarly for FH, δt is chosen as a fraction of T_h. This ensures that at least one cell is close to the peak of the (noiseless) correlation output of eqs. (5) or (6).

Once the nature and size of these cells have been determined, the next question is how to go about performing the search most successfully. Clearly, the strategy will depend on a variety of factors such as criteria of performance, degree of complexity and computational power available (directly related to cost), prior available information about the location of the correct cell, etc. A brute-force approach would try to create a bank of parallel correlation branches, each matched to a possible quantized value of the timing offset; it would then process the received waveform through all of them simultaneously, pick the largest and declare a candidate solution. Unless the uncertainty region (number of cells) is small, corresponding to either a small code period or a small initial uncertainty, such a solution (which we may call the "totally parallel solution") becomes obviously unwieldy in complexity very quickly. We note, however, that small uncertainty regions may be encountered in a nested design, whereby a multitude of different-period codes are combined for precisely the purpose of aiding acquisition. Furthermore, neural network structures are currently being explored for this purpose, where the neural network is trained for all possible such values. Such a scheme would emulate the spirit (if not the exact statistical processing) of the above solutions.

In current practice, total parallelism is out of the question when the number of cells is very large (although it appears doable for smaller uncertainty regions [MiGeDa85], [DaFl88], [SoGu90]) and simpler solutions are necessary. One of the most familiar such approaches is the simple technique of serial search, namely, one where the search starts from a specific cell and serially examines the remaining cells in some direction and in a pre specified order until the correct cell is found [Sage64], [PoWe84a]. Hence, serial search techniques do not account for any additional information gathered during the past search time, which could conceivably be used to alter the direction of search towards cells that show increased posterior likelihood of being the correct ones [HoWo78] (see also [Guma63], [Posn63] for related concepts in the radar target-search problem). A serial search starts from a cell which could be chosen totally arbitrarily (no prior information), or by some prior knowledge about a likely cell, and proceeds in a simple and easily implementable pre-directed manner. When the uncertainty space is two-dimensional as discussed above and searching all possible cells serially appears very time-consuming, then a speed-up may be achieved by employing a bank of filters, each matched to a possible Doppler offset [ChHuSt90]. The same idea can be applied to the one-dimensional case (no frequency uncertainty), where now a bank of correlators may be employed, each starting from a different point of the uncertainty region, which effectively amounts to dividing the search in many parallel sub-searches and therefore reducing the total search time by a proportional amount.

A note is due here: although it holds true that only one cell contains the exact delay and Doppler offsets of the incoming code, the set of desirable cells acceptable to the receiver includes a number of cells adjacent to the exact one. Indeed, the receiver will terminate acquisition and initiate tracking the first time a cell is reached (and correctly identified) which is close enough to the true synchronization so that the tracking loop can pull in and perform the remaining synchronization operation successfully. All these desirable cells are collectively called *hypothesis H_1*, and the remaining non-desirable "out-of-sync" cells comprise *hypothesis H_0* As an example, consider the case where the receiver examines the code delay uncertainty in steps of half a chip time ($\delta t = T_c / 2$) and there is no frequency uncertainty. Then all four cells located in the interval $(-T_c, T_c)$ around the true delay of the incoming code are included in hypothesis H_1,, since some amount of code correlation exists for each one of these cells, an amount that can initiate the code tracking loop.

The above definition of cells and hypotheses implies that each test does not pertain to a single value of the unknown parameter ζ, but rather to a range of values. It is straightforward to show that, under mild conditions and approximations pertaining to the pseudo-random nature of the code, this reformulated hypothesis testing results in a statistic (correlation) and threshold setting that do not depend on the given (tested) value of the unknown parameter (a Uniformly Most Powerful Test). This is because the threshold value is set by the desirable probability of false alarm per cell (see below) which is independent of ζ under H_0.

The generic code acquisition model of Fig. 1 below recapitulates the basic ideas discussed so far. The two-dimensional time/frequency code offset

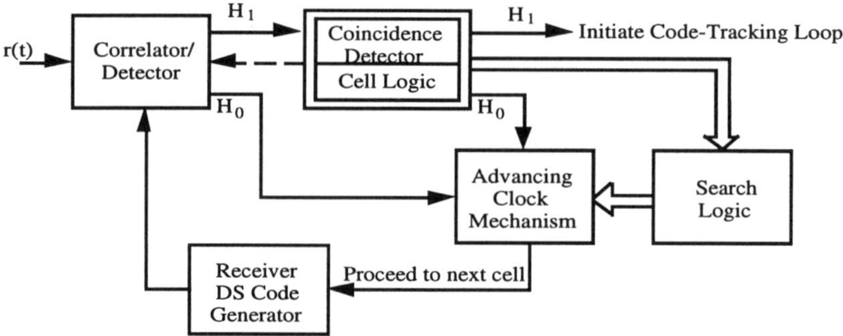

Fig.1 - Generic Model for the Serial-Search Likelihood-Ratio Code Acquisition Receiver

uncertainty within the noisy received waveform is quantized into a number of cells, which are typically searched in a serial fashion by a correlation receiver, although parallel multiple branches are also possible. Motivated by an ML argument, the receiver creates a cross-correlation between the incoming waveform and the local code at a specific offset, whose output is used to decide whether the currently examined cell is a desirable $(H_1,)$ one. The process continues until one such cell is correctly identified. At that point, acquisition is terminated and tracking is initiated.

We may use similar, likelihood-based arguments to introduce the tracking loop structures currently in use [Stif71], [Holm82], [ZiPe85]. Let us assume that the spreading code $c(t)$ is a string of pseudo-random symbols modulated by a *differentiable* chip pulse, i.e., $c(t) = \sum c_n p(t - nT_c)$ where $p(t)$ is confined in $(0, T_c)$ and is differentiable with respect to the time argument t. In commercial systems, the total spread bandwidth is a function of both the code rate <u>and</u> the pulse shape, so the latter is typically chosen with bandwidth efficiency in mind; in that case, pulse smoothness and differentiability is assured. In certain other, military and space applications where bandwidth restrictions are not severe, square pulses may be employed (which are non-differentiable), in which case the arguments below will only serve purposes of motivation, not implementation.

After successful code acquisition the receiver is interested in reducing the residual uncertainty about ζ down to zero. The receiver may view the envelope $Y(\zeta)$ of, say, eq. (5a) as a continuous function of the parameter ζ and wishes to maximize it over the choice of that parameter value. If the point of maximum,is interior to the uncertainty region associated with ζ, then $\hat{\zeta}_{ML}$ is obtained by setting the derivative of the above correlation with respect to ζ equal to zero:

$$\frac{d}{d\zeta} \int r(t)c(t+\zeta T_c)dt = \int r(t)c'(t+\zeta T_c)dt = 0 \tag{7}$$

where $c'(t) = dc(t)/dt$. In other words, the ML epoch estimate for the original code is the amount by which the <u>derivative</u> of the local code waveform must be delayed in order to produce zero correlation with the incoming waveform. This procedure can be viewed as an alternative to the direct method (or quantized versions thereof) of correlating with all possible offsets and selecting the maximum.

This argument brings out an important aspect of the estimation procedure, namely that the derivative of the local code waveform plays an important part in determining $\hat{\zeta}_{ML}$, at least in a localized sense. In practice, the extra complexity associated with differentiating this signal (or, in the case of square pulses, the difficulty with delta functions) is bypassed through a finite-difference approximation to the derivative:

$$\frac{d}{dt}c(t+\zeta T_c) \approx \frac{c(t+\zeta T_c - dT_c) - c(t+\zeta T_c + dT_c)}{2dT_c} \tag{8}$$

where dT_c is a small increment. Thus, the approximation (8) leads to utilizing an "early" and a "late" correlation between the received $r(t)$ with $c(t+dT_c)$ and $c(t-dT_c)$, respectively. This "early-late" concept is, in fact, the main building block embedded in a variety of signal tracking applications, despite the structural variations due to phase incoherence, presence of data, etc. It can be found in the theory and design of symbol synchronizers in data communications [Holm82], in early-late-gate range-tracking of radar systems [Skol80], and so on.

Our discussion so far has been facilitated by the assumption of a <u>fixed</u> (but unknown) delay parameter ζT_c over the whole observation interval, which is rarely true in practice; rather, it can typically be modeled as a time-varying function $\zeta(t)$ arising from transceiver dynamics, clock instabilities, etc. One possible approach for dealing with this reality would be to divide time into intervals over which $\zeta(t)$ is approximately constant (say, over the inverse bandwidth of this delay process), estimate the delay in each interval ζ with one of the above mentioned open-loop procedures and subsequently update a local reference. This, however, would amount to throwing away the useful correlation which exists between the delay values in successive intervals. A convenient (although not unique) solution which exploits such memory is to close the loop in a feedback fashion: a function of the estimated delay-error between the incoming waveform and the local signal, created through appropriate comparison in a *delay-lock discriminator* (i.e., delay detector), is fed back to the local code via a delay-controlling device in order to "pull" the code timing towards the direction which minimizes that error.

To be specific, let $\varepsilon(t) = (\tau(t) - \hat{\tau}(t))/T_c$ denote the normalized timing error between the incoming code $c(t - \tau(t))$ and the local version $c(t - \hat{\tau}(t))$, respectively[2]. The goal of the tracking loop is to force $\varepsilon(t)$ towards zero at all times by utilizing the following feedback mechanism: The incoming waveform, consisting of the code-modulated useful signal plus noise, is properly mixed (correlated) with the locally generated code in the delay-lock discriminator in order to produce a new waveform $e(t)$, called the *error voltage*, which is functionally related to the delay error $\varepsilon(t)$. This functional memoryless dependence between $\varepsilon(t)$ and $e(t)$ is called *the loop's S-curve*. The error voltage is then used to control the timing (epoch, phase) of the Voltage-Controlled Clock (VCC), which generates the local code, in the direction of reducing the phase error $\varepsilon(t)$ existing at a particular point in time. Such a generic closed-loop (feedback) system is shown in Fig. 2. Note

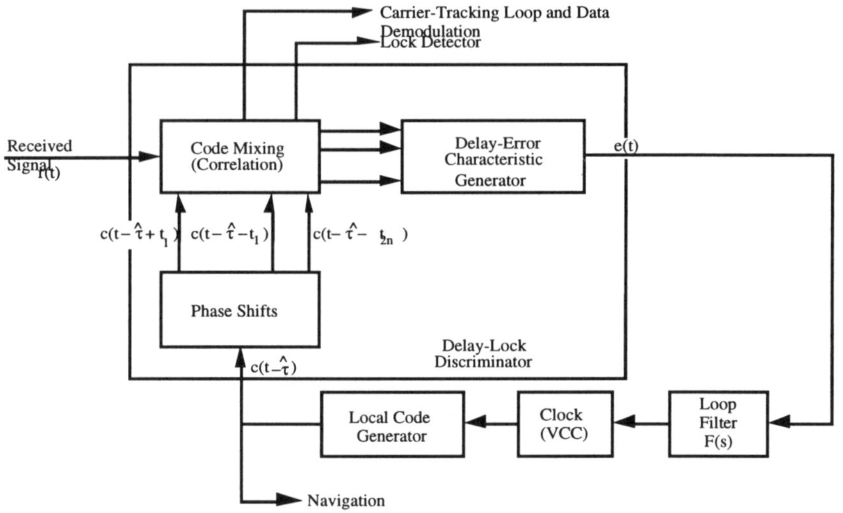

Fig.2 - Generic Model for a Code-Tracking Loop

that more than one phase shifts of the local code will always be required in the signal-mixing/phase-error generation process in order to produce $e(t)$. Also, a linear, time-invariant filter is sometimes inserted between $e(t)$ and the VCC in order to smooth out the statistical randomness of $e(t)$ through time averaging. The loop's final delay-estimate is then the output epoch of the local code generator.

We should note that eq. (8), upon which this discussion is based, is only an approximation, particularly for square pulses. Furthermore, once a feedback structure as in Fig. 2 is adopted *a priori*, the question arises as to the optimal type of the delay-lock discriminator (namely, the "right" function to mix the incoming signal with) which would minimize some function of the error (such as its variance). This issue has been discussed in [SpMa61], [Layl69], [Stif71], [Kosb88], etc., and the partial answer, when available, is signal-shape, criterion and Signal-to-Noise-Ratio (SNR) dependent. It resembles the Hilbert Transform of the incoming signal in certain cases (which is indeed the exact phase-locked-loop solution for tracking a pure sinusoid), whereas it may note even be a linear transformation of $c(t)$ in other cases [Kosb88]; the general answer is yet unavailable.

We have, thus far, introduced the basic concepts of LF-motivated code synchronization and shown the resulting general structures for the two tasks. The remaining parts of the paper deal with receiver structure categorization and performance measures in Sections 2.A and 2.B for acquisition and tracking, respectively, some brief exposure of performance analysis tools for the two cases in Sections 3.A and 3.B, respectively and, some design considerations in Section 4. Some specifics of code synchronization in fading channels are additionally discussed in Section 5 An extensive bibliography is included at the end.

2. Receiver Categorization and Performance Measures

We now present briefly the various structures employed in practice for the code acquisition and tracking tasks described above and provide a broad categorization of such structures according to some prevalent features or characteristics, which we also describe. Furthermore, we discuss the commonly accepted performance measures for these structures, according to which they are evaluated, ranked and optimized. It should be understood, however, that Spread-Spectrum System (S^3) receiver design (including the synchronizer part) is a rapidly evolving art, science and business due to the emergence of the S^3 technique as a competitive commercial tool in wireless communications; thus, the state of this art is shifting rapidly towards all-digital, massively parallel implementations that affect directly the menu of available receiver design options and their characteristics. For example, we may probably witness a shift away from sophisticated cell detector structures (such as elaborate multiple-dwell schemes---see below) towards multiple parallel search branches (recall the comments on parallelism above), for the same total computational burden. Similarly, for code tracking, multiple-offset extended-range loops [Bowl79] with appropriate delay-lock discriminators might be feasible and preferable for handling multiple rays (resolvable multipath fading) These problems are discussed in Section 5.Nonetheless, most of the concepts in this section should transcend the specificity of the design and should apply equally well to the new ones. We discuss code acquisition first (section 2.A) and code tracking next (section 2.B).

2.A. Code Acquisition: Categorization and Performance Measures

A number of sources contribute to the randomness of the acquisition process, including (a) initial uncertainty about the code-phase offset, (b) channel distortion (e.g., fading channels), (c) noise plus interference (narrowband, multi-user, etc.), (d) random data, (e) unknown carrier phase (noncoherent receivers) and Doppler offset, (f) partial code correlation, etc. The receiver's set of adjustables includes threshold settings, correlation times, number of tests per code chip, search strategy, cell logic (i.e., verification strategy), etc., as we describe below. For a well optimized design, implicit is the knowledge of important parameters such as the SNR, code rate, code length, code-uncertainty region, reset penalty-time and others.

Regarding performance measures for code acquisition, the dominant parameter of interest is the random time which elapses prior to acquisition, T_{acq}, and its statistics. We may distinguish two basic scenarios: (I) the case where, although the fastest possible acquisition is desired, no absolute time limit (stop time, T_{stop}) exists, a situation which arises when the transmitted waveform is always present as, for instance, in satellite-based position location systems; (II) in "push-to-talk" and other packetized or frame-based systems, data transmission starts after a certain time interval T_{stop}, which signifies the end of a preamble. Then, it is imperative that acquisition be performed in that specific interval with very high probability, for otherwise communication is impossible. To characterize both scenarios completely, knowledge of the Cumulative Probability Distribution Function (cpdf) $F_{T_{acq}}(t)$ of T_{acq} would be needed, which might be more than required in order to identify a good design (not to mention that $F_{T_{acq}}(t)$ is difficult to obtain except for simple cases). Hence, one typically settles for either the mean acquisition time $E\{T_{acq}\}$ and, possibly, the variance $var\{T_{acq}\}$ (case I). For case (II), however, the corresponding performance measure is the probability of prompt acquisition $Pr\{T_{acq} \leq T_{stop}\} = F_{T_{acq}}(T_{stop})$.

Under what circumstances will the code not be acquired? One possibility is the existence of the stop time as described above, and the inability of the search process to locate the correct cell in that interval of time due to noise, interference, fading, etc. Another is the case of a *catastrophic* False Alarm (FA), i.e., when the acquisition mechanism erroneously decides that code synchronization has occurred and a very prolonged tracking effort via a code-tracking loop is initiated. The *overall probability of missing the code*, P_M^{ov}, can be computed from total probability as

$$P_M^{ov} = Pr\{ FA \ before \ T_{stop} \}$$

$$+ Pr\{neither \ FA \ nor \ ACQ \ before \ T_{stop}\}$$

(9)

where ACQ denotes the correct acquisition event. The complementary *overall probability of detection*, P_D^{ov}, is many times cited as the key performance measure in such circumstances.

It is not necessarily true that all false alarms result in catastrophic delays. In fact, certain correlation-based logic can be employed (sometimes known as "lock detector logic") which detects the out-of-sync situation quickly and proceeds to discard the examined cell. In reality, the time required for indicating false lock is a random variable. In this case, we refer to the false-alarm state as a *returning state* associated with a random penalty time T_p, as opposed to the previous catastrophic case with an *absorbing* FA state [PoWe84a]. For computational ease, T_p can also be modeled as a fixed time [HoCh77], [DiCa79], [DiWe80].

The type of performance criterion comprises a first partition of acquisition receivers. Further distinctions concern the basic unit of any design, which is the per-cell decision-making device (*detector*). This can be classified as either *coherent* or *noncoherent* (see the Introduction above), and it can also be categorized according to the specific statistical testing philosophy employed for each cell test, i.e., *Bayes, Neyman-Pearson* or *other*. In a Neyman-Pearson type of test, the detector thresholds are set in a fixed position for all cells which, for stationary interference, result in a fixed *false alarm probability per cell*, $P_{FA,c}$. That is essentially equivalent to disregarding any available prior information about the correct cell position and treating all cells as equally likely candidates. This is simpler to implement than a receiver which attempts to make good use of any cell-position prior information; in the latter case, a Bayes test is more appropriate.

An acquisition detector will always consist of an initial testing level (called the *first dwell*), which may or may not be followed by a *verification mode* (logic), whose purpose is to secure the initial positive (sync) indication and prevent eventual false alarm. The verification mode is typically called a "multiple-dwell detector" [DiWe83], and consists of a series of successive tests of the same cell. Since a cell can potentially be rejected at any such stage, and since there is always the possibility (however remote) of a false alarm, it follows that the time spent by the detector in each cell is a random variable. This is true regardless of the nature of the individual dwell times, which are the times spent in each dwell described above. Those can be either *fixed*, in which case decisions are made at pre-arranged time intervals, or *random*, when a variable-time test is employed (the so-called Wald sequential test [Wald47]), arising from the random crossing time of appropriate thresholds [BuMi55], [Comp87], [SuWe90]. We note that the advent of digitized receivers and clocks tends to increasingly favor fixed-dwell tests, which can be designed to be very competitive performance-wise with sequential tests. Detectors can employ either *full-period* or *partial-period* correlation, the latter being favored at the early stages of multiple-dwell tests. Verification modes can employ *immediate-rejection logic* or *non immediate-rejection logic,* the terminology pertaining to the action taken by the supervising logic once a cell fails to pass an intermediate-dwell test [Hopk77], [Holm82]. Another esoteric distinction comes about because successive multiple-dwell decisions can be made either statistically *independent* (integrate-and-dump kind of operation) or *dependent* (continuous type of integration, with sampling but no dumping).

Finally, detectors are classified according to whether they employ *passive* or *active* correlation [PoWe84a], [PoWe84b]. The former term refers to a passive filter matched to a segment of the known spreading code. Ideally, this creates no correlation with the received waveform until the identical segment in the incoming code matches perfectly in time with the receiver filter impulse-response, at which moment maximum correlation is created. The alternative of active correlation is when the two codes (local and incoming) stay at a fixed phase offset, corresponding to the cell under examination, for the whole cell correlation period. This can be done, for example, by running the receiver code at the same rate with the incoming code while correlating. If the decision at the end of that time interval is negative, a clock pulse advances the local code forward to the next cell position and the process is repeated. We note that passive correlation inherently enjoys a *fast decision rate* (equal to or a multiple of the code rate), while active correlation is confined to slow decision rates. *Hybrids* of the above are also possible, whereby the first-stage decision is based on passive and the verification on active correlations.

The remaining important aspect of the acquisition process is the search strategy, i.e., the procedure adopted by the receiver in its search through the uncertainty region. A host of *serial-search* strategies was proposed and analyzed in [Poly82], [PoWe84a,b], [PoSi84]; see also [SiOmScLe85]. The main serial search types are the *straight search*, the *Z-search* and the *expanding-window search.* For further description and performance analysis, the reader may consult the above references, as well as [Ale77], [Hopk77], [HoCh77], [DiWe80], [ElTaGu80], [RaSc80], [Pand81], [PuRaSc81], [Brau82], [MePo83], [Wein83], [DaMiSc84], [MaBl84], [ZiPe85], [Chen88], [Jova88], [Su88], [WiRaVa88], [ChHuSt90], [PaDoKu90], [SoGu90], [SuWe90], [SoGu92], [Jova92], [LeTa92], etc.; the problem of analyzing performance under non-white noise (narrowband interference and jamming, mostly) has been addressed in [Kreb80], [PaPo85], [SiWe86], [MiLeFrTo92]; the impact of code acquisition on packet radio networks has been addressed in [MaPu93]; adaptive-array techniques have been studied in [DlSc89]. A fundamentally different class of search is *sequential state estimation* which refers to those techniques that try to estimate directly the state of the pseudo-noise code dictating the spreading code, and load that estimate onto the receiver shift register [Ward65], [Kilg73], [WaYi77]. Since these techniques are based on hard-decision chip estimates made in a spread-SNR environment (i.e., estimating the ± 1 value of each chip separately), and since a single chip error makes the whole state estimate useless, their utility is constrained to relatively benign, high chip-SNR environments where they can be very fast (the time it takes to estimate n chips for an n-stage shift register).

2.B. Code Tracking: Categorization and Performance Measures

What distinguishes code tracking loops into different classes is the way they create the error voltage *e(t)* by proper mixing of the input waveform with the various delayed versions of the local code. This is intimately related to the prevailing assumptions about the input parameters and the particular application. Thus, we distinguish between *coherent* (or *baseband*) [Spil63], [Gill66], [Stif71] and *noncoherent* loops [Simo77], [Holm82], [Poly82], [Ward85] depending on whether the carrier phase is available or not. Most of the time the spread-SNR is too low to allow the carrier-tracking loop to operate satisfactorily before the signal is despread; thus, carrier phase is typically unknown and noncoherent structures are more prevalent. Furthermore, in the coherent case, code correlation can be performed either at RF *(carrier-frequency correlation)* or at baseband *(video correlation)*, a distinction which is meaningful from the standpoint of implementation only (both structures are theoretically identical [Gill66]).

Depending on the application, loops can be employed for either *navigation* or *data despreading (demodulation)*. In navigation, measuring and tracking the two-way time delay enables accurate distance measurements. A type of loop well suited for such applications is the *extended-range characteristic loop* [Bowl79], where an increase in the pull-in range of the loop (as to accommodate higher dynamics) is traded against higher equivalent noise; this is accomplished by utilizing more than two time-shifted versions of the code in the phase detector, in contrast to most of the standard designs which employ exactly two (i.e., one early and one late shift).

Even when a specific loop structure makes theoretical sense, modifications are common for the purpose of reducing hardware and cost, to alleviate imbalance problems, component mismatch and other such reasons. Such alterations offer higher robustness to component variations at the expense of a somewhat worse performance in a certain sense (see below). Examples in this category are the *full-time* [Holm82], [PoWe85] versus *time-shared* (or *tau-dither*) loops [Hart74], [Hopk77], [WaPo85] and the *sum/difference* versus *reference/difference* [Lafl79], [YoBo82] loops. Full-time loops employ early and late code correlations simultaneously and on a continual time basis, while time-shared loops use a single correlator which alternates between the early and the late versions. Similarly, a sum/difference loop utilizes the product of the sum times the difference of the early and late correlations, while a reference/difference loop utilizes the product of the difference term times the *on-time* (or *reference*) correlation term. The tradeoff here involves gain and phase imbalances in the arm filters [Comp87]. Note that, historically, the full-time sum/difference loop is the one traditionally referred to as the "Delay-Lock Loop" (DLL) [Holm82], [ZiPe85], [SiOmScLe85], although the term is actually applicable to all such code-tracking loops. Finally, let us note that appropriate tracking loop structures can be configured for arbitrary data and code modulation schemes (such as CPM), for frequency-hopping systems, etc. For slow FH, combination of bit DLL and frame synchronization can be used to track FH patern [GlMi91],[GlMi91a].

In order to understand the appropriate nature of performance measures for code tracking, we must delve a bit further in to the operational details of such a loop. The main issue here is the <u>finite</u> range of values for the delay error within which the tracking loop can maintain "lock" (namely, it can keep the error small). Unlike carrier tracking or symbol tracking loops, whose S-curve is periodic, code tracking loops entail S-curves which are effectively non-zero only within a finite range. For example, the S-curve of a PLL contains an infinite number of stable operating points spaced at 2π rad apart, so that phase tracking continues indefinitely even after the PLL "slips cycles". On the other hand, a DLL will only track when there exists non-zero correlation and subsequent error voltage, meaning that the delay error must be within some limits $\varepsilon_{min} \leq \varepsilon \leq \varepsilon_{max}$, where ε_{min} and are loop-structure dependent. When the error process reaches any of these two boundaries, the loop "loses lock" and code re acquisition becomes necessary. The average time for the occurrence of such an event (called the *mean time to lose lock* or *average tracking time*) plays an important role in the loop operation and constitutes one of the standard design specifications of such loops (see [HoBi78], [PoWe85], [WeBo90]). From a practical standpoint, what is required is that this mean tracking time is much larger than the expected operating time. Note that the loop senses the loss of lock via the *lock detector*, which performs a continuous correlation between the incoming and local codes and decides whether the outcome is adequate. When the decision is "loss of lock" and re acquisition is initiated, the system undergoes a random time delay until the acquisition circuit, following some appropriate procedure as described previously, relocates the timing of the two codes with an error in the permissible range $\left(\varepsilon_{min}, \varepsilon_{max} \right)$. Tracking then takes over, and the combined process repeats at infinitum. Care must also be exercised to make sure that the lock detector does not interrupt the tracking process unnecessarily.

A typical sample trajectory of the combined process just described will consist of individual tracking portions, interspersed by code re acquisition portions in an alternating fashion. The individual tracking intervals τ_i^{tr} are independent and identically distributed r.v.'s, and so are the individual acquisition intervals τ_j^{acq}. The model without τ_j^{acq} (i.e., assuming they are negligible in comparison to τ_i^{tr}) was first introduced by Meyr [Mey75], [Mey76], [LiMe77] as the "renewal tracking process model" and was later generalized in [Poly82], [PoWe85]. All the "tracking" portions of the renewal process, which are the portions of interest, are generated by the same stochastic diffusion equation which governs the particular code tracking under consideration and thus form a Markovian continuous-time process when viewed separately from the "re acquisition" portions. Because of the regenerative nature of such a renewal process, a steady-state pdf for the delay error does not exist, but a stationary solution $p(\varepsilon)$ does and it forms the basis for the performance criteria discussed below.

Of interest in most practical systems is the *average* behavior, such as the overall error variance $\sigma_\varepsilon^2 = \int \varepsilon^2 p(\varepsilon) d\varepsilon$ or, for data demodulation, the unconditional average bit error rate $\sigma_\varepsilon^2 = \int \varepsilon^2 p(\varepsilon) d\varepsilon$ with $P_{error}(\varepsilon)$ the conditional error rate given a value ε of the error (which typically affects the prevailing SNR). The latter is part of the more general concept of digital communication system efficiency that takes into acount all level of synchronization (code, symbol, carrier, frame etc.) as well as the influence of coding and modulation [Glis87].It can be shown (see [PoWe85] for details) that either measure can be expressed in terms of (a) the stationary conditional pdf $p(\varepsilon|S)$, whose interpretation is that the differential quantity $p(\varepsilon|S)d\varepsilon$ represents the infinitesimal probability that the renewal error process, having been initiated an infinite time ago and being looked upon at an arbitrary time instant within one of the "tracking segments," will be in a differential neighborhood of ε, and (b) the mean tracking time $\overline{\tau}^{tr}$ for each such segment. Furthermore, it can be shown that the mean tracking time is directly derivable from $p(\varepsilon|S)$, which means that the latter contains all the useful tracking information. In conclusion, the standard measures of performance such as the tracking *variance* and the *mean time to lose lock* (a synonym for the mean tracking time) are all embedded in the knowledge of $p(\varepsilon|S)$, which thus becomes the right goal of analysis. We should note, however, that a simplified version of the tracking variance, namely that of the *linearized* loop (valid and useful for high SNR), do not require this pdf and can be derived directly from elementary linear-system theory. This is not possible for the mean tracking time.

3. Performance Analysis Tools

A significant amount of literature and effort has been devoted to analyzing the performance of both code acquisition and code tracking systems, the goal of the analysis being the maximization of some performance criteria discussed above (such as minimum acquisition time, maximum tracking time, minimum tracking error variance, etc.) by means of appropriate parameter selection such as, for instance, threshold values in acquisition detectors or gain values in closed-loop code trackers. The analysis is necessarily specific to the structure under consideration, but there exist few key tools and general concepts which recur in all analytical efforts. Our goal in this brief section is to highlight some of these concepts.

3.A. Code Acquisition Analysis

As we identified in Section 2.A, the two building blocks of any acquisition receiver is the detector and the search strategy. Accordingly, the purpose of analysis is to combine these two elements in order to derive the statistics of the acquisition time or the overall probability of detection. In fact, the impact of the specific detector structure on the performance measure can typically be summarized by few important parameters <u>per cell</u>, as we will illustrate with a simple example below. This implies that optimization of the detector structure and the choice of the search strategy can proceed independently.

There exist two general analytical directions for this type of analysis: the time-domain combinatorial technique and the transform-domain (or circular-state diagram) technique. The former proceeds along total-probability arguments and is occasionally faster and more insightful, whereas the latter seems more systematic and able to handle complicated search techniques; it is also more amenable to numerical evaluation via FFT techniques. Since a certain background in Markovian-chain state diagrams and their connection to signal flow-graph reduction techniques is required for understanding the latter, it will not be addressed in the limited space here; the interested reader is referred to [PoWe84a,b], [PoSi84] for a detailed exposure. Instead, we will discuss a simple example approached by the time-domain technique, in order to highlight the key parameters involved.

Let us consider a single-dwell serial search algorithm with non-absorbing false alarm, with a test (dwell) time equal to τ_d sec, a deterministic FA penalty time $T_p = K\tau_d$ sec (K a large integer) and unlimited search time (i.e., $T_{stop} = \infty$); thus, the receiver can revisit the q cells (which comprise the uncertainty region, equal to the code length for this example) as many times as required until acquisition is accomplished. Suppose we are interested in the average acquisition time $\overline{T}_{acq}$. Then, the following time-domain argument can be employed: with $P_{FA,c}$ the false alarm probability per cell, the average time spent per H_0 cell before testing proceeds to the next cell, denoted by $\overline{T}_0$, is given by

$$\overline{T}_0 = \left[\left(1 - P_{FA,c}\right) + (K+1)P_{FA,c}\right]\tau_d \tag{10}$$

Similar expressions can be derived for the average time it takes to correctly identify an H_1 cell, as well as the time to miss it; however, since the H_0 cells are overwhelmingly more than the H_1 cells, these times can be neglected without sacrificing accuracy.

Contrary to the time it takes to identify an H_1 cell (which can be easily neglected), the <u>probability</u> of such an event is of crucial importance, which we denote by $P_{D,c}$. Although it is very desirable to have $P_{D,c}$ arbitrarily large as to ensure acquisition on the first round, that cannot be achieved without also raising $P_{FA,c}$, an undesirable effect; hence, the trade-off in detector design. Note also that a run through the H_1 region will encounter the code more than once, hence the probability of detection *per run* $P_{D,r}$ is what counts, which can easily be derived in terms of $P_{D,c}$.

We can now return to the time-domain argument at hand: If the H_1 cell is correctly identified in the very first run through the uncertainty region and if the initial starting position is totally random with respect to the location of the H_1 cell, then the average number of H_0 cells visited will be $q/2$; furthermore, every time the H_1 cell is missed (with probability $1 - P_{D,r}$), another q tests will necessarily be added to the time cost. We can, therefore, write

$$\bar{T}_{acq} = \sum_{k=0}^{\infty} \left(\frac{q}{2} + kq \right) \bar{T}_0 \left(1 - P_{D,r} \right)^k P_{D,r} =$$

$$q \left(\frac{2 - P_{D,r}}{2 P_{D,r}} \right) \bar{T}_0 \tag{11}$$

Eq. (11) is a standard result in the acquisition literature [Hopk77], [ZiPe85] and shows that the mean acquisition time depends on the pair $\left(P_{D,c}, P_{FA,c} \right)$, the average time per cell and the size of the uncertainty region. If, one the other hand, the full statistical description of the acquisition process is desired (i.e., $F_{T_{acq}}(t)$), then we would need the conditional distributions of the random times of cell false-alarm, $F(t|FA)$, its complement $F(t|FA^c)$, cell-detection $F(t|D)$, its complement $F(t|M)$, where M denotes "miss", and their associated probabilities. These *transition statistics* can be obtained for any particular detector structure, although not necessarily in an easy way or in closed form (especially for dependent decisions, sequential tests, noncoherent environments, etc.). If, however, a simple radiometric or envelope detector is employed with fixed dwells and independent decisions, then these conditional distributions are simple step-like functions and the corresponding probabilities can be derived from Marcum's-Q type computations.

The above simple combinatorial argument has been extended to various other search techniques, but the results become cumbersome rather quickly. The transform-domain techniques mentioned previously help in this regard, although they themselves involve a good amount of computation and diagram reduction. It should be pointed out that all these techniques rely on the assumption of statistically identical and independent cells, at least as far as the computation of $F_{T_{acq}}(t)$ goes (that is not necessary for $\overline{T}_{acq}$). Since fading and other environments are expected to induce correlation between successive tests, some new theory will be required in this field.

3.B   Code Tracking Analysis

The target of analysis for each of the loops considered in the literature is to derive the Stochastic Integro-Differential Equation (SIDE) which governs the evolution of the delay-error process, and the procedure for doing so is standard: after a description of the loop structure under consideration the error signal $e(t)$ is developed analytically, typically under certain simplifications. Then, the SIDE (in operator-form) identifies the two fundamental quantities of interest: the loop's S-curve and the equivalent noise at the input of the loop filter. As mentioned, the S-curve is the nonlinear dependence between the normalized delay error $\varepsilon(t)$ and the resulting feedback voltage $e(t)$ which, in the absence of noise, would eventually drive the error to zero. The so-called linear theory consists of linearizing this memoryless nonlinearity around the origin $\varepsilon(t) = 0$. The equivalent noise represents the compound effect of the thermal noise, the signal-times-noise and the signal-times-signal (noisy part, also called "self-noise") terms on the loop operation; its computed level (i.e., its power spectral density) determines whether the linear theory is adequate at the considered operating point or the nonlinear renewal (threshold) study is required.

All the above steps are common to all loops and result in the same form of diffusion equation, since the basic block diagram is common to all (see Fig. 2); what differs from loop to loop is the exact mathematical form of the S-curve and the equivalent noise statistics. The standard form of the SIDE is (with p the Heaviside operator $d(.)dt$) [PoWe85]

$$p\varepsilon(t) = p\frac{\tau(t)}{T_c} - S'(0,\delta)K_L D_2 P_s F(p) \times$$

$$\left\{ S(\varepsilon,\delta) + \frac{n_{total}(t,\varepsilon)}{S'(0,\delta)D_2 P} \right\}$$

$$(12)$$

with $\tau(t)$ the input signal timing dynamics, $S(\varepsilon,\delta)$ the loop's normalized (by its derivative at the origin) S-curve as a function of the error ε and the chip-normalized offset δ of the early and late gates with respect to the on-time channel, $S'(0,\delta)$ its derivative with respect to ε evaluated at the origin, K_L the loop's total gain, P the signal power, D_2 the signal power loss due to arm filtering (present for noncoherent loops), F(p) the loop filter's Laplace Transform, and $n_{total}(t,\varepsilon)$ the total noise consisting of the aforementioned components.

Eq. (12) is the starting point for evaluating the performance measures mentioned in Section 2.B plus others. For example, by employing the linear approximation for the S-curve, a closed-loop transfer function for the tracking loop can be defined and the linearized variance derived (typically in the absence of dynamics) as a function of the level of incoming noise (noise-induced jitter). If the noise is dropped as insignificant (high-SNR approximation), eq. (12) can be used along with the exact S-curve to study (on an appropriate phase plane) the dynamic response of the loop due to various input timing dynamics, leading to such performance measures such as pull-in time and pull-in range (dynamics-induced jitter) [Spil63], [Niel75], [Holm82]. If the noise is significant, meaning that the loop operates near threshold and the nonlinear theory is required, eq. (12) can be employed within the renewal framework to derive the aforementioned stationary conditional pdf $p(\varepsilon|S)$ with its associated variance and mean time to lose lock $\overline{\tau}^{tr}$ (this is doable for a first order loop with $F(p)=1$; see [Meyr75], [PoWe85]). It is interesting to note that $\overline{\tau}^{tr}$ results as a by-product of solving for $p(\varepsilon|S)$ in this case, so that additional computations as in [HoBi78] are not necessary. For higher-order loops, the (approximate) computation of the mean tracking time involves other sophisticated mathematical tools, such as the singular perturbation approach [WeBo90]. The transient behavior of higher-order, nonlinear loops in noise is an open problem.

Recent trends in the theory of parameter estimation as applied to the epoch estimation problem tend to challenge the traditional closed-loop structure discussed above on at least two grounds. One pertains to their performance: the loop error variance due to noise is invariably shown to be inversely proportional to the prevailing SNR in the loop, even for non-differentiable (square) pulses, whereas the theory of square-edge ML epoch estimation in noise predicts an inverse-square relationship with SNR [Kosb88], [KoPo92]. Thus, significant jitter reduction might be possible with parallel architectures which imitate the ML estimator, assuming that complexity restrictions allow it. The other issue is the implementational desire to perform joint parameter and data estimation in a forward fashion (possibly on a single chip) without sacrificing accuracy [Ilti90], [Ilti94], [RaPoTz94]. This approach will be elaborated a bit in Section 5 of the paper.

4.   Code acquisition in fading channels

In general case the received signal given by Eq.(1) can be represented as r(t)=s(t,**A**)+n(t) where **A** is a vector of signal parameters to be estimated (e.g. code delay ζ, frequency Doppler ω_D, carrier phase θ_c etc.). So, the general form of the likelihood function given by Eq.(5) becomes

$$\Lambda(r|\mathbf{A}) = \int r(t)s(t,\mathbf{A})dt - \int s^2(t,\mathbf{A})dt$$

246

or the simplified form

$$\Lambda(r|\mathbf{A}) = \int r(t)s(t,\mathbf{A})dt$$

In the fading channel with impulse response h(t,$\mathbf{A}_1$), where $\mathbf{A}_1$ is the vector of channel parameters, the received signal becomes r(t,$\mathbf{A}$,$\mathbf{A}_1$))=s(t,$\mathbf{A}$)*h(t,$\mathbf{A}_1$)+n(t)=R(t,$\mathbf{B}$)+n(t) where (*) stands for convolution and $\mathbf{B}$ for the joint vector of signal and channel parameters ($\mathbf{A}$,$\mathbf{A}_1$). The likelihood function now becomes

$$\Lambda(r|\mathbf{B}) = \int r(t)R(t,\mathbf{B})dt \qquad (5c)$$

For the multipath fading h(t) can be represented as

$$h(t) = \sum_{l=1}^{L-1} \beta_l \delta(t - \zeta_l)$$

and vector $\mathbf{B}$ of the parameters to be estimated becomes B($\beta,\zeta,\theta,\omega_D$) where in the general case

$$\beta(\beta_0,\beta_1,...,\beta_{L-1}), \zeta(\zeta_0,\zeta_1,...,\zeta_{L-1})$$
$$\theta(\theta_0,\theta_1,...,\theta_{L-1}), \omega_D(\omega_{D0},\omega_{D1},...,\omega_{DL-1})$$

From this introductory observations we can see that synchronisation in fading channels becomes more and more joint channel and signal parameters estimation problem.

4A. Doppler effect:

The general theory of optimum synchronization of deterministic signals in fading dispersive channels was first addressed in [SolSc88] where a maximum-likelihood (ML) estimator for delay and Doppler was developed. The estimator technique is "open loop" and require a prior knowledge of channel scattering function. Due to the compexity of such an approach the most of the published work in the field of spread spectrum systems is trying to solve the problem from the point of view of the specific application.If Doppler effect is a dominant factor (LEO satellite communications) then ML-function presented in the previous section will be function of code delay and code and carrier Doppler.

The received complex DS signal in the case when there is no modulation can be represented as [Hur87],[ChHuSt90]

$$s(t) = Re\left\{\sqrt{2}S_0(t)e^{j\omega_c t}\right\}$$

$$S_0(t) = \sqrt{S}\,c\left(\frac{t}{1-\zeta'} - \zeta T_c\right)exp\left[jw_D t + \theta\right] \tag{13}$$

where we have used the following notation: $S_0(t)$ - the complex envelope of the signal, T_c - the PN code chip time, ζ - the received code-phase offset at t=o, ζ'/T_c - the received code frequency offset, ω_D - the received carrier radian frequency offset, θ - the carrier phase and ω_c - the carrier frequency.

In accordance with the general principles of ML parameter estimation we define the maximum likelihood estimates of ζ, ζ' and ω_D those values $\hat{\tau}, \hat{\tau}'$ and $\hat{\omega}_D$ that simultaneously maximize

$$\Lambda\left(\tau, \tau', \omega_D\right) =$$

$$\left|\int_{T_1}^{T_2} r_0(t)\,c\left(\frac{t}{1-\tau'} - \tau T_c\right)exp(-j\omega_D t)dt\right|^2 \tag{14}$$

where $r_o(t) = S_o(t) + n_o(t)$ and $n_o(t)$ is the complex envelope of the narrowband white Gaussian noise.

In practice Λ is maximized over a finite quantized set of the unknown parameters $\tau = \{\tau_1, \tau_2,.. \tau_{M1}\}, \tau = \{\tau'_1, \tau'_2,..., \tau'_{M2}\}$ and $\omega_D = \{\omega_1, \omega_2,.. \omega_{M3}\}..$ The quantization step size of ω_d, denoted by $\Delta\omega$, is determined by the length of the coherent integration time and the quantization step size of $\tau'$is determined by the allowable amount of code chip slip during the correlation process.

The likelihood function $\Lambda(\tau, \tau', \omega)$ is computed for each set (combination) of parameters $\left(\tau_i, \tau'_j, \omega_k\right)$ and the set that maximizes its value is chosen for the estimate.

Details regarding hardware implementation of the algorithm can be found in [Hur87],[Aft92],[Thom93]. Hardware descried in [Aft92] is suggested for spread spectrum demodulator for data relay systems. The structure is based on so called fast correlator. The principle of the fast correlator is that the convolution between two signals may be formed by transforming to the frequency domain, multiplying the transforms together and then transforming back to the time domain. To perform the correlation it is necessary to reverse one sequence in the time domain prior to convolution, or equivalently complex conjugating one sequence in the frequency domain

248

before multiplying the transforms together. If the transforms are performing using the FFT (Fast Fourier Transform) the technique is called "Fast Correlation". Thus correlation of the received signal with locally generated replica is carried out by multiplying the fourier transform of the signal by the reference.A bank of such correlators is used to span the entire doppler range. An increased search time, but reduced complexity is achieved by searching possible combinations (τ, τ', ω_D) serially. For this approach these parameters should be constant within the search period. The same principle based on neural network applications is described in [Thom93].

If Doppler is compensated prior to code acquisition the residual doppler can be ignored and three dimensional parallel search described above is now reduced to one dimensional (τ) parallel search analyzed in reference [DaFl88]. The effect of the residual Doppler frequency and data modulation decorrelation is assessed through an effective loss in signal power.

In the case when data modulation is present the modified Eq(13) must be used where $S_o(t)$ is now replaced by

$$S_o(t) = \sqrt{S} c\left(\frac{t}{1-\zeta'} - \zeta T_c\right) \times$$

$$d\left(\frac{t-T_1}{1-\zeta'} - \xi T_s\right) Exp(\omega_D t + \theta) \tag{13a}$$

where in addition to the previously defined parameters ξT_s is the receiver data-symbol time offset relative to the beginning of an integration subinterval i.e. relative to T_1. The analysis of the effects of data modulation on the acquisition performance is given in [ChHuSt90].

Although conceptually simple, ML approach is hardware intensive and in many applications (like mobile communications) not acceptable.

Application of matched filters would provide fast acquisition but unfortunately presence of data modulation, frequency offsets and code doppler limits the length (integration time) of these filters. At the same time, in low chip signal energy to noise power density ratio situations, long correlation periods are necessary to get desired statistics. One way to resolve this conflict is to use relatively short matched filters and noncoherent integration, or postdetection integration. Four different algorithms based on this idea are presented in [Su88].

In the first case uncertainly region (τ, ω_D) is searched serially with the fixed dwell time, the decision statistics is stored for each pair of (τ_i, ω_j) and the maximum value is then chosen with which the verification mode is initiated.The second algorithm initiates verification mode as soon as the decision statistics is larger than a threshold.The remaining two algorithms don´t use fixed dwell but rather variable dwell time and exploit advantages of sequential detection.

Further simplification of these schemes leeds to combining the PNMF´s outputs to generate feedback control signal to compensate Doppler. A number of such schemes is presented in [Gri80],[Suz86].

4.B. Multipath effect:

Personal/mobile communication systems usually operate in the indoor or urban environments where they suffer from severe multipath effects. Single dwell and multi dwell PN code acquisition in two ray Rayleigh fading channel is analysed in [ChCh93].

In general case of multipath channel model when N-path RAKE receiver is used [LePu87] the acquisition procedure is supposed to initially determine delays of N signal components. Of course the practical approach to the problem depends very much on the specific values of the delays and Dopplers in each of the N paths. One such approach is described in [Gr90]. The acquisition is based on the measuring of the magnitude of the complex envelope of the channel impulse response $|x(n)|$ shown in fig. 3b

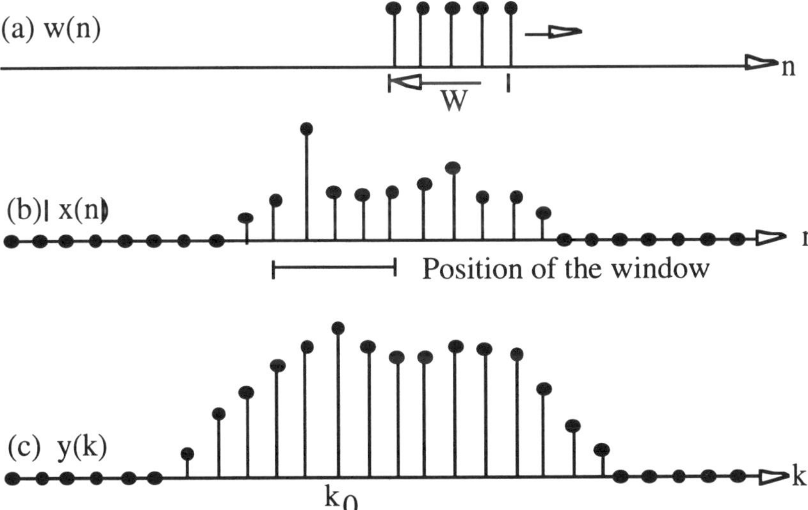

Fig.3 - Acquisition Algorithm. (a) Receiver window. (b) magnitude of the complex envelope of the channel impulse response. (c) Distribution of the summed signal magnitudes that fall into the receiver window

If the multipath delay spread is not to large (indoor communications) the RAKE-type receiver would process any signal component that falls into a window of width W and uses it for data demodulation. If the receiver has an odd number of N despreading correlator paths with a delay of one code chip

250

between them then the receiver window w(n) has a normalized width of W=N-1 as shown in Fig. 3a for N=5. The distribution y(k) of the summed signal magnitudes that fall into the receiver window is plotted as a function of the window position relative to the impulse response in Fig.3c. This function can be measured by sliding the window alone the impulse response x(n) and measuring (calculating) y(k) at every location k of the window. After sliding over one period of the PN code, a maximum at some position k_0 will be found. The receiver window is then centered around this position by shifting the local code, and the code tracking loop is switched on.The acquisition algorithm can be described mathematically by a correlation between the receiver window w(n)=rect(n/w) and the magnitude of the complex envelope of the channel impulse response $|x(n)|$.

$$y(k) = \sum_{n=-\infty}^{\infty} |x(n)| rect\left(\frac{n+k}{w}\right)$$

$$= \sum_{n=k-w/2}^{k+w/2} |x(n)| \qquad (15)$$

$$y(k)\Big|_{max} \Rightarrow k_0$$

For large delay spread where significant multipath components are not close to each other different strategy would provide better performance . Fig.4 demonstrates such an example So for the M-tap matched filter and N-path

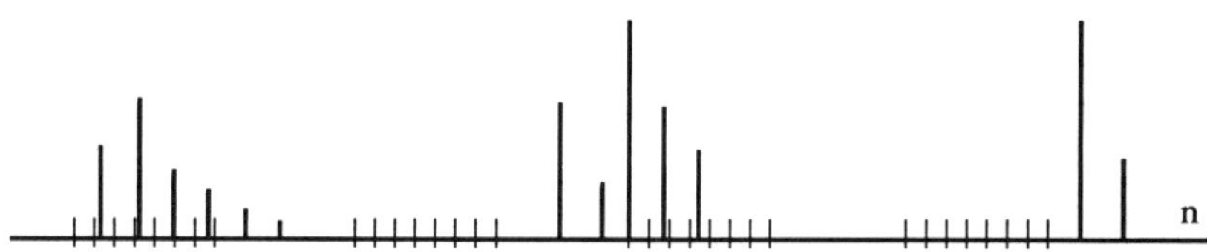

Fig.4 - Magnitude of the complex envelope of the channel impulse response with large delay spread.

Rake receiver the acquisition procedure can be defined as

$$\mathbf{y}(M, N) = Max(M, N, x(n)) \qquad (16)$$

Which means nothing more but choosing N the largest out of M available samples of the channel impulse responce x(n).

4.C   Rician fast fading channel:

When the transmission medium is such that there is a strong stable path and a number of weak paths we talk about Rician fading channel. Typical practical situation for these channels are the aircraft/satellite [Miy83] and line of sight (LOS) communications. Parallel acquisiton schemes for such channels are discussed in [SoGu90],[SoGu92]. The acquisition system itself employs a bank of parallel I-Q noncoherent pseudonoise matched filters and can be considered as an extension of the approach already described in the previous section [PoWe84b] and in [MiGeDa85] too. The only difference is that now in the presence of Rice fading probabilities of false alarm and signal detection have to be modified accordingly.

4.D   Fading channel characterized by Gilbert model.

Two level scheme for coarse code acquisition of FH signal was described first in [RaSc80]. The scheme uses a bank of passive correlators followed by a bank of active correlators. Signal structure includes short sync prefixes and the passive correlators matched to these prefixes are supposed to locate their possitions. These passive devices correlate over sliding time windows providing a rapid search capabilities. When the output of a passive correlator exceeds a threshold an active correlator is engaged in verification mode. The multiple sync prefxes reduce the probability of missing the signal in a fading enviroment. In order not to miss the signal during the period when an active correlator is enganged by a false alarm a multiple active correlator structure is necessary. In the analysis of such a system a model based on queueing and detection theory is used. The behaviour of this system in the fading channel characterized by Gilbert model [Gi60],[Kan78] is described in [WiRaVa88].

5.   Code tracking in fading channel

If Doppler effect is a dominant factor in the signal degradation then after signal acquisition the residual delay error can be further redused and trached by using delay lock-loop. One should be aware of the fact that for doppler and doppler rate the second and the third order loops respectively must be used in order to eliminate steady state tracking errors. In the Rician fading with the strong specular component DLL can still be used although now the random (Rayligh) component of the signal will act as an additional interference.

In multipath enviroment, when Rake receiver is used, synchronization process becomes rather complex issue. For N-paths Rake, N delayed replicas of the signal have to be tracked. A simpler version of this problem is when these N components are concentrated in N successive chip interrals so that the problem reduces to finding delay τ and signal amplitudes in N successive chip interrals. So for the channel with the impulse response.

$$h(\tau,t) = \sum_{n=0}^{N-1} f_n(t)\delta(t - nT_s) \tag{17}$$

and transmitted signal S(t) the received sampled signal is

252

$$r(k) = \sum_{l=0}^{N-1} f_l(k) S\big[(k-l)T_s + \tau(k)\big] + n(k) \tag{18}$$

and from the standpoint of the Rake receiver design that will use optimum ratio diversity combiner the task is to obtain minimum variance estimate of the multipath and delay parameters. If extended Kalman filter is used [Ilti90] then these estimates denoted as

$$\hat{f}_l(k|k) = E\big\{ f_l(k) | r(k), r(k-1)..r(0) \big\}$$

$$\hat{\tau}(k,k) = E\big\{ \tau(k) | r(k), r(k-1)...r(0) \big\} \tag{19}$$

are obtained by using the following dynamic model of the channel

$$f_l(k+1) = \alpha_l f_l(k) + W_l(k)$$

$$\tau(k+1) = \gamma \tau(k) + W_\tau(k) \tag{20}$$

for l=0,1,...N and where the $W_l(k)$ and $W_\tau(k)$ are mutually independent white Gaussian processes.

The use of higher order models for the channel coefficients, an example being N-th order AR model that can be represented as

$$f_l(k+1) = \sum_{n=0}^{N-1} \alpha_l(n) f_l(k-n) + W_l(k), \tag{21}$$

requires spectral analysis of the scattering function to determine coefficients $\alpha_l(n)$.

In the delay tracking problem, the state model is linear, while the measurement model is nonlinear.By using the following notation [Ilti90]

$$\mathbf{x}(k) = \big[\tau(k), f_0(k), f_1(k), ... f_{N-1}(k) \big]^T$$

$$\mathbf{F} = \begin{vmatrix} \gamma & 0 & ... & 0 \\ 0 & \alpha_0 & ... & 0 \\ ... & ... & ... & ... \\ 0 & 0 & ... & \alpha_{N-1} \end{vmatrix}$$

$$\mathbf{G} = \mathbf{I}$$

$$\mathbf{W}(k) = \left[W_\tau(k), W_0(k), W_1(k) \dots W_{N-1}(k) \right]^T$$

$$\mathbf{H}\left[x(k)\right] = \sum_{l=0}^{N-1} f_l(k) S\left[(k-l)T_s + \tau(k)\right] \tag{22}$$

$$Z(k) = r(k) = H\left[x(k)\right] + n(k)$$

the parameter estimation updating process can be represented as [AnMo79]

$$\mathbf{x}(k+1) = \mathbf{F}\mathbf{x}(k) + \mathbf{G}\mathbf{W}(k)$$

$$z(k) = H\left[x(k)\right] + n(k) \tag{23}$$

The scalar measurement z(k) is a nonlinear function of the state $\mathbf{x}(k)$. By using the first two termes in a Taulor´s series expansion about prediction $\hat{x}(k|k-1)$ we have

$$z(k) = H\left[x(k)\right] \cong H\left[\hat{x}(k|k-1)\right] +$$

$$+ \sum_{i=1}^{N+1} \left[x_i(k) - \hat{x}_i(k|k-1)\right] \frac{\partial}{\partial x_i} H(x)\Big|_{x=\hat{x}(k|k-1)} \tag{24}$$

After this linearization we have

$$\hat{\mathbf{x}}(k|k) = \hat{\mathbf{x}}(k|k-1) + \mathbf{K}(k)\left[z(k) - H\left(\hat{x}(k|k-1)\right)\right]$$

$$\mathbf{K}(k) = \mathbf{P}(k|k-1)\mathbf{H}'(k) \cdot$$

$$\left[\mathbf{H}'(k)^H \mathbf{P}(k|k-1)\mathbf{H}'(k) + \sigma_n^2\right]^{-1} \tag{25}$$

$$\mathbf{P}(k|k) = \left[\mathbf{I} - \mathbf{K}(k)\mathbf{H}'(k)^H\right]\mathbf{P}(k|k-1)$$

The matrix $H'(k)$ represents the time-varying gradient of the observation scalar with respect to the one-step prediction vector.

$$\mathbf{H}'(k) = \begin{bmatrix} \dfrac{\partial}{\partial x_1} H\left(\hat{x}(k|k-1)\right) \\[2ex] \dfrac{\partial}{\partial x_2} H\left(\hat{x}(k|k-1)\right) \\[2ex] , \dots , \\[2ex] \dfrac{\partial}{\partial x_{N_{f+1}}} H\left(\hat{x}(k|k-1)\right) \end{bmatrix} \tag{26}$$

254

Finally, the one-step predictions of the state vector and error covariance matrix follow the usual Kalman filter form:

$$\hat{\mathbf{x}}(k+1|k) = \mathbf{F}\hat{\mathbf{x}}(k|k)$$
$$\mathbf{P}(k+1|k) = \mathbf{F}\mathbf{P}(k|k)F^H + \mathbf{G}\mathbf{Q}\mathbf{G}^T \tag{27}$$

where

$$\mathbf{Q} = diag\left[\, \sigma_\tau^2, \sigma_{w_0}^2, \cdots, \sigma_{w_2}^2, \cdots, \sigma_{w_{N_f-1}}^2 \,\right] \tag{28}$$

After substituting the explicit forms of $x(k)$ and $H(x(k))$ into the above equations, the following measurement update equation is obtained for the joint delay/multipath estimator:

$$
\begin{bmatrix} \hat{\tau}(k|k) \\ \hat{f}_0(k|k) \\ \hat{f}_1(k|k) \\ \cdots \\ \hat{f}_{N_f-1}(k|k) \end{bmatrix}
=
\begin{bmatrix} \hat{\tau}(k|k-1) \\ \hat{f}_0(k|k-1) \\ \hat{f}_1(k|k-1) \\ \cdots \\ \hat{f}_{N_f-1}(k|k-1) \end{bmatrix}
+ \frac{1}{\sigma^2(k|k-1)} \mathbf{P}(k|k-1)
$$

$$
\cdot
\begin{bmatrix}
\frac{\partial}{\partial \hat{\tau}} S_f^*\!\left(kT_s + \hat{\tau}(k|k-1)\right) \\
S^*\!\left(kT_s + \hat{\tau}(k|k-1)\right) \\
S^*\!\left((k-1)T_s + \hat{\tau}(k|k-1)\right) \\
\cdots \\
S^*\!\left((k-N_f+1)T_s + \hat{\tau}(k|k-1)\right)
\end{bmatrix}
\tag{29}
$$

$$
\cdot\left[r(k) - S_{\hat{f}}\!\left(kT_s + \hat{\tau}(k|k-1)\right) \right]
$$

where $S_{\hat{f}}(t)$ denotes the estimate of the multipath distorted signal using the one-step predictions of $f_l(k)$.

$$S_{\hat{f}}(t) = \sum_{n=0}^{N_{f-1}} \hat{f}_n(k|k-1)S(t-nT_s) \tag{30}$$

The innovations variance $\sigma^2(k|k-1)$ is given by

$$\sigma^2(k|k-1) = \left[\mathbf{H}'(k)^H \mathbf{P}(k|k-1)\mathbf{H}'(k) + \sigma_n^2 \right]. \tag{31}$$

By using this line of reasoning PN code synchronization using the EKF joint estimator is analyzed in [Ilti90]. If in addition to this Doppler effect is present the size of the vector of the parameters to be estimated is further increased. For the channel dynamic model

$$\tau(k+1) = \alpha_\tau \tau(k) + W_\tau(k)$$
$$v_r(k+1) = \alpha_v v_r(k) + W_v(k) \tag{32}$$
$$f_n(k+1) = \alpha_f f_n(k) + W_n(k)$$

where v_r is the Doppler velocity the analysis is presented in [Ilti91]. The analysis can be further extended to include interference suppression. The analysis of such a case where interference is modeled as the N_α-th order AR process is presented in [Ilti94].

For the better understanding of code delay estimation and tracking the interested reader should also study the more general problem of time delay estimation. Special list of references covering this problem is enclosed at the end of the paper.

REFERENCES: Code acquisition

[Aft92] A. Aftelak et al - Design and Implementation of Spread Spectrum Demodulator for Data-Relay Systems, IAF - 92 - 0416 - 1992

[Alem77] W. K. Alem, "Advanced Techniques for Direct Sequence Spread Spectrum Acquisition," Ph.D. Dissertation, Dept. of Electrical Engineering, University of Southern California, February 1977.

[AlHuHoUd78] W. K. Alem, G. K. Huth, J. K. Holmes, and S. Udalov, "Spread Spectrum Acquisition and Tracking Performance for Shuttle Communication Links," *IEEE Trans Comm.*, Vol. COM-26, pp. 1689 – 1702, November 1978.

[AnMo79] B. Anderson and I. Moore - Optimal Filtering ,Englewood Cliffs, NY: Prentice-Hall, 1979

[Brau82] W. R. Braun, "Performance Analysis for the Expanding Search PN Acquisition Algorithm," *IEEE Trans. Comm.*, Vol. 30, No. 3, pp. 424-435, March 1982.

[BuMi55] J. J. Bussgang and D. Middleton, "Optimum Sequential Detection of Signals in Noise," *Trans. IRE*, Vol. IT-1, pp. 5 – 18, December 1955.

[Cahn71] C. R. Cahn, "Performance of Digital Matched Filter Correlator With Unknown Interference," *IEEE Trans. on Comm.*, Vol. 19, No. 6, pp. 1163-1172, December 1971.

[Chen88] U. Cheng, "Performance of a Class of Parallel Spread-Spectrum Code Acquisition Schemes in the Presence of Data Modulation," *IEEE Trans. on Comm.* Vol. 36, No. 5, pp. 596-604, May 1988.

[ChCh93] T. CHENG et al - Single Dwell and Multi Dwell PN Code Acquisition in Multipath Rayleigh Fading Channel, *PIMRC'93*,Yokohama, Japan, September 8-11, 1993, pp.276-283.

[ChHuSt90] U. Cheng, W. Hurd and J. Statman, "Spread Spectrum Code Acquisition in the Presence of Doppler Shifts and Data Modulation," *IEEE Trans. on Comm.*, Vol. 38, No. 2, pp. 241-250, February 1990.

[Comp87] G. M. Comparetto, "A General Analysis for a Dual Threshold Sequential Detection PN Acquisition Receiver," *IEEE Trans. Comm.*, Vol. 35, No. 9, pp. 956-960, September 1987.

[DaMiSc84] S. Davidovici, L. B. Milstein, D. L. Schilling, "New Rapid Acquisition Technique for Direct Sequence Spread-Spectrum Communications," *IEEE Trans. Comm.*, Vol. 32, No. 11, pp. 1161-1168, November 1984.

[DaFl88] L. D. Davisson and P. G. Flikkema, "Fast Single-Element PN Acquisition for the TDRSS MA System," *IEEE Trans. Comm.*, Vol. 36, No. 11, pp. 1226-1235, November 1988.

[Dica79] D. M. Di Carlo, "Multiple Dwell Serial Synchronization of Pseudonoise Signals," Ph.D. Dissertation, Dept. of Electrical Engineering, University of Southern California, May 1979.

[DiWe80] D. M. Di Carlo and C. L. Weber, "Statistical Performance of Single Dwell Serial Synchronization Systems," *IEEE Trans. Comm.*, Vol. 28, No. 8, pp. 1382-1388, August 1980.

[DiWe83] D. M. DiCarlo and C. L. Weber, "Multiple Dwell Serial Search: Performance and Application to Direct Sequence Code Acquisition," *IEEE Trans. Comm.*, Vol. 31, No. 5, pp. 650-659, May 1983.

[DlSc89] D. M. Dlugos and R. A. Scholtz, "Acquisition of Spread Spectrum Signals by an Adaptive Array," *IEEE Trans. on Acoustics, Speech and Signal Processing*, Vol. 37, No. 8, pp. 1253-1270, August 1989.

[ElTaGu80] A. K. Elhakeem, G. S. Takbar, and S. C. Gupta, "New Code Acquisition Techniques in Spread Spectrum Communications," *IEEE Trans. Comm.* , Vol. 28, pp. 249-257, February 1980.

[Gi60] E. Gilbert - Capacity of a Burst-Noise Channel *Bell System Tech. J,* 1960, Vol.39, pp. 1253-1265

[Glis88] G. S. Glisic, "Automatic Decision Threshold Level Control (ADTLC) in Direct-Sequence Spread-Spectrum Systems Based on Matching Filtering," *IEEE Trans. Comm.*, Vol. 36, No. 4, pp. 519-527, April 1988.

[Glis91] G. S. Glisic, "Automatic Decision Threshold Level Control in Direct-Sequence Spread-Spectrum Systems," *IEEE Trans. Comm.*, Vol. 39, No. 2, pp. 187-192, February 1991.

[Gr90] M. GROB et al - Microcellular Direct Sequence Spread Spectrum Radio System Using N-Path RAKE Receiver - *IEEE Journal on Selected Areas in Commun.*, Vol.8, No.5, June 1990, pp. 772-780

[Gri80] DeGrieco - The Application of Charge Coupled Devices to Spread Spectrum Systems - *IEEE Trans. on Comm.* Vol. COM-28, No.3, Sep. 1980

[Guma63] C. Gumacos, "Analysis of an Optimum Sync Search Procedure," *IRE Trans. Comm. Systems*, Vol. 11, pp. 89-99, March 1963.

[HoCh77] J. K. Holmes and C. C. Chen, "Acquisition Time Performance of PN Spread Spectrum Systems," *IEEE Trans. Comm.*, Special Issue on Spread Spectrum Communications, Vol. 25, pp. 778 – 784, August 1977.

[HoWo78] J. K. Holmes and K. T. Woo, "An Optimum PN Code Search Technique for a Given A Priori Signal Location Density," *NTC 78 Conference Record*, Birmingham, Alabama; Section 18.6, pp. 18.6.1 – 18.6.5, December 3 – 6, 1978.

[Hopk77] P. M. Hopkins, "A Unified Analysis of Pseudonoise Synchronization by Envelope Correlation," *IEEE Trans. Comm.*, Vol. 25, pp. 770 – 778, August 1977.

[Hur87] W Hurd et al - High Dynamic GPS Receiver Using Maximum Likelihood Estimation and Frequency Tracking - *IEEE Trans. Aerospace* Vol.23, Sept.1987

[Jova88] V. M. Jovanovic, "Analysis of Strategies for Serial Search Spread-Spectrum Code Acquisition-Direct Approach," *IEEE Trans. Comm.*, Vol. 36, pp. 1208-1220, November 1988.

[Jova92] V. M. Jovanovic, "On the Distribution Function of the Spread-Spectrum Code Acquisition Time," *IEEE J. Select. Areas Comm.*, Vol. 10, No. 4, pp. 760-769, May 1992.

[Kan78] L. Kanal et al - Models for Channels with Memory and their Applications to Error Control - *Proc. IEEE, 1978,* vol.66, pp. 724-744

[Kilg73] C. C. Kilgus, "Pseudonoise Code Acquisition Using Majority Logic Decoding," *IEEE Trans. Comm.*, Vol. 21, pp. 772 –774, June 1973.

[Kreb80] J. Krebser, "Performance of FH-Synchronizers with Constant Search Rate in the Presence of Partial Band Noise," *Proceedings of the 1980 International Zurich Seminar on Communications*.

[LePu87] I. Lehnert and M. Pursley - Multipath Diversity Reception of Spread Spectrum Multiple Access Communications - *IEEE Transactions on Communications,* Vol. COM-35, No.11, November 1987, pp. 1189-1198

[LeTa92] Y. H. Lee and S. Tantaratana, "Sequential Acquisition of PN Sequences for DS/SS Communications: Design and Performance," *IEEE J. Select. Areas Comm.*, Vol. 10, No. 4, pp. 750-759, May 1992.

[MaPu93] U. Madhow and M. B. Pursley, "Acquisition in Direct-Sequence Spread-Spectrum Communication Networks: An Asymptotic Analysis," *IEEE Trans. Inform. Theory*, Vol. 39, No. 3, pp. 903-912, May 1993.

[MaBl84] J. W. Mark and I. F. Blake, "Rapid Acquisition Techniques in CDMA Spread-Spectrum Systems," *IEE Proceedings*, Vol. 131, Part F, No. 2, pp. 223-232, April 1984.

[MePo83] H. Meyr and G. Polzer, " Performance Analysis for General PN-Spread Spectrum Acquisition Techniques," *IEEE Trans. Comm.*, Vol. 31, No. 12, pp. 1317-1319, Dec. 1983.

[MiLeFrTo92] L. E. Miller, J. S. Lee, R. H. French and D. J. Torrieri, "Analysis of an Antijam FH Acquisition Scheme, "*IEEE Trans. Comm.*, Vol. 40, No. 1, pp. 160-170, January 1992.

[MiGeDa85] L. B. Milstein, J. Gevargis and P. K. Das, "Rapid Acquisition for Direct Sequence Spread Spectrum Communications Using Parallel SAW Convolvers," *IEEE Trans. Comm.*, Vol. 33, No. 7, pp. 593-600, July 1985.

[Miy83] Y. Miyagaki et al - Double Symbol Error Rates of M-arg DPSK in a Satellite - Aircraft Multipath Channels - *IEEE Trans. Comm.* Vol. COM-31, pp. 1285-1289, Dec. 1983

[PaDoKu90] S. M. Pan, D. E. Dodds and S. Kumar, "Acquisition Time Distribution for Spread-Spectrum Receiver," *IEEE J. Select. Areas Comm.*, Vol. 8, No. 5, pp. 800-808, June 1990.

[Pand81] M. Pandit, "Mean Acquisition Time of Active and Passive-Correlation Acquisition Systems for Spread-Spectrum Communication Systems," *IEE Proc.*, Vol. 128, Part F, No. 4, pp. 211-214, August 1981.

[PaPo85] P. Pawlowski and A. Polydoros, "Optimization of a Matched Filter Receiver for FH Code Acquisition in Jamming," *Proc. IEEE 1985 Conf. Military Commun.*, pp. 1.1.1-1.1.7, Oct. 1985.

[Poly82] A. Polydoros, "On the Synchronization Aspects of Direct Sequence Spread Spectrum Systems," Ph.D. Dissertation, Dept. of Electrical Engineering, University of Southern California, August 1982.

[PoSi84] A. Polydoros and M. Simon, "Generalized Serial Search Code Acquisition: The Equivalent Circular State Diagram Approach," *IEEE Trans. on Comm.*, Vol. 32, No. 12, pp. 1260-1268, December 1984.

[PoWe84a] A. Polydoros and C. L. Weber, "A Unified Approach to Serial Search Spread-Spectrum Code Acquisition-Part I: General Theory," *IEEE Trans. Comm.*, Vol. 32, No. 5, pp. 542-549, May 1984.

[PoWe84b] A. Polydoros and C. L. Weber, "A Unified Approach to Serial Search Spread-Spectrum Code Acquisition-Part II: A Matched-Filter Receiver," *IEEE Trans. Comm.*, Vol. 32, No. 5, pp. 550-560, May 1984.

[Posn63] E. C. Posner, "Optimal Search Procedures," *IEEE Trans. Inform.Theory*, Vol. IT-11, pp. 157 – 160, July 1963.

[PuRaSc81] C. A. Putman, S. S. Rappaport, and D. L. Schilling, "A Comparison of Schemes for Coarse Acquisition of Frequency Hopped Spread Spectrum Signals," *Proc. ICC '81*, Denver, Colorado, pp. 34.2.1 – 34.2.5, June 1981.

[RaGr84] S. S. Rappaport and D. M. Grieco, "Spread-Spectrum Signal Acquisition: Methods and Technology," *IEEE Comm. Magazine*, Vol. 22, No. 6, pp. 6-21, June 1984.

[RaSc80] S. Rappaport and D. Schilling, "A Two Level Coarse Code Acquisition Scheme for Spread Spectrum Radio," *IEEE Trans. Comm.*, Vol. 28, pp. 1739-1742, 1980.

[Sage64] G. F. Sage, "Serial Synchronization of Pseudonoise Systems," *IEEE Trans. Comm.*, Vol. 12, pp. 123 – 127, December 1964.

[SiOmScLe85] M. K. Simon, J. K. Omura, R. A. Scholtz, B. K. Levitt, *Spread Spectrum Communication III*, Rockville, MD: Computer Science, 1985.

[SiWe86] E. W. Siess and C. L. Weber, "Acquisition of Direct Sequence Signals with Modulation and Jamming," *IEEE Journal on Selected Areas in Comm.*, Vol. 4, No. 2, pp. 254-272, March 1986.

260

[SoGu90] E. A. Sourour and S. C. Gupta, "Direct-Sequence Spread-Spectrum Parallel Acquisition in a Fading Mobile Channel," *IEEE Trans. Comm.*, Vol. 38, No. 7, pp. 992-998, July 1990.

[SoGu92] E. A. Sourour and S. C. Gupta, "Direct-Sequence Spread-Spectrum Parallel Acquisition in Nonselective and Frequency-Selective Rician Fading Channels," *IEEE Trans. Comm.*, Vol. 10, No. 3, pp. 535-544, April 1992.

[SolSc88] S. Soliman and R. Scholtz - Synchronization over Fading Dispersive Channels - *IEEE Transcations on Comm.*, Vol.36, No.4, 1988, pp. 499-505

[Stif68] J. J. Stiffler, "Rapid Acquisition Sequences," *IEEE Trans. Inform. Theory*, Vol. IT-14, No. 2, pp. 221-225, March 1968.

[Stif71] J. J. Stiffler, *Theory of Synchronous Communications*, Prentice-Hall, Englewood Cliffs, New Jersey, 1971.

[SuWe90] Y. T. Su and C. L. Weber, "A Class of Sequential Tests and Its Applications," *IEEE Trans. Comm.*, Vol. 38, No. 2, pp. 165-171, February 1990.

[Su88] Y. T. Su, "Rapid Code Acquisition Algorithm Employing PN Matched Filters," *IEEE Trans. Comm.*, Vol. 36., No. 6, pp. 724-733, June 1988.

[Suz86] R. Suzuki et al - Spread spectrum satellite communication terminal with coherent matched filter - Conference record, *GLOBECOM '86*, Vol.2, pp. 728-732

[Thom93] M. Thompson et al - Non-Coherent PN Code Acquisition in Direct Sequence Spread Spectrum Systems Using a Neural Network, *Milcom '93*, Conference Record, Vol.1, pp. 30-34

[VaEi87] I. Vajda and G. Einarsson, "Code Acquisition for a Frequency-Hopping System," *IEEE Trans. Comm.*, Vol. COM-35, No. 5, pp. 566-568, May 1987.

[VanT68] H. L. Van Trees, *Detection, Estimation and Modulation Theory, Part I*, Wiley, New York, 1968.

[Wald47] A. Wald, *Sequential Analysis*, Wiley, New York, 1947.

[Ward65] R. B. Ward, "Acquisition of Pseudonoise Signals by Sequential Estimation," *IEEE Trans. Comm.*, Vol. 13, pp. 475 – 483, December 1965.

[WaYi77] R. B. Ward and K. P. Yiu, "Acquisition of Pseudonoise Signals by Recursion Aided Sequential Estimation," *IEEE Trans. Comm.*, Vol. 25, pp. 784 – 794, August 1977.

[Wein83] A. Weinberg, "Generalized Analysis for the Evaluation of Search Strategy Effects on PN Acquisition Performance," *IEEE Trans. Comm.*, Vol. 31, pp. 37-49, January 1983.

[WiRaVa88] N. D. Wilson, S. S. Rappaport, M. M. Vasudevan, "Rapid Acquisition Scheme for Spread-Spectrum Radio in a Fading Environment," *IEE Proceedings*, Vol. 135, Part F, No. 1, February 1988.

[ZiPe85] R. E. Ziemer and and R. L. Peterson, *Digital Communications and Spread Spectrum Systems,* McMillan, Inc., 1985.

Code tracking

[Bowl79] W. M. Bowles, "GPS CodeTracking and Acquisition Using Extended-Range Detectors," Charles Stark Draper Lab., Inc., April 1979; see also *Proc. NTC'80*, Houston, Texas, pp. 24.1.1–24.1.5, December 1980.

[Comp87] G. Comparetto, "A Noncoherent Delay-Locked Loop Using the Exit-Time Criterion," *IEEE Trans. Comm.*, Vol. 35, No. 11, pp. 1240-1244, November 1987.

[Gill66] W. J. Gill, "A Comparison of Binary Delay-Lock Loop Implementations," *IEEE Trans. Aerospace Electr. Sys.*, Vol. 2, pp. 415–424, July 1966.

[GlMi91] S.Glisic and L.Milstein, Discrete Tracking System for Slow FH:Part I: Algorithms with Distributed Synchronization Group, *IEEE Transactions on Comm.*,Vol.39,No.2, pp304-314, February 1991.

[GlMi91a] S.Glisic , L.Milstein, Discrete Tracking System for Slow FH: Part II: Algorithms

with Concentrated Synchronization Group, *IEEE Transactions on Comm.*,Vol.39,No.2, pp314-324, February 1991.

[Glis87] S.Glisic et al, Efficiency of Digital Communication Systems, IEEE Transactions on Communications, Vol.COM-35,No.6,pp679-684, June 1987.

[Hart74] H. P. Hartmann, "Analysis of a Dithering Loop for PN Code Tracking," *IEEE Trans. Aerospace Electr. Sys.*, Vol. 10, pp. 2–9, January 1974.

[HoBi78] J. K. Holmes and L. Biederman, "Delay-Lock-Loop Mean Time to Lose Lock," *IEEE Trans. Comm.* , Vol. 26, pp. 1549–1556, November 1978.

[Holm82] J. K. Holmes, *Coherent Spread Spectrum Systems*, John Wiley, New York, 1982.

[**Hopk77**] P. M. Hopkins, "Double Dither Loop for Pseudonoise Code Tracking," *IEEE Trans. Aerospace Electr. Sys.*, Vol. 13, pp. 644–650, November 1977.

[**Ilti90**] R. Iltis, "Joint Estimation of PN Code Delay and Multipath using the Extended Kalman Filter," *IEEE Trans. Comm.*, Vol. 38, No. 10, pp. 1677-1685, October 1990.

[**Ilti94**] R. Iltis, "An EKF-based Joint Estimator for Interference, Multipath and Code Delay in a DS Spread Spectrum Receiver," to appear in *IEEE Trans. Comm.*,

[**Kosb88**] K. Kosbar, "Open and Closed Loop Delay Estimation with Applications to Pseudo-Noise Code Tracking," Ph.D. Dissertation, Dept. of Electrical Engineering, University of Southern California, July 1988.

[**KoPo92**] K. Kosbar and A. Polydoros, "A Lower-Bounding Technique for the Delay Estimation of Discontinuous Signals," *IEEE Trans. Inform Theory*, Vol. 38, No. 2, pp. 451-457, March 1992,

[**Lafl79**] D. T. LaFlame, "A Delay-Lock Loop Implementation Which Is Insensitive to Arm Gain Imbalance," *IEEE Trans. Comm.*, Vol. 27, pp. 1632–1633, October 1979.

[**Layl69**] J. W. Layland, "On Optimal Signals for Phase-Locked Loops," *IEEE Trans. Comm. Technology*, Vol. 17, No. 5, pp. 526-531, October 1969.

[**LiMe77**] W. C. Lindsey and H. Meyr, "Complete Statistical Description of the Phase-Error Process Generated by Correlative Tracking Systems," *IEEE Trans. Inform. Theory*, Vol. 23, pp. 194–202, March 1977.

[**Lind72**] W. C. Lindsey, *Synchronization Systems in Communication and Control*, Prentice Hall, Englewood Cliffs, New Jersey, 1972.

[**Meyr75**] H. Meyr, "Nonlinear Analysis of Correlative Tracking Systems Using Renewal Process Theory," *IEEE Trans. Comm.*, Vol. 23, pp. 192–203, February 1975.

[**Meyr76**] H. Meyr, "Delay-Lock Tracking of Stochastic Signals," *IEEE Trans. Comm.*, Vol. 24, pp. 331–339, March 1976.

[**Niel75**] P. T. Neilson, "On the Acquisition Behavior of Binary Delay-Lock Loops," *IEEE Trans. Aerospace Electr. Sys.*, Vol. 11, pp. 415–418, May 1975.

[**Poly82**] A. Polydoros, "On the Synchronization Aspects of Direct-Sequence Spread Spectrum Systems," Ph.D. Dissertation, Dept. of Electrical Engineering, University of Southern California, August 1982.

[PoWe85] A. Polydoros and C. L. Weber, "Analysis and Optimization of Correlative Code-Tracking Loops in Spread Spectrum," *IEEE Trans. Comm.*, Vol. 33, pp. 30–43, January 1985.

[RaPoTz94] R. Raheli, A. Polydoros and C-K. Tzou, "Per-Survivor Processing: A General Approach to MLSE in Uncertain Environments," accepted for publication in *IEEE Trans. Comm.*, 1994.

[Simo77] M. K. Simon, "Noncoherent Pseudonoise Code-Tracking Performance of Spread Spectrum Receivers," *IEEE Trans. Comm.*, Vol. 25, pp. 327–345, March 1977.

[SiOmScLe85] M. K. Simon, J. K. Omura, R. A. Scholtz, and B. K. Levitt, *Spread Spectrum Communications, Vol III*, Computer Science Press, 1985.

[SpMa61] J.J. Spilker, Jr. and D. T. Magill, "The Delay-Lock Discriminator—An Optimum Tracking Device," *Proc. IRE*, Vol. 49, pp. 1–8, September 1961.

[Spil63] J. J. Spilker, Jr., "Delay-Lock Tracking of Binary Signals," *IEEE Trans. Space. Electron. Telem.*, Vol 9, pp. 1–8, March 1963.

[Stif71] J. J. Stiffler,*Theory of Synchronous Communications*, Prentice Hall, Englewood Cliffs, New Jersey, 1971.

[WaPo85] R. Ward and A. Polydoros, "Optimization of Full-Time and Time-Shared Noncoherent Code Tracking Loops," *Proceedings of MILCOM '85*, Boston, Mass., pp. 1.6.1–5, October 1985.

[Ward85] R. Ward, "Optimization of Full-Time and Time-Shared Noncoherent Code Tracking Loops," Ph.D. Dissertation, Dept. of Electrical Engineering, University of Southern California, August 1985.

[WeBo90] A. L. Welti and B. Z. Bobrovsky, "Mean Time to Lose Lock for a Coherent Second-Order PN-Code Tracking Loop - The Singular Perturbation Approach," *IEEE Journal on Selected Areas in Comm.,,* Vol. 8, No. 5, pp. 809-818, June 1985.

[YoBo] R. A. Yost and R. W. Boyd, "A Modified PN Code Tracking Loop: Its Performance Analysis and Comparative Evaluation," *IEEE Trans. Comm.*, Vol. 30, pp. 1027–1036, May 1982.

[ZiPe85] R. E. Ziemer and R. L. Peterson, *Digital Communications and Spread Spectrum Systems*, McMillan, Inc., 1985.

General problem of time delay estimation

[BeRa90] R. E. Bethel, R. G. Rahikka, Optimum time delay detection and tracking, *IEEE Transaction on Aerospace and Electronic Systems,* Vol.26, No.5, Sept. 1990, pp. 700-712/4.5

[BeRa92] R. E. Bethel, R. G. Rahikka, Multisignal time delay detection and tracking, IEEE Transaction on Aerospace and Electronic Systems, Vol.28, No.3, July 1992, p. 675-696/4.5

[BoMe86] J. Bohmann, H. Meyr, An all-digital realization of a baseband DLL implemented as a dynamical state estimator, *IEEE Transaction on Acoustics, Speech, and Signal Processing,* Vol. ASSP-34, No.3, June 1986, pp. 535-545/4.5

[BoKa93] D. Bourdreau, P. Kabal, Joint time-delay estimation and adaptive recursive least squares filtering, *IEEE Transactions on Signal Processing,* Vol.41, No.2, Feb. 1993, pp. 592-601/4.5

[ChEiPa91] B. Champagne, M. Eizenman, S. Pasupathy, Exact Maximum likelihood time delay estimation for short observation intervals, *IEEE Transactions on Signal Processing,* Vol. 39, No.6, June 1991, pp. 1245-1257/4.5

[ChRi81] Y. T. Chan, J. M. F: Riley, J. B. Plant, Modeling of time delay and its applications to estimation of nonstationary delays, *IEEE Transactions on Acoustics, Speech, and Signal Processing,* Vol. ASSP-29, No.3, June 1981, pp. 577-581/4.5

[ChNi90] H-I Chiang, C. L. Nikias, A new method for adaptive time delay estimation for nongaussian signals, *IEEE Transactions on Acoustics, Speech, and Signal Processing,* Vol.38, No.2, Feb. 1990, pp. 209-217/4.5

[DiHeAg93] P. Diaz, D. Henche, R. Agusti, A PN code delay estimator based on the extended kalman filter for a DS/CDMA cellular system, *The Fourth International Symposium on Personal, Indoor and Mobile Radio Communications,* Yokohama, Japan, Sebtember 8-11, 1993/4.5

[EtSt81] D. M. Etter, S. D. Stearns, Adaptive estimation of time delays in sampled data systems, *IEEE Transactions on Acoustics, Speech, and Signal Processing,* Vol ASSP-29, No.3, June 1981, pp. 582-587/4.5

[FeBeRe81] P. L. Feintuch, N. J. Bershad, F. A. Reed, Time delay estimation using the LMS adaptive filter- dynamic behavior, *IEEE Transactions on Acoustics, Speech, nad Signal Processing,* Vol. ASSP-29, No.3, June 1981, pp. 571-576/4.5

[HaLi81] W. H. Haas, C. S. Lindquist, A synthesis of frequency domain filters for time delay estimation, *IEEE Transactions on Acoustics, Speech, and Signal Processing,* Vol. ASSP-29, No.3, June 1981, pp. 540-548/4.5

[HoChCh93] K. C. Ho, Y. T. Chan, P. C. Ching, Adaptive time-delay estimation in nonstationary signal and/or noise power enviroments, *IEEE Transactions on Signal Processing,* Vol.41, No.7, July 1993, pp. 2289-2299/4.5

[JaSc93] G. Jacovitti, G. Scarano, Discrete time techniques for time delay estimation, *IEEE Transactions on Signal Processing,* Vol.41, No.2, Feb. 1993, pp. 525-533/4.5

[KoZa93]. K. L. Kosbar, J. L. Zaninovich, Periodic PN sequence delay estimation using phase spectrum data, *GLOBECOM '93,* pp. 1665-1669/4.5

[LoMo91] I. M. G. Lourtie, J. M. F. Moura, Multisource delay estimation:nonstationary signals, *IEEE Transactions on Signals Processing,* Vol.39, No.5, May 1991, pp.1033-1048/4.5

[MeSp84] H. Meyr, G. Spies, The structure and performance of estimetors for real-time estimation of randomly varying time delay, IEEE Transactions on Acoustics, Speech, and Signal Processing, Vol. ASSP-32, No.1, Feb. 1984, pp. 81-94/4.5

[MiLe81] L. E. Miller, J. S. Lee, Error analysisof time delay estimation using a finite integration time correlator, *IEEE Transactions on Acoustics, Speech, and Signal Processing,* Vol. ASSP-29, No.3, June 1981, pp. 490-496/4.5

[PaJo91] M-A Pallas, G. Jourdain, Active high resolution time delay estimation for large BT signals, *IEEE Transaction on Acoustics, Speech, and Signal Processing,* Vol.39, No.4, April 1991, pp. 781-788/4.5

[ReFe81] F. A. Reed, P. L. Feintuch, N. J. Bershad, Time delay estimation using the LMS adaptive filter - static behavior, *IEEE Transactions on Acoustics, Speech, and Signal Processing,* Vol. ASSP-29, No.3, June 1981, pp. 561-571/4.5

[RoWi81] M. A. Rodriques, R. H. Williams, T. J. Carlow, Signal delay and waveform estimation using unwrapped phase averaging, *IEEE Transaction on Acoustics, Speech, and Signal Processing,* Vol. ASSP-29, No.3, June 1981, pp. 508-513/4.5

[ScAhCa81] K. Scaraborough, N. Ahmed, G. C. Carter, On the simulation of a class of time delay estimation algorithms, *IEEE Transactions on Acoustics, Speech, and Signal Processing,* Vol. ASSP-29, No.3, June 1981, pp. 534-540/4.5

[SeWeMu91] M. Segal, E. Weinstein, B. R. Musicus, Estimate-maximize algorithms for multichannel time delay and signal estimation, *IEEE Transaction on Acoustics, Speech, and Signal Processing,* Vol.39, No.1, Jan. 1991, pp.1-16/4.5

[St81] S. Stein, Algorithms for ambiguity function processing, *IEEE Transactions on Acoustics, Speech, and Signal Processing,* Vol. ASSP-29, No.3, June 1981, pp.588-599/4.5

[Tu91] **J**. K. Tugnait, On time delay estimation with unknoen spatially correlated gaussian noise using fourth-order cumulants and cross cumulants, *IEEE Transactions on Signal Processing,* Vol.39, No.6, June 1991, pp. 1258-1267/4.5

[Tu93] J. K. Tugnait, Time delay estimation with unknown spatially correlated gaussian noise, *IEEE Transactions on Signal Processing,* Vol.41, No.2, Feb. 1993, pp. 549-558/4.5

[Wax81] M. Wax, The estimate of time delay between two signals with random relative phase shift, *IEEE Transactions on Acoustics, Speech, and Signal Processing,* Vol. ASSP-29, No.3, June 1981, pp. 497-501/4.5

Chapter 5

CDMA Applications

Design Aspects of a CDMA Cellular Radio Network

W. C. Y. Lee

Abstract

This paper has identified the key elements in designing a CDMA system in which two scenarios are considered: non-uniform capacity but uniform cell-size scenario; and nonuniform capacity and non-uniform cell-size scenario. Due to the many variables, the number of channels in each cell has to meet certain conditions from the reverse channel formula. Then the total transmitted power of all the forward-link channels at each different cell site can be determined based on the worst-case interference conditions.

I. System Design Philosophy

Deploying CDMA systems is like tuning a sophisticated automobile engine. When proper tuning is done, the engine runs very smoothly. But a sophisticated automobile engine needs a sophisticated. computer-aided tuning device, just like a sophisticated CDMA system needs a computer-aided designing tool. As we know, in analog and TDMA systems, capacity increases are due to the elimination of interference from the desired signal. The signal level of a desired signal is always much stronger than the interference level, say 18 dB or better, for AMPS. However, in a CDMA system, the capacity increase is based on how much interference the desired signal can tolerate. The signal level of a desired signal is always below the interference level. Also, all the users have to share the same radio channel. If one user takes more power than it needs, then the others will suffer and system capacity will be reduced. This scenario is the same as dining in a formal restaurant. The volume of the conversations at every table is low. Therefore, no walls are needed between tables. The guests never feel their conversations are being interrupted by the next table. Therefore, many conversations can occur in the same dining room. This is the concept of CDMA that all the voice channels are sharing one big radio channel. If one table starts to raise their voices, the rest of the tables have to either leave or raise their voices too. The former case destroys CDMA. The latter is the so-called cocktail party syndrome which reduces the capacity of CDMA. Neither one is desired. In order to tolerate interference, this section addresses how to tune the CDMA cellular radio network.

Designing a uniform CDMA system is comparatively simple. Uniform CDMA means all the cells will be assigned the same number of channels. However, in reality, CDMA systems are not uniform. The voice channels of each cell in a CDMA system are not the same. Due to the demographical

S.G. Glisic and P.A. Leppänen (eds.), Code Division Multiple Access Communications, 269-282.
© 1995 *Kluwer Academic Publishers. Printed in the Netherlands.*

needs, some cells need more voice channels and some need less voice channels. Since CDMA has only one radio channel, to generate different voice channels on demand from a single CDMA radio is a big challenge. We would like to present the challenges by illustrating the design aspect of the CDMA system.

II.   Key Elements in Designing a CDMA System

The design of a CDMA system is much more sophisticated than the design of a TDMA system. In analog and TDMA systems, the most important key element is C/I. There are two different kinds of C/I. One is the measured (C/I) which is used to indicate the voice quality in the system. The higher the measured value is the better. The other is called the specified $(C/I)_S$ which is a specified value for a specified cellular system. For example, the $(C/I)_S$ in the AMPS system is 18 dB. Since in analog and TDMA systems, due to the spectral and geographical separations, the interference (I) is much lower than the received signal (C), sometimes we can utilize field strength meters to measure C to determine the coverage of each cell. The field strength meter therefore becomes a useful tool in designing the TDMA system. In CDMA all the traffic channels are served solely by a single radio channel in every cell[1,2,3,4]. Therefore, in an m-voice channel cell, one of the m traffic channels is the desired channel and the remaining m-l traffic channels are the interference channels. In this case, the interference is much stronger than the desired channel. Then C/I is hard to obtain by using the signal strength meter which will receive more interference than the desired signal. Thus, the key elements in designing a CDMA system are different from the key element in designing a TDMA system.

Relationship between C/I and FER

In CDMA, the key element is E_b/I_0 (energy per bit/power per Hz) which is related to the frame error rate (FER). An acceptable speech quality of a specified vocoder would determine the FER which is related to a E_b/I_0 at a given vehicle speed. From a system design aspect, we consider the system performance with all the vehicle speeds and environmental conditions and come up with a specified E_b/I_0. Now we can design the CDMA system based on the specified E_b/I_0

The following equation is used:

$$\frac{C}{I} = \left(\frac{E_b}{I_0}\right) \cdot \left(\frac{R_b}{B}\right) \cdot \eta \qquad (1)$$

where R_b is the bits per second and B is the CDMA channel bandwidth. η is the speech activity cycle in percent. From Eq. (1), B/R_b is the processing gain (P.G.) which is known in a given CDMA system. E_b/I_0 and η are also known in the system, then the C/I of each CDMA channel is obtained. Each coded channel in CDMA can be treated as a frequency channel in FDMA or

TDMA. If the coded channels are sent over a cable transmission medium, the interference among the coded channels can be treated as adjacent channel interference. Due to the nature of channel orthogonality, the interference should be very small. But in the mobile radio environment, due to the creation of the multipath wave phenomenon, the orthogonality among the channels cannot be held. Therefore, the processing gain is the only interference protection among the channels.

E_b/I_0 always varies in order to meet a specified FER under different conditions. From Eq. (1) we can find a required $(C/I)_s$ from a specified $(E_b/I_0)_s$ in a worst-case scenario for designing the system. However, the values of $(E_b/I_0)_s$ for the forward link channels and for the reverse-link channels are different due to the different modulation schemes. Therefore, we may have two different requirements for C/I. One $(C/I)_F$ for the forward link channels and the other $(C/I)_R$ for the reverse link channels.

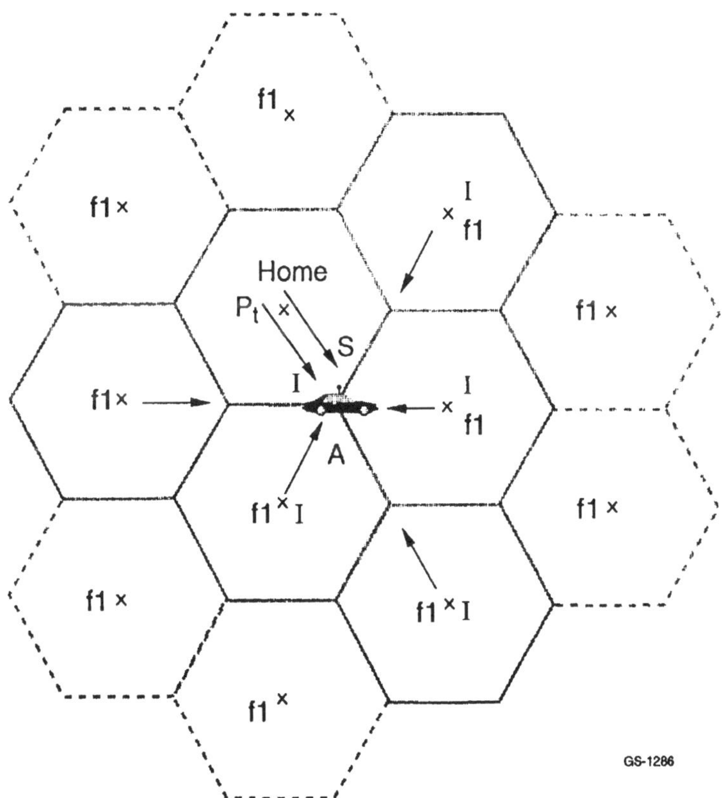

Fig. 1 CDMA System and its Interference (From a Forward Link)

III. Calculation of Design Parameters in Uniform Cell-Size Scenario

Now we are trying to find the design parameters of each cell for the forward link and the reverse link in a uniformed capacity condition which is a reality.

(A) For the Forward Link

A worst-case scenario is used to find the relation among the transmitted powers of all cell sites. First we form an equation which relates the C/I received at a mobile location A to the transmitted powers of all cell sites. The mobile location A is as shown in Figure 1.

$$C/I = \frac{\alpha_1 \cdot R^{-4}}{\underbrace{\alpha_1(m_1 - 1) \cdot R^{-4}}_{\text{self cell}} + \underbrace{(\alpha_2 m_2 + \alpha_3 m_3) \cdot R^{-4}}_{\text{2 adjacent cells}}}$$

(2)

$$\overline{+\underbrace{\beta \cdot (2R)^{-4}}_{\text{3 intermediate cells}} + \underbrace{\gamma(2.633R)^{-4}}_{\text{6 distant cells}}}$$

where α_i (i = 1, 3) is the transmitted power of each voice channel in the cell and m_i is the number of channels per cell. β and γ are the transmitted powers of the combined adjacent cells at a distance 2R and 2.633R, respectively. By solving Eq. (2) we can determine m_i as follows:

$$m_1 = \left(\frac{1}{C/I} + 1\right) - \left[\frac{\alpha_2 m_2 + \alpha_3 m_3}{\alpha_1}\right]$$

$$- \frac{\beta}{\alpha_1}(2)^{-4} - \frac{\gamma}{\alpha_1}(2.633)^{-4}$$

(3)

<u>Case A</u> - No adjacent cell interference

Let $\alpha_2 = \alpha_3 = \beta = \gamma = 0$ in Eq. (3), then

$$m_1 = \frac{1}{(C/I)} + 1$$

(4)

If the value of C/I obtained from Eq. (1) is C/I = -17 dB, then $m_1 = 51$ which is the maximum voice channels in a cell.

<u>Case B</u> - No interference other than from the two close-in interfering cells.

In Eq. (3), the values of the third and fourth terms are much smaller as compared with that of the first two terms and therefore can be neglected. Then

$$\alpha_1 = \frac{\alpha_2 m_2 + \alpha_3 m_3}{\dfrac{1}{C/I} + 1 - m_1} \qquad (5)$$

If C/I = -17 dB, and the assigned voice channels at three cells are $m_1 = 30$, $m_2 = 25$, and $m_3 = 15$, respectively, then Eq. (5) becomes:

$$\alpha_1 = \frac{25\alpha_2 + 15\alpha_3}{51 - 30} = 1.19\alpha_2 + 0.714\alpha_3 \qquad (6)$$

Eq. (6) expresses the relationship among α_1, α_2, and α_3.

The total transmitted power P in each cell site is $P_1 = \alpha_1 m_1$, $P_2 = \alpha_1 m_2$, $P_3 = \alpha_3 m_3$, When m_1, m_2, m_3, are given, thus P_1, P_2 and P_3 are the maximum transmitted powers of the three cells. Then Eq. (5) can be simplified as:

$$\left(\frac{1}{(C/I)} + 1 \right) \frac{P_1}{m_1} = P_1 + P_2 + P_3 \qquad (7)$$

Following the same derivation steps, we can obtain the following equations:

$$\left(\frac{1}{(C/I)} + 1 \right) \frac{P_2}{m_2} = P_1 + P_2 + P_3 \qquad (8)$$

$$\left(\frac{1}{(C/I)} + 1 \right) \frac{P_3}{m_3} = P_1 + P_2 + P_3 \qquad (9)$$

The relationship of three maximum transmitted powers of three cells are:

$$\frac{P_1}{m_1} = \frac{P_2}{m_2} = \frac{P_3}{m_3} \qquad (10)$$

Deduced from Eq. (10), a design criterion which we will be used in general for a CDMA system of N cells can be expressed as:

$$\frac{P_i}{m_i} = \frac{P_j}{m_j} = \text{cons} \tan t \qquad (11)$$

274

where i indicates the i cell and j indicates the j cell. Eq. (16.4-11) indicates that the more voice channels generated, the more transmit power is needed. Therefore, either applying power control to the voice channels in a cell such that more voice channels can be provided with a given transmit power, or using less number of channels in a cell such that the transmit power P will reduce, as a result, interference is reduced in both cases.

(B)   For the Reverse Link

The worst-case scenario (shown in Figure 2) is also used in the reverse link analysis. Assume that all the mobile units traveling in the two adjacent cells will be located at the cell boundary of the home cell. From the reverse link, the powers of the ml voice signals received at the home site are the same, due to the power control implementation to overcome the near-to-far interference.

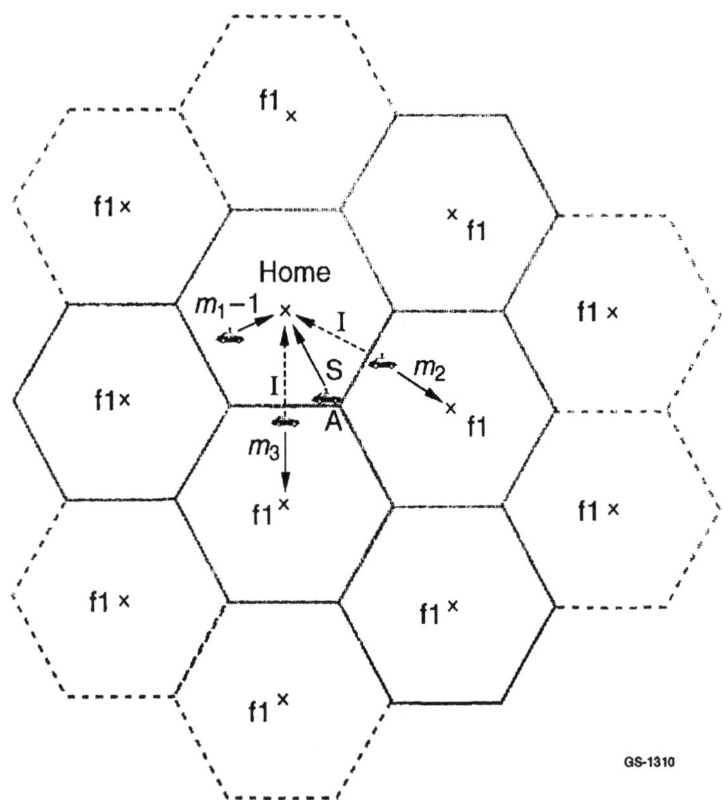

Fig. 2 CDMA System and its Interference (From a Reverse Link Scenario)

Let the received signal from a desired mobile unit at the home cell site be C. Assume that each signal of other m_1 channels received at the home site is also C as shown in Figure 2. Also, assume that the interference of certain mobile units, say r $\cdot$ m_1, from the two adjacent cells comes from the cell boundary. Because of the power control in each adjacent cell, the interference coming from the adjacent cell for each voice channel would roughly be C as received by the home cell site. The received C/I at the desired voice channel can be expressed as:

$$\frac{C}{I} = \frac{C}{(m_1 - 1) \cdot C + r_{12} \cdot m_2 C + r_{13} \cdot m_3 C}$$
$$= \frac{1}{m_1 - 1 + r_{12} m_2 + r_{13} m_3} \tag{12}$$

where r_{12} and r_{13} are a portion of the total number of voice channels in adjacent cells that will interfere with the desired signal at the home cell which is Cell 1.

From Eq. (12), the worst-case scenario is when:

$$m_1 + r_{12} \cdot m_2 + r_{13} m_3 \leq \frac{1}{\frac{C}{I}} + 1 \tag{13}$$

Following the same steps, we find:

$$r_{21} m_1 + m_2 + r_{23} m_3 \leq \frac{1}{\frac{C}{I}} + 1 \tag{14}$$

$$r_{31} m_1 + r_{32} m_2 + m_3 \leq \frac{1}{\frac{C}{I}} + 1 \tag{15}$$

The value of r depends on the size of the overlapped region in the adjacent cell, and can be reasonably assumed as 1/6 (which is 0.166) if the system is properly designed.

If C/I = -17 dB which is 50, and r_{12} =r_{13} = 0.166, then Eq. (13) becomes:

$$m_1 + 0.166 \cdot (m_2 + m_3) \leq 51 \tag{13a}$$

The relationships among the number of voice channels in each cell, m_1, m_2, and m_3 are expressed in Eq. (13), Eq. (14) and Eq. (15).

C.   Designing a CDMA System

From the reverse-link scenario, we can check to see whether all the conditions expressed in Eq. (13), Eq. (14) and Eq. (15) can be met. The

276

unknowns in these conditions come from the demanded voice channels, m_1, m_2, and m_3.

Then based on the forward link equations, Eq. (7) to Eq. (10), we can determine the maximum transmitted power of each cell.

IV. Calculation of Design Parameters in Non-Uniform Cell Scenario

A. Transmit Power on the Forward-Link Channels

We may first assign the number of voice channels m in each cell due to requirements from demographical data. Then we may calculate the total transmit power on the forward link channels in each cell from a worst-case scenario as shown in Figure 3.

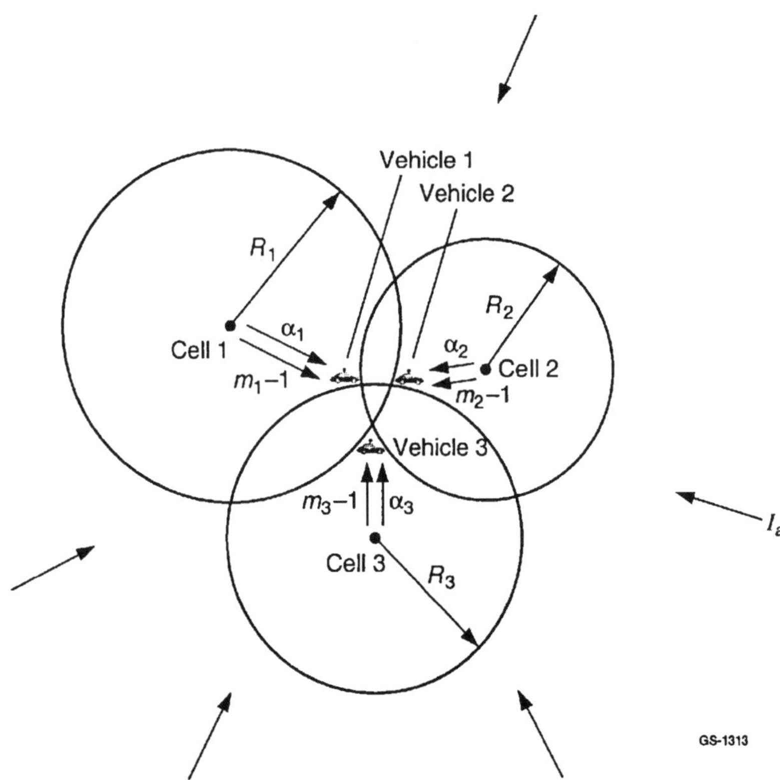

Fig. 3 The Worst-Case Scenario on a Forward-Link Channel
Reception in a Non-Uniform CDMA System

All the cell sizes are not the same in a non-uniformed capacity CDMA system. Among them, we only consider the three most effected cells due to the worst locations of the three vehicles as shown in Figure 3. The locations of the three vehicles, one belonging to each cell, are the most interfering locations.

In this case, the $(C/I)_F$ received at vehicle 1 is

$$\left(\frac{C_1}{I_1}\right)_F = \frac{\alpha_1 R_1^{-4}}{(m_1-1)\alpha_1 R_1^{-4} + \alpha_2 m_2 R_2^{-4} + \alpha_3 m_3 R_3^{-4} + I_{a1}} \tag{16}$$

where I_a is the interference coming from other interfering cells besides these three cells. I_a is usually very small as compared to the two terms (second and third in the denominator of Eq. (16)) and can be neglected.

From the received $(C_2/I)_F$ from vehicle 2.

$$\left(\frac{C_2}{I_2}\right)_F = \tag{17}$$

$$\frac{\alpha_2 R_2^{-4}}{(m_2-1)\alpha_2 R_2^{-4} + \alpha_1 m_1 R_1^{-4} + \alpha_3 m_3 R_3^{-4} + I_{a2}}$$

From the received $(C_3/I)_F$ from vehicle 3

$$\left(\frac{C_3}{I_3}\right)_F = \tag{18}$$

$$\frac{\alpha_2 R_2^{-4}}{(m_3-1)\alpha_3 R_3^{-4} + \alpha_1 m_1 R_1^{-4} + \alpha_2 m_2 R_2^{-4} + I_{a3}}$$

Let

$$\left(\frac{C_1}{I_1}\right)_F = \left(\frac{C_2}{I_2}\right)_F = \left(\frac{C_3}{I_3}\right)_F = \left(\frac{C}{I}\right)_F$$

and

$$I_{a1} = I_{a2} = I_{a3} = 0$$

Simplifying Eqs. (16-18), respectively, we obtain:

$$\alpha_1 m_1 + \alpha_2 m_2 \left(\frac{R_2}{R_1}\right)^{-4} + \alpha_3 m_3 \left(\frac{R_3}{R_1}\right)^{-4} \tag{19}$$

$$= \alpha_1 \left[\frac{1}{(C/I)_F} + 1\right] = \alpha_1 \cdot G$$

$$\alpha_1 m_1 \left(\frac{R_1}{R_2}\right)^{-4} + \alpha_2 m_2 + \alpha_3 m_3 \left(\frac{R_3}{R_2}\right)^{-4} = \alpha_2 \cdot G \tag{20}$$

$$\alpha_1 m_1 \left(\frac{R_1}{R_3}\right)^{-4} + \alpha_2 m_2 \left(\frac{R_2}{R_3}\right)^{-4} + \alpha_3 m_3 = \alpha_2 \cdot G \tag{21}$$

Solving Eqs. (19), (20) and (21) we come up with the following relation:

$$\alpha_1 R_1^{-4} = \alpha_2 R_2^{-4} = \alpha_3 R_3^{-4} \tag{22}$$

Also, assume that the minimum values of α_1, α_2, and α_3 will be α_1^0, α_2^0 and α_3^0, respectively, which are purely based on the received level, C_0, of individual voice channels at the vehicle locations as:

$$\alpha_1 \geq \alpha_1^0 = C_0 R_1^{+4} / k_1$$
$$\alpha_2 \geq \alpha_2^0 = C_0 R_2^{+4} / k_2 \tag{23}$$
$$\alpha_3 \geq \alpha_3^0 = C_0 R_3^{+4} / k_3$$

where C_0 is the required signal received level at the vehicle location, and k_i is a constant gain related to the antenna heights at the cell sites.

Now the total transmit power of each cell site will be:

$$P_1 = m_1 \alpha_1$$
$$P_2 = m_2 \alpha_2 \tag{24}$$
$$P_3 = m_3 \alpha_3$$

B.   Transmit Power on the Reverse-Link Channels

On the reverse-link channels, we also use the same worst-case scenario as shown in Figure 4. Based on the power control algorithm, all the signals will be the same when reaching the cell site. Then the vehicle 1 signal received at the cell site 1 is C_1, the rest of the signals are considered interference.

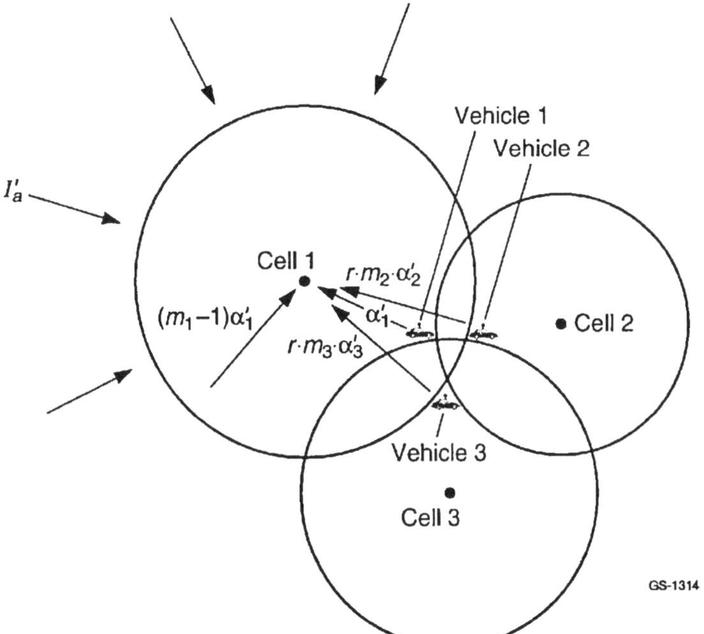

Fig. 4 The Worst-Case Scenario on a Reverse-Link Channel
Reception in a Non-Uniform CDMA System

$$\left(\frac{C_1}{I_1}\right)_R \geq$$

$$\frac{\alpha'_1 \cdot R_1^{-4}}{(m_1-1)\alpha'_1 R_1^{-4} + \eta_2 m_2 \alpha'_2 R_1^{-4} + \eta_3 m_3 \alpha'_3 R_1^{-4} + I'_{a1}} \tag{25}$$

where α'_1, α'_2 and α'_3 are the power of individual channels transmitted back to their corresponding cell sites. r_{12} and r_{13} are the portion of the total number of voice channels in adjacent cells that will interfere with the desired signal at Cell 1. I_{a1} is the interference coming from other vehicles in other cells which is not cell 2 and cell 3. I_{a1} is a relatively small value and can be neglected. Utilizing the same steps, we can derive the following two equations for the two other cases; Case 2: Cell Site 2 receives the vehicle 2 signal, and Case 3: Cell Site 3 receives the vehicle 3 signal.

280

$$\left(\frac{C_2}{I_2}\right)_R \geq$$

$$\frac{\alpha_2' R_2^{-4}}{r_{21} \cdot m_1 \alpha_1' R_2^{-4} + (m_2 - 1)\alpha_2' \cdot R_2^{-4} + r_{23} \cdot m_3 \alpha_3' R_2^{-4}} \tag{26}$$

$$\left(\frac{C_3}{I_3}\right)_R \geq$$

$$\frac{\alpha_3' R_3^{-4}}{r_{31} m_1 \alpha_1' R_3^{-4} + r_{32} m_2 \alpha_2' R_3^{-4} + (m_3 - 1)\alpha_3' \cdot R_3^{-4}} \tag{27}$$

where r is the percentage of total channels from the interfering cell and would be received by the home cell site. Simplifying Eqs. (25), (26) and (27) we obtain:

$$\left(\frac{I}{C}\right)_R \geq (m_1 - 1) + r_{12} m_2 \frac{\alpha_2'}{\alpha_1'} + r_{13} m_3 \frac{\alpha_3'}{\alpha_1'} \tag{28}$$

$$\left(\frac{I}{C}\right)_R \geq r_{21} m_1 \frac{\alpha_1'}{\alpha_2'} + (m_2 - 1) + r_{23} m_3 \frac{\alpha_3'}{\alpha_2'} \tag{29}$$

$$\left(\frac{I}{C}\right)_R \geq r_{31} m_1 \frac{\alpha_1'}{\alpha_3'} + r_{32} m_2 \frac{\alpha_2'}{\alpha_3'} + (m_3 - 1) \tag{30}$$

where

$$\left(\frac{C}{I}\right)_R = \left(\frac{C_1}{I_1}\right)_R = \left(\frac{C_2}{I_2}\right)_R = \left(\frac{C_3}{I_3}\right)_R$$

All the coefficients in Eq. (28) to Eq. (30) involve the transmit power of a voice channel in each of the three cells, α_1', α_2' and α_3'.

For the same reason as stated for obtaining Eq. (23), the minimum values of α_1', α_2' and α_3' can be defined as follows:

$$\alpha_1' \geq \alpha_1^o = C_0 R_1^4 / k_1$$

$$\alpha_2' \geq \alpha_2^o = C_0 R_2^4 / k_2$$

$$\alpha_3' \geq \alpha_3^o = C_0 R_3^4 / k_3 \tag{31}$$

where R_1, R_2 and R_3 are the radii of the three cells, respectively. k is a constant gain related to the antenna heights at the cell sites.

We may replace all the $\alpha's$ with $\alpha^o s$ shown in Eq. (31) in the coefficients of Eq. (28) and (30), then,

$$\left(\frac{I}{C}\right)_R \geq (m_1 - 1) + \eta_{12} m_2 \left(\frac{R_2}{R_1}\right)^4 + \eta_{13} m_3 \left(\frac{R_3}{R_1}\right)^4 \tag{32}$$

$$\left(\frac{I}{C}\right)_R \geq \eta_{21} m_1 \left(\frac{R_1}{R_2}\right)^4 + (m_2 - 1) + \eta_{23} m_3 \left(\frac{R_3}{R_2}\right)^4 \tag{33}$$

$$\left(\frac{I}{C}\right)_R \geq \eta_{31} m_1 \left(\frac{R_1}{R_3}\right)^4 + \eta_{32} m_2 \left(\frac{R_2}{R_3}\right)^4 + m_3 - 1 \tag{34}$$

Under the physical condition, the following relationships have to be held. The three values m_1, m_2 and m_3 have to be:

$$m_1, m_2, \text{ or } m_3 < \frac{1}{\left(\dfrac{C}{I}\right)_R} + 1 \tag{35}$$

which has been derived in Eq. (4).

C.   Designing a CDMA System

We first have to check whether all the requirements expressed in Eq. (32) to Eq. (35) are met with our given conditions. If they are met, then we can find the transmit power P_1, P_2, and P_3 from Eq. (24). Usually, among the three equations, Eq. (32) to Eq. (35), only one of them is dominating. If that one meets the given conditions, the other two will meet them also. The following example addresses this point:

V.　Conclusion

This paper has identified the key elements in designing a CDMA system in which two scenarios are considered: non-uniform capacity but uniform cell-size scenario and nonuniform capacity and cell-size scenario. Due to the many variables, the number of channels in each cell has to meet certain conditions from the reverse channel formula Then the total transmitted power of all the forward-link channels at each different cell site can be determined based on the worst-case interference condition.

In designing a CDMA system we do not have the frequency management problem but we have to properly distribute the interference in each cell such that the desired number of voice channels in each cell can be achieved. This paper has highlighted the methodology of designing a CDMA system.

References:

1. Lee, W. C. Y., "Overview of Cellular CDMA,"IEEE Transactions on Vehicular Technology, Vol. 40, May 1991, pp. 291-302. Also, "Mobile Communications Design Fundamentals," John Wiley & Sons, 1993, Chapter 9.

2. Gilhousen, K.S.; L M. Jacobs; R. Padovani; A. J. Viterbi; L. A. Weaver; and C. E. Wheatley, "On the Capacity of a Cellular CDMA System," IEEE Transactions on Vehicular Technology, Vol. 40, May 1991, pp. 303-312.

3. Pickholtz, RL; L. B. Mulstein; and D. L. Schilling, "Spread Spectrum for Mobile Communications," IEEE Transactions on Vehicular Technology, Volume 40, May 1991, pp. 313-322.

4. Simon, M. K; J. K. Omira; R. A. Scholtz; and B. D. Levin, "Spread Spectrum Communications," Vol. I, II, III, McGraw Hill, 1994, Chapter 6.

Consumer Communications Based on Spread Spectrum Techniques

Masao Nakagawa

Abstract- This paper shows consumer applications of Spread Spectrum Techniques; Power Line Communication, Data Carrier, Radio Remote Control, ISM Wireless LAN, Vehicle to Vehicle Communication, Digital TV Broadcasting, and Devices and Systems.

1.   Introduction

Spread Spectrum Techniques had been regarded as complicated and special ones for long years. Therefore, nobody believed that they would be applied to consumer applications. However, history sometimes gives us ironic conclusions, as PCM audio drove analog audio out no matter how PCM system is much more complicated than analog one. Spread Spectrum Techniques are being applied not only to Public Communications, but also to Consumer Communications.

Consumer Communication is a kind of non-public communications including LAN, WLAN, indoor communication, cordless phone, communication for robots, remote control, data carrier, home communication, office communication, factory communication and vehicle to vehicle communication. Fig.1 shows consumer communication networks and public communication networks, where personal communication networks are included in public communication networks. The predicted progress of Consumer Communications is attributed to low cost, unregulated design, and advantage to local and small-zone communication.

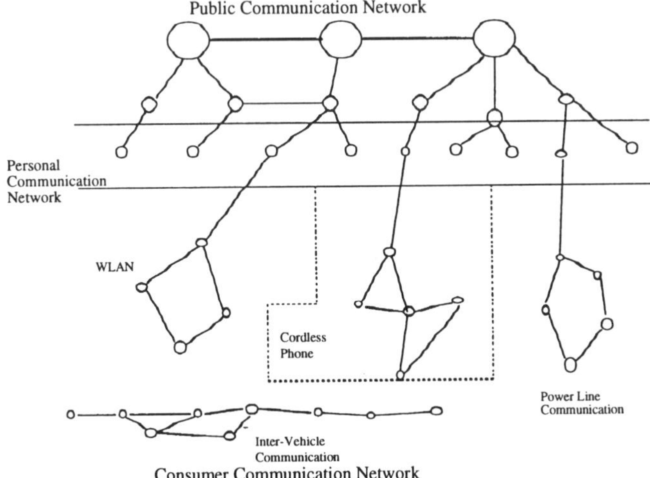

Fig.1 Consumer Communications and Public Communications networks

283

S.G. Glisic and P.A. Leppänen (eds.), Code Division Multiple Access Communications, 283-297.
© 1995 *Kluwer Academic Publishers. Printed in the Netherlands.*

However, several problems must first be solved, transmission line problems as well as device design problems, since Consumer Communications networks often have to employ poor transmission lines suffering from fading, low power transmission, interference, and interception. The key technology for overcoming the poor transmission lines is Spread Spectrum because of its anti-interference, anti-interception, high frequency efficiency and anti-fading properties.

This paper describes the growing applications of spread spectrum techniques to Consumer Communications in Japan.

2.   Concept of Spread Spectrum Communication

Here the concept of spread spectrum communication is shown. Fig. 2 shows the block diagram of a spread spectrum communication system. The data is applied to a narrowband modulator and the resulting narrowband signal is widely spread using the spreading code at the transmitter. The transmitted spread signal with interfering signals and noise is despread by the same spreading code and the resulting narrowband signal is demodulated at the receiver. Fig. 3 shows a spectra schematic of the frequency transformation at the transmitter and receiver. Because the spread signal can be demodulated even in the presence of noise and interfering signals with higher spectrum densities than the signal means that the spread signal has resistance to interfering signals and interception.

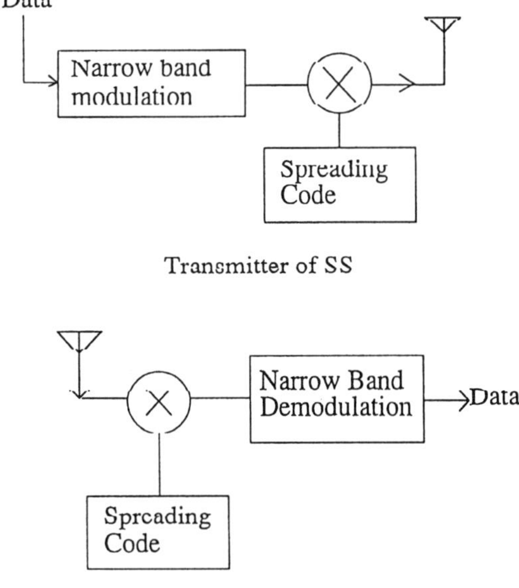

Fig. 2 Spread Spectrum Communication System

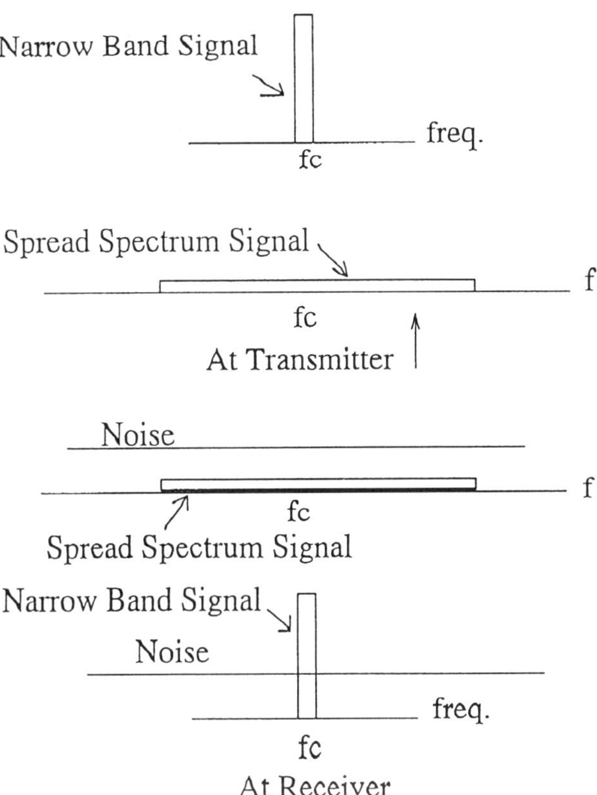

Fig. 3 Spectra schematic at the transmitter and receiver

3. Spread Spectrum Consumer Applications

Consumer Communication is classified into two categories; wire communication and wireless communication.   Here we discuss power line spread spectrum transmission as a wire consumer communication system and data carrier, remote control, ISM wireless LAN and vehicle to vehicle communication and ranging as wireless consumer communication systems.

3.1 Power Line Communication

Power lines which supply electrical power can be used as communication transmission lines without any extra cable construction. However, they have problems in communication performance. i. e., fluctuations of frequency characteristics due to load changes and heavy noises generated by motors and regulators. They are similar to those in fading channels in mobile communication.

The spread spectrum technique has been adopted to solve the above problems in power line communication. The first proposal using SS method was put forward by NEC in 1983[1]. A power line is utilized as a home bus connected with communication terminals, as shown in Fig.4. NEC Home Electronics Co., Ltd. has developed the spread spectrum power line home communication system whose specifications are shown in Table 1.

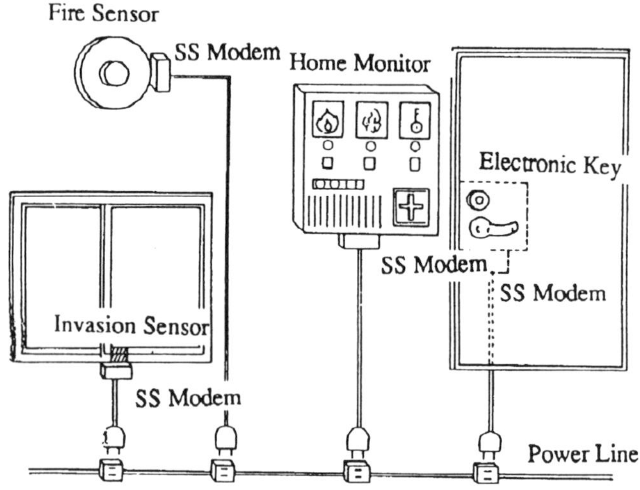

Fig. 4 Home bus system using spread spectrum power line transmission

Table 1 Specifications of a Spread Spectrum power line communication system

Term	Specification
Modulation	DS/SS
Processing Gain	31
Frequency Band	10 kHz - 450 kHz
Bit Rate	9.6 kbps
Acquisition Rate	13 msec Max

Power line communication suffers from severe bandwidth limitation, where the spreading gain has to be reduced to increase data bit rate in the limited bandwidth. A parallel spread spectrum communication system was proposed to increase the bit rate without large reduction of the spreading gain[2]. Fig.5 shows the block diagram and the autocorrelation function of the parallel spread spectrum system.

Although the narrow bandwidth(about 400KHz) limits the number of CDMA(Code Division Multiple Access) users, a CDMA method using a crosscorrelation canceler is proposed to combat with the limitation[3] .

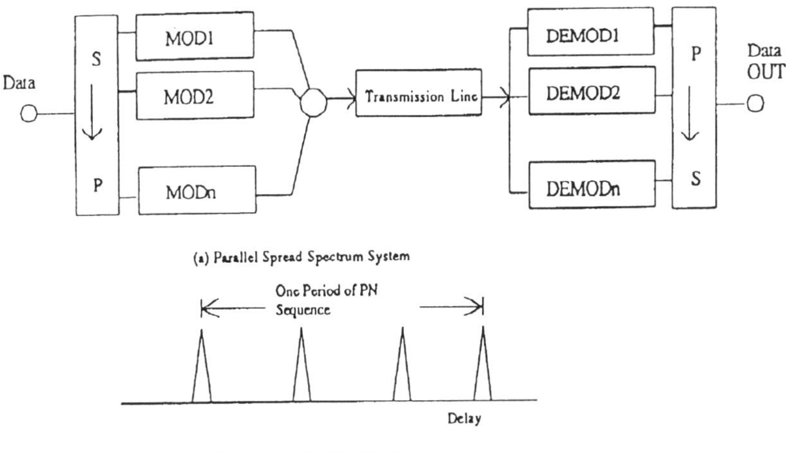

Fig.5 The block diagram and the autocorrelation function of the parallel spread spectrum system

3.2   Data Carrier

Data carrier systems[4] are mainly used in factories where manufactured objects need to be automatically identified. As shown in Fig.6, they usually consist of a controller unit, a read/write(R/W) head unit, and data carrier unit(tag) as illustrated in Fig. 7.

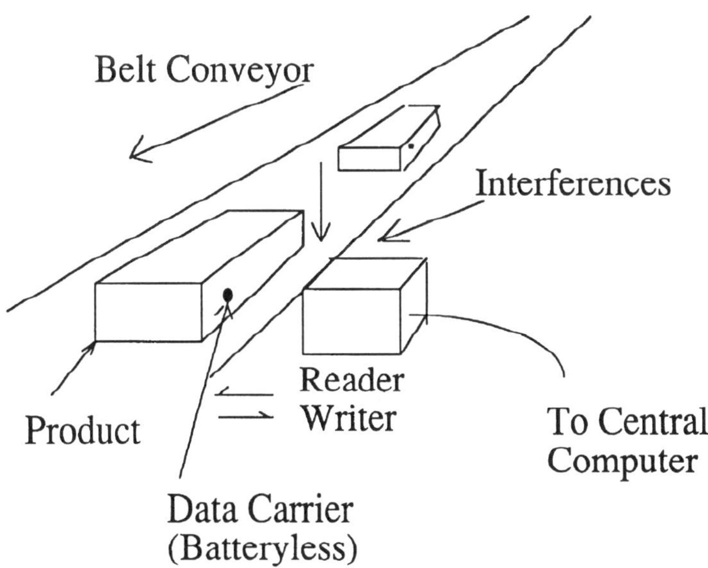

Fig. 6 Data Carrier System in Factory

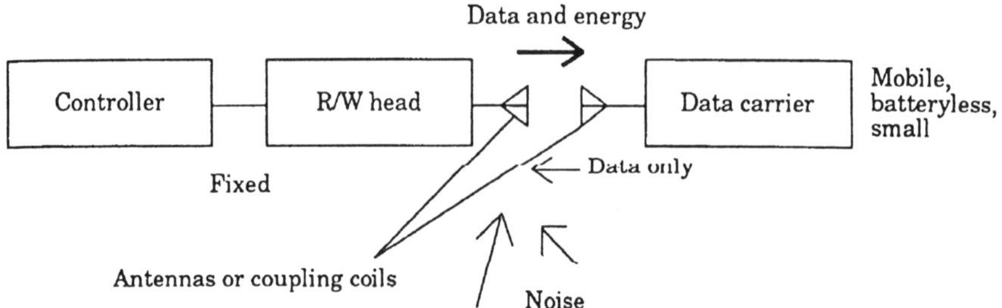

Fig.7 Blockdiagram of Data Carrier System

The controller and R/W units are fixed while the data carrier unit is mobile since it is attached to an object carried on a conveyor belt. Electrical power for transmitting data to the fixed R/W head unit is sent from the fixed head unit through coupling coils. The transmission power is so small that the data transmission is apt to be susceptible to noise such as clock noises generated by processors in the controller and R/W units. Clock noise does not have a uniform spectrum, but a spectrum consisting of lines. Spread spectrum modulation is robust against such noises.

3.3   Radio Remote Control

Radio remote control systems are widely used in factories, construction sites, offices and homes to control robots, cranes, and computers. These systems must be reliable enough, because a data transmission error may cause an accident. Since fading, interference, and noise degrade the transmitted signal of radio remote control systems, a radio remote control system using a DS/FH(Direct Sequence/Frequency Hopping) hybrid modulation was proposed[5]. Table 2 shows the specifications.

Table 2 Specifications of SS Radio Remote Control System

Term	Specification
Trans.Power	Low Power Regulation
FH Bandwidth	5 MHz
Number of Hopping	15(RS Code)
Hopping Frequency	4.88kHz
DS Processing Gain	32
Modulation	FSK(m=1)
Chip Rate	6.4 micro sec/chip
Error Control	1/2 convolutional code Viterbi decoding CRC
DS synch	64 steps matched filter
FH synch.	using DS signal

3.4   ISM Wireless LAN

Spread spectrum communication has been allowed on industrial scientific and medical(ISM) bands (900MHz, 2.4GHz and 5GHz)in the USA since 1985. One of them (2.4GHz) has also been used for spread spectrum communication in Japan since the end of 1992. The maximum transmitted power is 260mW, while the maximum bandwidth is 26MHz. Specifications of an ISM WLAN transceiver are shown in Table 3. This system is designed for factory automation uses.

Table 3 Specifications of ISM WLAN

Modulation	DS-Spread Spectrum
Radio Frequency	2.4 GHz ISM
Output Power	1 mW/MHz
Modulation speed	32 kbps
Bandwidth	less than 26 MHz
Error Control	ARQ + CRC
Size and Weight	210/148/50 mm, 1.5 kg

3.5   Vehicle to Vehicle Communication and Ranging

To avoid accidents and to gain the ability to communicatewith others while driving, a simultaneous communication and ranging system using spread spectrum signals has been proposed [6]). Fig. 8 shows the concept of a system called the "boomerang transmission system". The transmitted spread spectrum signal from one car returns carrying the data modulated by another car. These spread spectrum signals are not only useful for communications, but also for ranging.

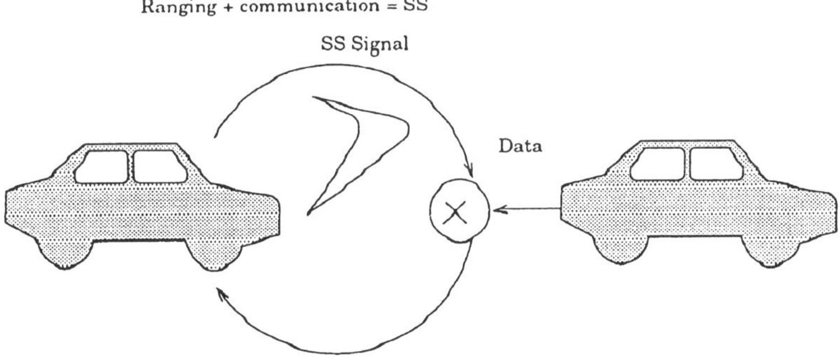

Fig. 8 Boomerang Transmission System Using Spread Spectrum Signal

3.6   TV Broadcasting Spread Spectrum Modulation

NHK(Japan Broadcasting Cooperation) proposes a spread spectrum digital modulation method for TV broadcasting[7]. The image source is encoded into three components, low frequency, middle frequency and high frequency ones. Although the middle and low frequency components are spread by some spreading codes, the high frequency code is not spread by a spreading code for the bandwidth limitation. These three components can be well decoded in a good receiving condition, however, they can not be equally decoded in a bad receiving condition. At least the low frequency component, the most impotant one, survives multipath fading and noise in a bad receiving condition. A graceful degradation with high quality can be done. The block diagram of the method is shown in Fig. 9.

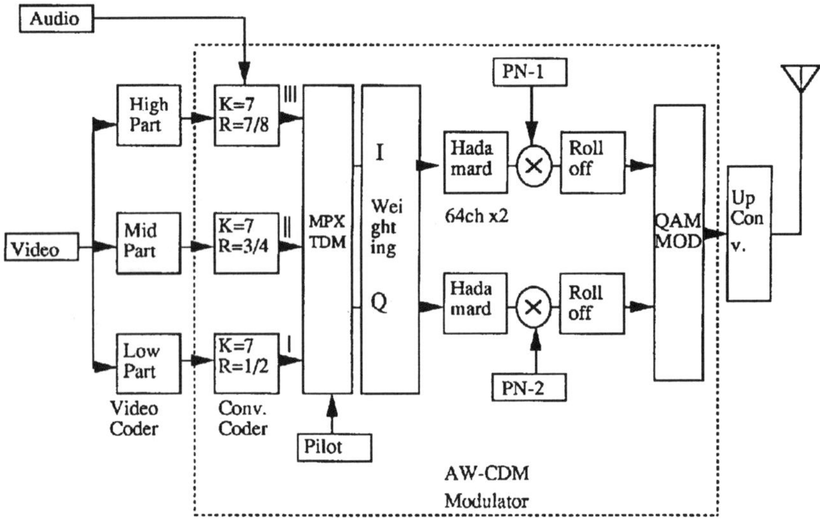

Fig.9 SS TV Modulator

4.   Devices and Systems

4.1   SAW device applications

One of the most difficult problems for a spread spectrum communication system is how to synchronize the receiver to the transmitted signal. The sliding correlator method is the typical and simplest, but needs a long acquisition time to synchronize. Matched filter and convolver methods need much shorter acquisition time and SAW devices are adapted to them.

4.1.1   Matched Filters

One of matched filters sold today shows the specifications in Table 4[8].

Table 4 SAW matched filter specifications

Modulation	DS/SS
PN code	127 chip Msequence
Chip rate	16Mcps
Centre freq.	144MHz
Bit rate	126kbps

4.1.2   Convolver

A SAW matched filter mentioned above utilizes the convolutional effect between the impulse response of the filter and the received pseudo noise (PN) code. However, it is difficult to change one PN code to another PN code because the impulse response depends on the pattern of the electrodes on the SAW filter.   A convolver whose structure with a received input f(t) and external reference input g(t) is shown in Fig. 10 overcome this problem. A received PN code can be changed to another PN code by changing the external reference PN code without any internal structural change. Table 5 shows the SAW convolver specifications.

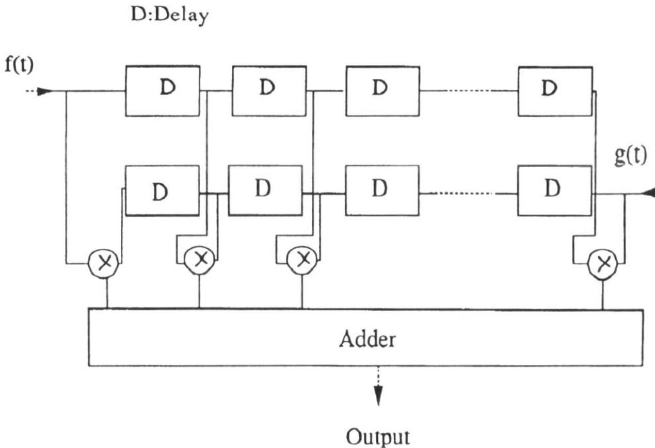

Fig. 10 SAW convolver structure

Table 5 SAW convolver specifications

Input 3 dB Bandwidth	24 MHz
Processing Time	9.0μsec
Time Bandwidth Product	216
Input Centre Freq.	215 MHz
Output Centre Freq.	430 MHz
Terminal Efficiency	-50 dBm
DC Voltage	+12v, 12v

4.2 Digital Signal Processing Applications

Digital signal processors can be adopted in spread spectrum communication systems because of the stability, the variability of parameters, the flexibility, and the expansibility to new functions.

4.2.1 Spread Spectrum Block Demodulator for Packet SS Transmission

Wireless LAN, one of consumer communication, is based on SS packet communication whose packet has to have a long preamble for the initial synchronization as shown in Figure11. The long preamble degrades the efficiency of the packet data transmission if conventional demodulators without memory function are employed.

SS Packet

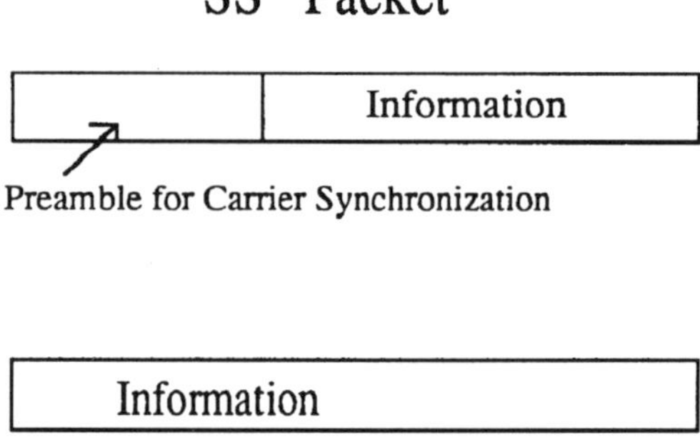

Fig. 11 Preamble in SS Packet

By using an SS block demodulator with a memory block to store a quasi-coherent demodulated packet as shown in Figure 12, the received SS packet can be synchronized without using a preamble[9].

4.2.2 Crosscorrelation Canceller in SS Packet Communication

Crosscorrelations between multiuser limit the multiple user capacity in SS communications. By using a crosscorrelation canceller at a receiver the number of the users can be increased. One of cancellers based on SS block demodulation[10] can increase the number of multiple access packet users in Wireless LAN. The basic concept of the canceller is illustrated in Figure 13.

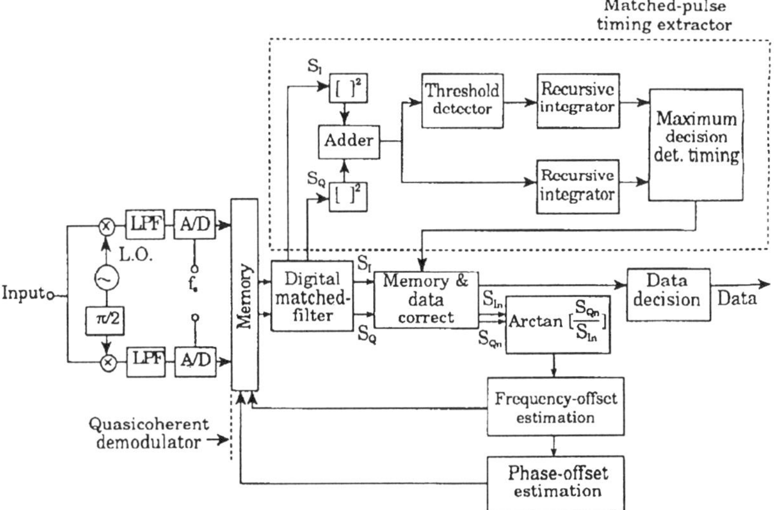

Fig. 12 SS Block Demodulator

Crosscorrelation Cancellation to Increase Capacity

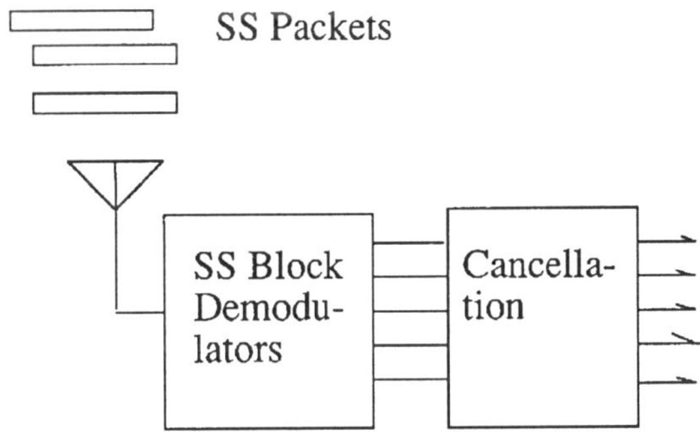

Cancellation at Base Station

Fig. 13 Crosscorrelation Canceller based on SS block Demodulation

4.2.3 Canceller with Unknown PN Code Estimator for Near Far Problem

Near far problem in Direct Sequence spread spectrum communication can be solved by an interference canceller the spreading PN code of a strong near signal is known or the base station power control is given. However, if

294

it is unknown or no power control is given, the problem has not been solved. Such a condition may be often found in consumer communication. A system to cancel an unknown strong near SS signal is proposed. The system first demodulates the strong signal as a wideband PSK signal and gives a baseband signal of the PN code multiplied by data. It can be divided into the PN code and data by an unknown PN code estimator. The spreading PN code is no longer unknown, then we can cancel the strong interfering signal as a usual canceller does. Figure 14 illustrates the system block diagram.

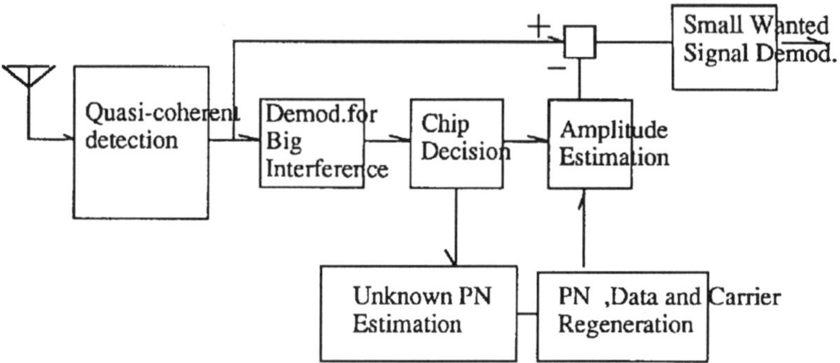

Fig. 14 Canceller with Unknown PN Code Estimator for Near Far Problem

4.2.4 Frequency Synthesizer for Frequency Hopping

It is well known that FH is the best way to avoid near far problem, however, a synthesizer in an FH system is a bottle neck for the system realization due to the hardware complexity. Frequency synthesizers are classified into three types, Multi-Oscillator (MO), DDS(Direct Digital Synthesizer) and PLL (Phase Locked Loop). MO gives fast hopping performance, but the system is too complicated. DDS will be the best solution both for fast hopping and simplicity in the future, but it can not provide high frequencies as GHz today. Although a PLL gives a simplest system, its hopping speed is slow due to its loop filter and loop delay. A new Hybrid (Analog and Digital) PLL synthesizer without loop filter with fast hopping is proposed [12]. The block diagram of the PLL synthesizer is shown in Fig. 15.

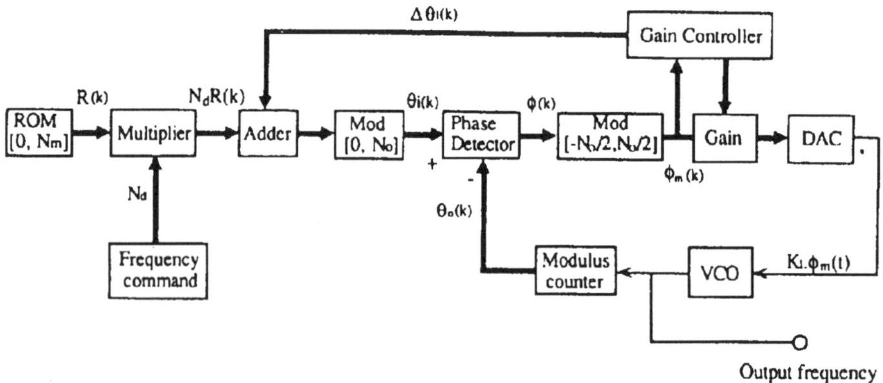

Fig. 15 The block diagram of the fast hopping PLL synthesizer

4.2.5 Time Division Duplex in SS Cellular System

The system concept of cordless telephones is similar to that of cellular mobile phones; base station(BS) and mobile station(MS). Duplex methods for cordless phones are FDD(Frequency Division Duplex) which conventional analog cordless phones employ and TDD(Time Division Duplex). Only FDD can be employed in analog mobile phones because voice data cannot be compressed. TDD will be employed in digital systems whose forward and reverse links take the same frequency band as shown in Fig.16. TDD gives not only simple system structure, but also power control which is important for DS CDMA transmission and Pre-RAKE diversity as shown below.

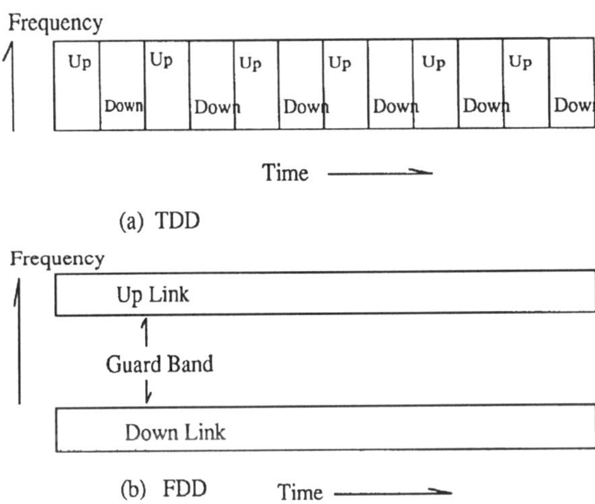

Fig. 16 Frequency bands for TDD(a) and FDD(b)

(a)Power Control using TDD[13]
Received signal powers from mobile stations have to be same at a base station, because different received powers reduce channel capacity in DS CDMA. Reverse link power control can be done in TDD making use of the same frequency band in the forward and reverse lines and short burst signals.

(b)Pre-RAKE using TDD[14]
A Mobile unit hates a complicated receiver like RAKE receiver. Pre-RAKE diversity method prevents the mobile unit from equipping RAKE diversity combination, but makes the base unit to equip RAKE diversity combination at the transmitter. Pre-RAKE diversity at the transmitter have to know the forward link multi-path information in advance which is given by the reverse link multi-path information. Fig. 17 shows the concept of Pre-RAKE in TDD DS CDMA.

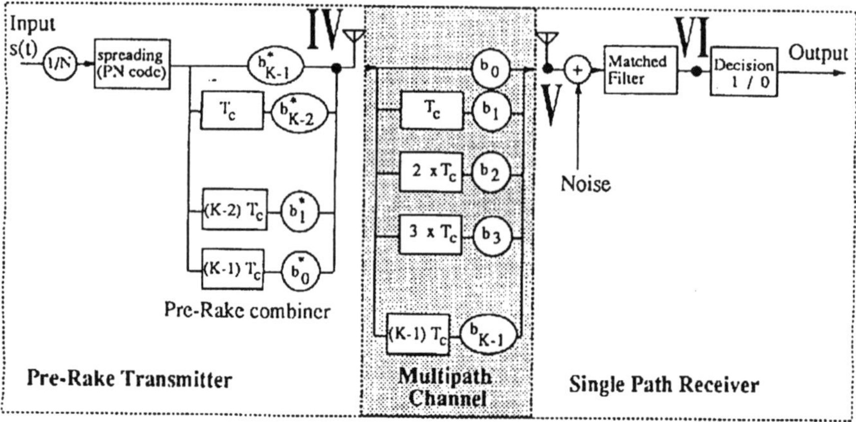

Fig.17 Pre-RAKE in TDD DS CDMA

5.   Conclusion

Consumer Communication will play an important future role and the key will be Spread Spectrum method.   Several examples were shown in this paper. Advanced communication technology will be passed from the military to the commercial and from the public to the individual.

References

[1]   Hirosaki B., Hasegawa S. and Endo K.,"A Power Line Home Bus System using SS Communication," IEEE ICCE(International Conference on Consumer Electronics"(June 1985)

[2]   Kadoi Y.,Sasaki S. and Marubayashi G.,"A Study of Parallel Spread Spectrum Communication System,"IEICE(Institute of Electronics, Information and Communication Engineers) Technical Report(Spread Spectrum), SSTA90-16(March 1990)

[3]   Kohno R., Imai H., Hatori M. and Pasupathy S.,"An Adaptive Canceller of Cochannel Interference for Spread Spectrum Multiple Access Communication Network in a Power Line," IEEE JSAC, Vol.8, No.4, pp.691-699(1990)

[4]   Harada K., Kajiwara A., Takeuchi K. and Nakagawa M.,"Characteristic of Data-Carrier using Spread Spectrum Communication,"IEEE ICCE'90(June 1990)

[5]   Yamamoto M., Hoshikuki A. Kohno R. and Imai H.," An Implementation of R/C System Using DS/FH Hybrid Spread Spectrum Technique," IEICE Technical Report(Spread Spectrum), SSTA90-24(June 1990)

[6]   Mizui K., Uchida M. and Nakagawa M.,"Vehicle to Vehicle Communication and Ranging System Using Spread Spectrum Technique-Proposal of Boomerang Transmission System," IEEE VTC(Vehicular Technology Conference) Report, pp. 335-338(May 1993)

[7]   Hamazumi H., Ito Y. and Miyazawa H.,"An Application of Code Division Multiplexing(CDM) to Digital Broadcasting," IEICE Technical Report(Spread Spectrum) SST 93-100(March 1994)

[8]   Takehara K.,"DS-SS Communication Demodulator using SAW Matched Filter," IEICE Technical Report(Spread Spectrum),SSTA89-30(Aug. 1989)

[9]   Kajiwara A. and Nakagawa M.,"Spread Spectrum Block Demodulator," International Conference on Spread Spectrum Techniques and Applications, London(Sept. 1990)

[10] Kajiwara A. and Nakagawa M.,"Spread Spectrum Block Demodulator with High-Capacity Crosscorrelation Canceller," IEEE Globecom' 91, 25 A4.1-4.5, pp.851-855(Dec. 1991)

[11] Ohkubo M. and Nakagawa M.," Countermeasure against SSMA Near Far Problem Using Estimator for Unknown Sequence," IEEE ISSSTA'94 (Internationa Symposium on Spread Spectrum Techniques and Applications," 1.4(July 1994)

[12] Kajiwara A. and Nakagawa M.,"A New PLL Frequency Synthesizer with High Switching Speed," IEEE Transaction on Vehicular Technology, Vol. 41, No.4, pp.407-413(November 1992) or "A fast PLL synthesizer for fast SS-FH communications," IEEE Globecom'89, Vol. 45.2, pp.1602-1606(Nov. 1989)

[13] Esmailzadeh R. and Nakagawa M.,"Time Division Duplex Method of Transmission of Direct Sequence Spread Spectrum Signals for Power Control Implementation," IEICE Trans. Communications, Vol.E76-B, No.8,pp.1030-1038(August 1993) or "Direct Sequence Spread Spectrum Communication in Selection Diversity Channels by Time Division Duplex Technique," IEEE ISSSTA'92(Nov. 1992)

[14] Esmailzadeh R. and Nakagawa M.,"Pre-RAKE Diversity Combination for Direct Sequence Spread Spectrum Mobile Communications Systems," IEICE Trans. Communications, Vol. E76-B, No. 8pp.1008-1015(Aug. 1993) or "Pre-RAKE Diversity Combination for Direct Sequence Spread Spectrum Communications," Proc. IEEE ICC'93(May 1993)

OPTIMAL POLICIES FOR MULTI-MEDIA INTEGRATION IN CDMA NETWORKS

E. Geraniotis, Y.-W. Chang, and W.-B. Yang,

Abstract

In this paper we describe two policies for efficient use of the spectrum by multi-media users in direct-sequence code-division multiple-access (DS/CDMA) networks. The first policy pertains to CDMA networks with voice traffic and lower priority data traffic; voice and data have different bit error rate (BER) requirements. In this scheme, data traffic can only use the capacity of the CDMA system that is left unused by voice users and gets buffered whenever there are insufficient resources (CDMA codes). The second policy pertains to CDMA networks with voice traffic and two types of data traffic: high priority data traffic with the same priority as voice traffic that requires real-time delivery and lower priority data traffic that can tolerate delay and thus can be queued. In this scheme a movable boundary policy in the CDMA code domain is used for the voice and high priority data, whereas a small number of CDMA codes are reserved for the lower priority data which also utilizes any CDMA codes left unused by the other two traffic types.

Optimal policies for the two above formulations are obtained by minimizing cost functions consisting of the rejection rate of voice traffic for the former case and of the weighted sum of the rejection rates of voice and high priority data traffic for the latter case, subject to performance requirements on the BERs of all traffic types. The queueing delay of the lower priority data traffic is also evaluated. A semi-Markov decision process (SMDP) with guaranteed BERs for voice and data traffic is used for formulating the dynamic code assignment problem. Value-iteration algorithms are applied to this SMDP to derive the optimal policies.

1 Introduction

Code-division multiple-access (CDMA) techniques find today many commercial applications beyond their traditional use in military communications. Cellular systems, mobile satellite networks, and personal communication networks (PCN) that use CDMA have been proposed and are currently under design, construction, or deployment [1]-[3]. Moreover, networks of LEO (Low Earth Orbit) and MEO (Medium Earth Orbit) satellites for world-wide (global) communications such as Loral/Qualcom's

S.G. Glisic and P.A. Leppänen (eds.), Code Division Multiple Access Communications, 299-329.
© 1995 *Kluwer Academic Publishers. Printed in the Netherlands.*

Globalstar and TRW's Odyssey that use CDMA have been proposed and are under design [5]-[6].

In our recent work of [7] we described and analyzed schemes for voice and data integration in CDMA networks that use for voice traffic a threshold-based admission policy and for data traffic the ALOHA protocol with retransmissions based on feedback about the network state (number of ongoing voice calls and data messages). In [8] we modified and extended the schemes of [7] to a hybrid satellite/terrestrial network using framed ALOHA with reservations on-board the satellite and the CDMA scheme of [7] for the terrestrial network. In both of these schemes voice has priority over data and data traffic is using whatever resources (CDMA codes or slots in framed ALOHA) were left unused by voice traffic. However, in the work of [7]-[8] there is neither any attempt to allocate the resources (CDMA codes) to the voice and data users in an optimal manner nor any attempt for a fair treatment of voice and data users.

According to the first policy, voice always has priority over data; thus data users can only transmit when some CDMA codes have been left unused by the voice users. Moreover, data packets are buffered whenever no CDMA codes are available. The voice and data bit error rate (BER) requirements can be different. This scheme is suitable for accommodating some data traffic in a CDMA network offering primarily voice service, provided that the data traffic does not require real-time delivery.

According to the second policy, voice and two types of data traffic can be accommodated. The first type of data traffic has the same priority as the voice users, that is, it requires real-time delivery and it also requires lower BER than voice. The second type of data has lower priority, it can tolerate delays and thus it can be buffered; the required BER is larger or equal to that of the high priority data but definitely smaller than the one required for the voice traffic. This second scheme is suitable for a truly multi-media CDMA network.

For the first policy we present an optimal allocation scheme that determines the number of newly arrived voice calls that are accepted in the network so that the long-term blocking (rejection) rate of voice calls is minimized and the packet error probabilities of voice traffic remains within acceptable limits. The unused CDMA capacity is be used by data traffic and the reamining data traffic is queued. We consider two models for the effects of other-user interference: the threshold model and the graceful degradation model. Although we refer to our schemes as code allocation schemes, they are is actually admission policies, since they determine how many voice (or data) users can be admitted in the system (and allocated CDMA codes) from the total number of users arriving at the system (and requesting services).

For the second policy we derive an optimal code allocation scheme that determines the number of newly arrived voice calls and data users with high priority that are accepted in the network so that the long-term weighted blocking rates of voice calls and data traffic is minimized and the packet error probabilities of these two traffic types are within acceptable limits. The activity characteristic of voice traffic is considered for increasing bandwidth efficiency. If the new arrivals are not accepted, they are blocked and there are no CDMA codes assigned to these arrivals. For the lower priority data we consider two policies. According to the first policy there are no CDMA codes reserved for these data, they get assigned CDMA codes only when the combined voice and high priority data traffic leaves certain codes unused; this is done in a pre-emptive manner, when voice or high priority data requests arrive the codes are assigned back to them and the low priority data are queued. The second policy operates like the first except that there is also a small number of CDMA codes that are always assigned to low priority traffic; if there is a large demand from low priority data traffic the reserved codes are assigned first and the remaining low priority data packets use the codes left unused by the voice or high priority data traffic or, if there is not a sufficient number of them, they are queued. For both schemes the BER requirement for the low priority data traffic is met.

The performance measures are the average blocking rates and average throughputs of the voice calls and all data messages as functions of the offered voice and data traffic loads under the proposed optimal code allocation policy. The queueing delay and the packet loss probability of the low priority data traffic is also evaluated. A semi-Markov decision process (SMDP) with guaranteed BERs for voice and data traffic is used for formulating the system operation as a dynamic code assignment problem. A value-iteration algorithm is applied to this SMDP to derive the optimal policy.

The paper is organized as follows. In Section 2, the system and CDMA other-user interference models are described. In Section 3, the performance analysis of a direct (nonoptimal) admission policy for voice is carried out. Then in Section 4, the optimal code allocation policies for the voice-only scenario as well as for voice and high priority data traffic are derived. In Section 5, performance analyses of the optimal admission policies for voice and high priority data traffic are conducted followed by the performance analysis of the low priority data traffic in Section 6. Finally, numerical results and conclusions are presented in Sections 7 and 8, respectively.

2   System Model

The CDMA network may have any generic architecture. The basic ingredients of our work are applicable to terrestrial (cellular, PCN) or satellite networks. Figure 1 shows an example of a CDMA network with voice and data traffic.

The voice call or data message of each user admitted in the network is packetized with the same fixed length packet. Time is divided into slots of duration equal to the transmission of one packet. In our models, packet transmissions start at common clock instances and packets have constant length. The typical packet length is 1000 to a few thousand bits.

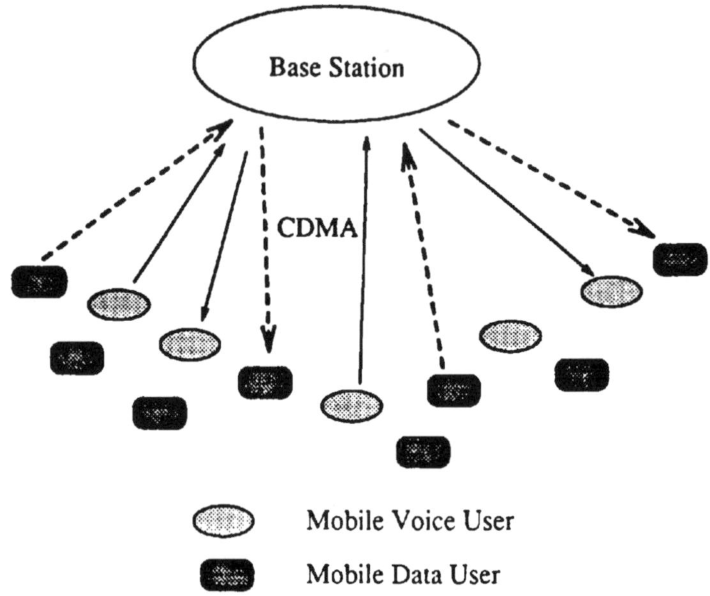

Figure 1. Model for Mobile CDMA System

2.1   Model for Voice Traffic

Let N_v be the total population of voice traffic in the CDMA system. The traffic generated by each voice user is modeled as a three-state (idle, silent, and talkspurt) discrete-time Markov processes each with transition probabilities p_{01}^v, p_{10}^v, p_{12}^v and p_{21}^v, as shown in Figure 2a. Figure 2b is a two-state Markov chain, which is the voice model for the system with different priorities of voice and data users. Thus, the steady-state probability of k active voice users is

$$P\{k \text{ active voice users}\} = \binom{N_v}{k}\left(p_{active}^v\right)^k\left(1 - p_{active}^v\right)^{N_v - k} \tag{1}$$

where

$$p_{active}^v = \frac{p_{01}^v}{p_{01}^v + p_{10}^v}, \, p_{idle}^v = 1 - p_{active}^v \tag{2}$$

The mean duration of the idle and active periods of voice traffic are $1/p_{01}^v$ and $1/p_{10}^v$.

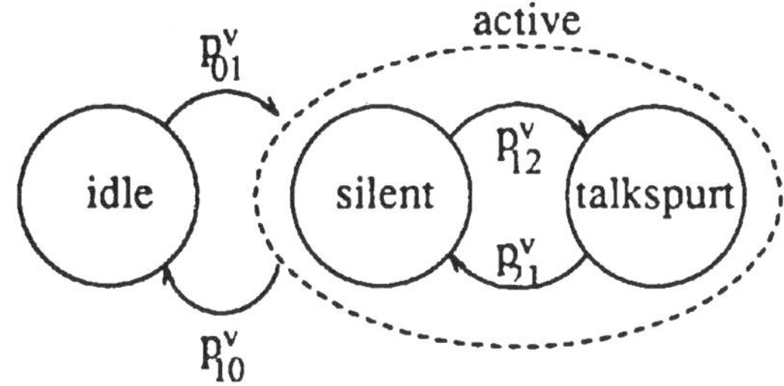

Figure 2a. Model for Voice Users

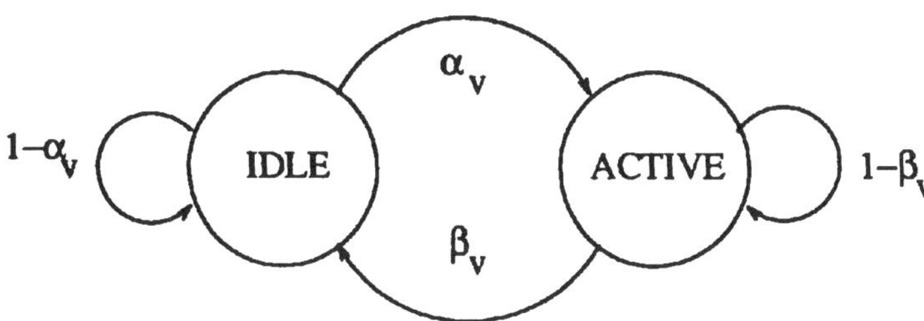

Figure 2b. Model for Voice Users

2.2 Model for Data Traffic

Let N_{d_1} and N_{d_2} be the total population of two types of data traffic. Each data user of high priority is modeled by a two-state (OFF/ON or idle/active) Bernoulli process, shown in Figure 3 with transition probabilities p_{01}^d and p_{10}^d. Similarly, the mean duration of the idle and active periods of data users are $1/p_{01}^d$ and $1/p_{10}^d$. For the low priority data traffic we assume the number of arriving data packets follows a Poisson distribution with mean rate λ_d. There is some justification for using different models for the high and low priority data users. The high priority data users are treated in the same manner as voice users and the finite population assumption is essential in deriving the optimal allocation strategy. On the other hand the low priority data users are assigned only the codes left unused by all other traffic (voice and data), no optimization takes place, and a Poisson population model simplifies our analysis and admission policy; we could also use a Bernoulli model for the low priority users, but that will complicate matters. Further, when the population of data users is large or when they have packets to transmit with small probability, the binomial model approaches the Poisson model. Finally, we assume that the data streams of all voice and data users are statistically independent.

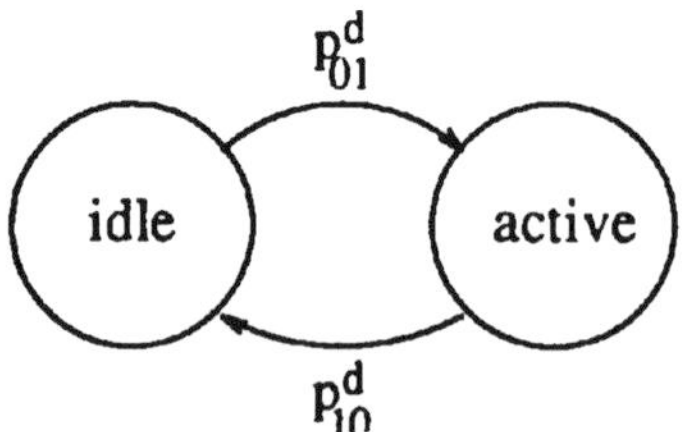

Figure 3. Model for Data Users

2.3 CDMA System and Interference Models

Direct-sequence code-division multiple-access (DS/CDMA) is employed by all users in the network. The same frequency band is shared by voice and data traffic. Voice and data traffic have the same data rate; thus the same pool of CDMA codes is used. Each user (data or voice node) employs a distinct code for the transmission of its packets.

We define the multiple-access capability (MAC) index K_v as the number of voice users that can be accommodated simultaneously, such that the expected packet error probability of voice traffic remains below a specified threshold. Similarly, the MAC index K_{d_1} and K_{d_2} for data users is the number of data users that can transmit simultaneously with a tolerable packet error. Thus we have

$$P_E(k) \le P_E^v, \forall\, k \le K_v \tag{3}$$

$$P_E(k) \le P_E^{d_1}, \forall\, k \le K_{d_1} \tag{4}$$

$$P_E(k) \le P_E^{d_2}, \forall\, k \le K_{d_2} \tag{5}$$

where $P_E^v, P_E^{d_1}$ and $P_E^{d_2}$ are the maximum tolerable voice and data packer error probabilities, respectively, and $P_E(k)$ is the packet error probability in the presence of k simultaneous packet transmissions [?], where k includes both voice users and active data users. In practice, $P_E^v > \max\left\{ P_E^{d_1}, P_E^{d_2} \right\}$, and, therefore, $K_v > \max\left\{ K_{d_1}, K_{d_2} \right\}$. Here we also assume $P_E^{d_1} \le P_E^{d_2}$, so that $K_{d_1} \le K_{d_2}$; this assumption can be easily relaxed and any relationship between $P_E^{d_1}$ and $P_E^{d_2}$ can be accommodated. If the total number of simultaneous users is $k \le K_{d_1}$ then all k data or voice packets are received with acceptable error probabilities. If $K_{d_1} < k \le K_{d_2}$, then, among the k packets, the voice packets and the data packets with lower priority are received with acceptable error probabilities, whereas the data packets with higher priority are received with error probabilities higher than the acceptable value $P_E^{d_1}$ If $K_{d_2} < k \le K_v$ then only the voice packets are received with acceptable error probabilities. Finally, if $K_v < k$, then all voice or data packets are received with unacceptable error probabilities. The model described above is referred to as the threshold model; its usefulness and simplicity lies in characterizing acceptable operating conditions in terms of maximum allowable numbers of simultaneous users. Shown in Figure 4 are the boundaries (moving) for allocation of CDMA codes to voice and low priority data users.

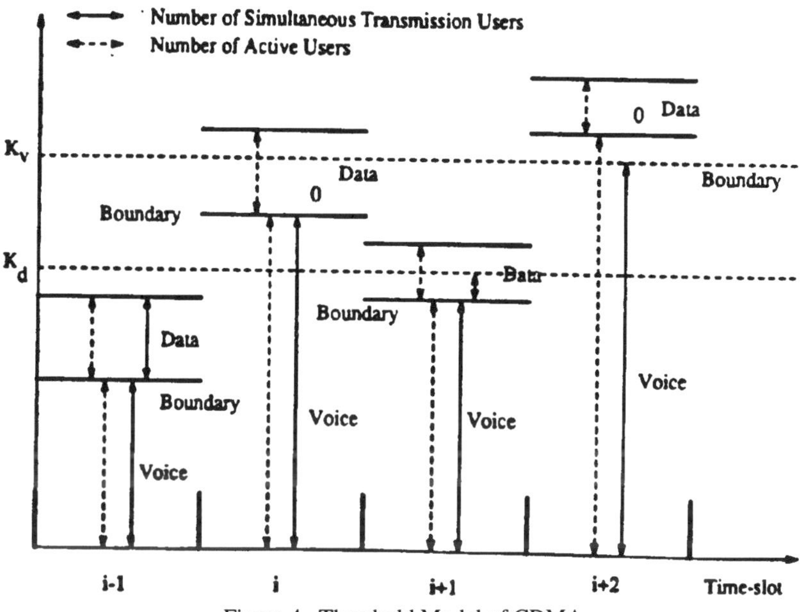

Figure 4. Threshold Model of CDMA

For the multiple-reception of data packets both the threshold model and the graceful degradation model can be used. According to the latter model, there is a non-zero probability of correct reception for any arbitrary number of packets (even for numbers exceeding the MAC index), which depends on the total number of simultaneously transmitted packets (data and voice) in the CDMA network. This takes the form

$$P_{gdm}\left(l \text{ succesful data packets} \,|\, m \text{ voice and } n \text{ data packets are transmitted}\right)$$

$$\approx \begin{cases} \binom{n}{l}\left[1 - P_E(m+n)\right]^l \left[P_E(m+n)\right]^{n-l}, & \text{if } 0 \le l \le n \\ 0, & \text{elsewhere} \end{cases}$$

$$= P_{gdm}\left(n, l, P_E(m+n)\right)$$

(6)

where P_E is the standard probability of packet error for a direct-sequence CDMA system. The above approximation has been justified in our work of [9] for frequency-hopped spread-spectrum systems (FH/SSMA) and [10] for direct sequence DS/SSMA systems.

If binary phase-shift keying (BPSK) with coherent demodulation and binary convolutional codes with rate 1/2 and constraint length 7 and Viterbi decoding and hard decisions is used, one can evaluate an upper bound on the packet error probability of a single-receiver system $P_E(k)$. If N is the number of chips per bit and k the number of simultaneous DS/CDMA signals that are both time- and phase-synchronous; then the symbol error probability ρ under a Gaussian noise channel from [11]

$$\rho \approx Q\left(\left[\frac{N_0}{E_b} + \frac{(K-1)}{N}\right]^{-1/2}\right) \tag{7}$$

where E_b is the energy per information bit in the received signal, $N_0/2$ is the two-sided spectral density of Gaussian noise, and

$$Q(x) = \frac{1}{\sqrt{2\pi}} \int_x^\infty exp\left(-\frac{u^2}{2}\right) du. \tag{8}$$

¿ From ρ we can obtain an upper bound of packet error probability $P_E(k)$ from (3.29) and (4.1) of [11] and Table 1 of [12]. Then the upper bound for $P_E(k)$ is plugged in the inequalities involving $P_E^v, P_E^{d_1}$ and $P_E^{d_2}$ above to obtain the MAC indices K_v, K_{d_1} and K_{d_2}, when the BER (and thus the packet error rate) specifications for voice and the two data traffic types are given.

Since each active voice user is in either silent or talkspurt mode, the number of codes which are assigned to active users can be larger than K_v while still guaranteeing the same voice-packet error probability. This is because, even though the number of total active voice users may be larger than K_v, if the number of voice users in talkspurt is less than K_v, the voice packet error probability will still be less than P_E^v.

For the model of two priority data, voice and high priority data packets share all but a small number $D2$ (where $D2 < K_{d_2}$) of the available CDMA codes according to a dynamic movable boundary allocation scheme. $D2$ codes are always allocated to low priority data users and the remaining low priority data users either use the CDMA codes (if any) left unused by voice and high priority data traffic or their packets are queued.

Given the MAC indices K_v and K_{d_1}, we may compute the maximum number of acceptable voice calls and data users as follows. Let V_t and $D_t^{(1)}$ be the number of active voice and data users at the t-th time-slot. Then the values of V_t and $D_t^{(1)}$ must satisfy the following inequalities

$$\begin{cases} 0 \le D_t^{(1)} \le K_{d_1} - D_2 \\ 0 \le V_t \le M_1\left(D_t^{(1)}\right) \end{cases} \tag{9}$$

where $M_1\left(D_t^{(1)}\right)$ is the maximum number of active voice calls when the number of active data users is $D_t^{(1)}$. The reason that we determine the bound on active voice calls from the current state of high priority data traffic, is that the data have more stringent requirements on BER (recall that $P_E^{d_1}, P_E^{d_2} < P_E^v$). The value of $M_1\left(D_t^{(1)}\right)$ is obtained by ensuring that the probability that the number of active calls in talkspurt (out of $M_1\left(D_t^{(1)}\right)$)) is larger than $K_{d_1} - D_t^{(1)}$ - $D2$ remains smaller than a prespecified acceptable value δ, namely,

$$M_1\left(D_t^{(1)}\right) = \arg \max_M \left\{ \sum_{i=K_{d_1}-D_t^{(1)}-D2+1}^{M} \binom{M}{i} [p_v]^i [1-p_v]^{M-i} < \delta \right\} \tag{10}$$

This is equivalent to stating that under the conditions **??** and 10 the data packet error probability remains smaller than $P_E^{d_1}$ and the voice packet error probability remains smaller than P_E^v with confidence (probability) $1-\delta$.

In 10 p_v denotes the probability that each active voice user is in talkspurt mode, it turns out that

$$p_v = \frac{p_{12}^v}{p_{12}^v + p_{21}^v} \tag{11}$$

Typical values of the transition probabilities are [13]

$$p_{21}=0.00833, \quad p_{12}=0.00556 \tag{12}$$

Therefore, a typical value of p_v is 0.4.

The function $M_1\left(D_t^{(1)}\right)$ of 10 is non-increasing and the maximum number of active voice calls or active data users is given by $M_1\left(D_t^{(1)} = 0\right)$. Thus, the maximum number of CDMA codes to be assigned is $M_1\left(D_t^{(1)} = 0\right)$. However, as the value of $D_t^{(1)}$ increases, $M_1\left(D_t^{(1)}\right)$ decreases dramatically. Consequently, the number of available codes for voice traffic must decrease to meet the performance requirement $P_E^{d_1}$ of data. If $M_1\left(D_t^{(1)} = 0\right) = K_v / p_v$, which is generally used for calculating the maximum number of active voice users in the absence of data traffic, then the packet error probability of data exceeds the prespecified value with unacceptably high probability.

3 Voice Traffic Analysis with Direct Admission Policy

According to the direct admission policy analyzed in this section, if a new voice call arrives and there is an available code, then the new voice call is accepted immediately and can transmit its first packet in the beginning of the next time-slot; otherwise, the new voice call is rejected (blocked). That is, if there are i active voice calls and the MAC index is K_v, we accepted only up to $K_v - i$ new voice arrivals; thus if there are more than $K_v - i$ new arrivals, the surplus (above $K_v - i$) will be blocked (denied access to the system).

Let p_{ij}^v be the transition probability from the current state $s^v = i$ to the next state $s^v = j$ and π^v be the steady-state probability of being at state i, where $0 \le i, j \le K_v$. Thus, we have a set of linear equations

$$\begin{cases} \pi_j^v = \sum_{i=0}^{K_v} \pi_i^v p_{ij}^v \\ \sum_{i=0}^{K_v} \pi_i^v = 1 \end{cases} \tag{13}$$

for which the transition probabilities, if we use the above admission scheme to manage the voice traffic, are given by

$$p_{ij}^v = \sum_{l=0}^{N_v - i} b\left(i, i + \min\{l, K_v - i\} - j, p_{10}^v\right) \cdot b\left(N_v - i, l, p_{01}^v\right) \tag{14}$$

where $b(M,m,p)$ denotes the binomial distribution with parameters M and p, where $0 \le p \le 1$.

$$b(M, m, p) = \begin{cases} \binom{M}{m} p^m (1-p)^{M-m}, & \text{if } 0 \leq m \leq M \\ 0, & \text{elsewhere} \end{cases} \quad (15)$$

where $b(0, 0, p) = 1$. This equation is a direct consequence of the Markov model of voice traffic in Figure 1, the MAC index K_v representing the upper bound for the number of calls in the system, and the direct admission policy.

After we obtain the steady-state probability distribution by solving the above set of linear equations, we can compute the probability of blocking voice calls under the direct admission policy as follows:

$$P_B^v = \frac{\sum_{i=0}^{K_v} \sum_{k=K_v-i+1}^{N_v-i} (k - K_v + i) \pi_i^v b\left(N_v - i, k, p_{01}^v\right)}{\sum_{i=0}^{K_v} \sum_{k=1}^{M_v-i} k\, \pi_i^v b\left(N_v - i, k, p_{01}^v\right)} \quad (16)$$

4 An SMDP Approach to Optimal Admission Control Policies

The goal of the optimal call admission policy minimize the long-run average cost per unit time. A policy that rejects certain voice calls at some instance of time (which would otherwise be accepted) may admit more calls on the average than a policy that always accepts calls, whenever there are available channels.

In order to find an optimal policy (in some practical sense) for the admission control problem we introduce a semi-Markov decision process (SMDP). A Markov decision process does not suffice here because the times between consecutive decision epochs are not identical but random.

4.1 Optimal Admission Policy for Voice Traffic

Let the state space of the CDMA network be

$$I = \left\{ x_t = (V_t, v_t) \mid 0 \leq V_t \leq K_v, 0 \leq v_t \leq N_v - V_t \right\} \quad (17)$$

where V_t is the number of the currently active voice calls, whereas v_t is the number of new voice calls. The action space is $A^{(v)} = I \times A_t^{(v)}$, where $A_t^{(v)} = \left\{0, 1, 2, \ldots, \min\left(v_t, K_v - V_t\right)\right\}$. It is assumed that a central controller has access to this information about the currently active (ON) and the arriving voice calls.

Let the action vector be a_t then $a_t \in A_t^{(v)}$ which corresponds to the state x_t, where $x_t \in I$. Here the service completion epochs are introduced as fictitious decision epochs in addition to the real decision epochs, which are the arrival epochs of voice calls. Clearly, the actions a_t is equal to 0 on fictitious decision epochs, if there are no new arrivals at the epoch slot.

The transition probabilities that at the next epoch the system will be in state j, if action a_t is chosen at the present state i, are the following:

$$P_{ij}(a_t) = P\left[j = \left(\overline{V}_t, \overline{v}_t\right) \mid i = \left(V_t, v_t\right), a_t \right]$$
$$= b\left(V_t, V_t + a_t - \overline{V}_t, p_{10}^v\right) \cdot b\left(N_v - V_t - a_t, \overline{v}_t, p_{01}^v\right) \tag{18}$$

The relationship follows from the fact that the events of the various voice users becoming active or turning idle are independent.

Let $C(x_t, a_t)$ be the expected cost incurred until the next decision epoch, if action at is selected at the present state x_t. The cost (the number of rejected calls) at the state x_t is defined as follows:

$$C(x_t, a_t) = v_t - a_t \tag{19}$$

In addition, the expected time until the next decision epoch, if action a_t is selected at the present state x_t, is $\tau(x_t, a_t)$ in packet-slots, where

$$\tau(x_t, a_t) = \left[\left\{ 1 - \left(1 - p_{01}^v\right)^{N_v - V_t - a_t} \cdot \left(1 - p_{10}^v\right)^{V_t + a_t} \right\}^{-1} \right] \tag{20}$$

Here it is assumed that $\tau(x_t, a_t) > 0$, for all states and actions. In the above expression the entity in { } represents the probability of no new arrival and no new service completion at the current system state (x_t, a_t) in any packet-slot; then 1/{ } stands for the mean of the geometrically distributed number of time-slots that the system remains unchanged (holding time).

The value-iteration algorithm [14] of a semi-Markov decision model is applied to derive the optimal admission policy as described below.

<u>step 0:</u> Choose $V_0(x_t)$ and τ such that $0 \leq V_0(x_t) \leq \min_{a_t}$
$\{C(x_t, a_t) / \tau(x_t, a_t)\}$, for all x_t and $0 \leq \tau \leq \min_{x_t, a_t}(x_t, a_t)$, for all x_t and a_t. Let $n := 1$.

step 1: Compute the recursive function of $V_n(x_t)$ for all x_t, from

$$V_n(x_t) = \min_{a_t \in A_t} \left\{ \frac{C(x_t, a_t)}{\tau(x_t, a_t)} + \frac{\tau}{\tau(x_t, a_t)} \cdot \sum_{\bar{x}_t} P_{x_t \bar{x}_t}(a_t) V_{n-1}(\bar{x}_t) \right.$$

$$\left. + [1 - \frac{\tau}{\tau(x_t, a_t)}] V_{n-1}(x_t) \right\}$$

and determine $R(n)$ as a stationary policy, actions maximize the right-hand side of the equation.

step 2: Compute the bounds

$$l_n = \min_{x_t} \left\{ V_n(x_t) - V_{n-1}(x_t) \right\} \tag{21}$$

and

$$L_n = \max_{x_t} \left\{ V_n(x_t) - V_{n-1}(x_t) \right\} \tag{22}$$

The algorithm terminates and outputs policy $R^v(n)$, (depending on V_t, v_t) when $0 \le (L_n - l_n) \le \varepsilon l_n$ where ε is a prespecified bound on the relative error accuracy). Otherwise, go to *step 3*.

step 3: $n := n+1$ and go to *step 1*.

¿From this algorithm, we obtain an optimal control scheme R^v for managing voice traffic. The optimal action a_t corresponding to the state x_t is obtained according to the policy R^v.

4.2   Optimal Admission Policy for Voice and High Priority Data

In this section, we find an optimal policy for code allocation, that is a sequence of allocation functions

$$\begin{cases} F*(V_t, D_t^{(1)}, v_t, d_t^{(1)}, t), \ t = 0,1,2,\ldots, \\ G*(V_t, D_t^{(1)}, v_t, d_t^{(1)}, t), \ t = 0,1,2,\ldots, \end{cases} \tag{23}$$

which minimizes an average cost under the constraints that the mean packet error probabilities of voice and data traffic remain below prespecified values. The average cost is defined as the weighted sum of the rejection (or blocking) rates of voice and high priority data traffic. By calibrating the weights in this sum we can assign any desirable priority to voice and data traffic. The value of $F*(\cdot,\cdot,\cdot,\cdot,\cdot)$ is the action a_t^v for voice traffic which represents the number of codes which are assigned to newly active voice calls. Similarly, the value of $G*(\cdot,\cdot,\cdot,\cdot,\cdot)$ is the action $a_t^{d_1}$ for data traffic which represents the number of codes which are assigned to newly active data users.

Let S be the state space of the code allocation problem, then we have

$$S = \left\{ z_t = (V_t, D_t^{(1)}, v_t, d_t^{(1)}) \mid 0 \leq D_t^{(1)} \leq K_{d_1} - D2, \ 0 \leq V_t \leq M_1(D_t^{(1)}), \right.$$
$$\left. 0 \leq v_t \leq N_v - V_t, 0 \leq d_t^{(1)} \leq N_{d_1} - D_t^{(1)} \right\} \tag{24}$$

where v_t and $d_t^{(1)}$ are the number of newly active voice and data users at the t-th time-slot, respectively.

The cost function is

$$C(z_t, a_t) = w_v \left(v_t - a_t^v \right) + w_d \left(d_t^{(1)} - a_t^{d_1} \right), \tag{25}$$

where w_v and w_d are relative weighting constants. The action subspace is given by

$$A(z_t) = \left\{ \left(a_t^v, a_t^{d_1} \right) \mid 0 \leq a_t^v \leq v_t, \ 0 \leq a_t^{d_1} \leq d_t^{(1)}, 0 \leq D_t^{(1)} + a_t^{d_1} \leq K_{d_1} - D2, \right.$$
$$\left. 0 \leq V_t + a_t^v \leq M_1 \left(D_t^{(1)} + a_t^{d_1} \right) \right\} \tag{26}$$

The transition probabilities that at the next epoch the system will be in state $\bar{z}_t$, if action a_t is selected at the present state $\bar{z}_t$ are the following:

$$P_{z_t \bar{z}_t}(a_t) = P\left\{ \bar{z}_t = (\overline{V}_t, \overline{D}_t^{(1)}, \overline{v}_t, \overline{d}_t^{(1)}) \mid z_t = (V_t, D_t^{(1)}, v_t, d_t^{(1)}), a_t = \left(a_t^v, a_t^{d_1} \right) \right\}$$
$$= b\left(V_t, V_t + a_t^v - \overline{V}_t, p_{10}^v \right) \cdot b\left(D_t^{(1)}, D_t^{(1)} + a_t^{d_1} - \overline{D}_t^{(1)}, p_{10}^d \right)$$
$$\cdot b\left(N_v - V_t - a_t^v, \overline{v}_t, p_{01}^v \right) \cdot b\left(N_{d_1} - D_t^{(1)} - a_t^{d_1}, \overline{d}_t^{(1)}, p_{01}^d \right), \tag{27}$$

In addition, the expected time until the next decision epoch, given that action a_t is followed at the present state $\bar{z}_t$ is

$$\tau(z_t, a_t) = \left\lceil \left\{ 1 - \left(1 - p_{01}^v\right)^{N_v - V_t - a_t^v} \cdot \left(1 - p_{10}^v\right)^{V_t + a_t^v} \cdot \right.\right.$$
$$\left.\left. \left(1 - p_{01}^d\right)^{N_{d_1} - D_t^{(1)} - a_t^{d_1}} \cdot \left(1 - p_{10}^d\right)^{D_t^{(1)} + a_t^{d_1}} \right\} - 1 \right\rceil \tag{28}$$

Similarly, the value-iteration algorithm is applied to derive the optimal policy.

Implementation Issues

The optimal code allocation policy derived in this section is fairly complicated. It appears to lack a simple structure; for example we can not determine the optimal action by comparing the components of the state $\bar{z}_t$ to proper thresholds and/or to each other. Moreover, the derivation requires multi-dimensional optimization and is rather time-consuming to perform on-line. We are pursuing two solutions to this problem. The first is to construct a look-up table describing the optimal policy; that is, a table listing all possible states $z_t = \left(V_t, D_t^{(1)}, v_t, d_t^{(1)}\right)$ and the corresponding optimal actions $a_t = \left(a_t^v, a_t^{d_1}\right)$. This table is constructed by performing the necessary optimization computations off-line for the most likely set of system parameters. Excessive memory size will be the problem here for large systems. The alternative is to derive near-optimal or sub-optimal policies that possess structure and approximate closely the performance of the optimal policy; these will still use some parameters and small-size look-up tables derived off-line but a substantial fraction of the computations for determining the proper action given the network state will take place on-line.

5 Performance Analysis of Optimal Admission Policies

In this section we analyze the performance of the above two optimal policies. For the optimal voice admission policy, the blocking probability is derived and compared with the direct admission policy. The performance measures of voice and high priority data traffic include the blocking rates and throughputs of voice and data traffic.

To analyze the performance, we have to solve the following set of linear equations

314

$$\begin{cases} \pi\left(\overline{x}_t = \left(\overline{V}_t, \overline{D}_t^{(1)}\right)\right) = \sum_{V_t} \sum_{D_t^{(1)}} \pi\left(x_t = \left(V_t, D_t^{(1)}\right)\right) P_{\overline{x}_t|x_t} \\ \sum_{V_t} \sum_{D_t^{(1)}} \pi\left(x_t = \left(V_t, D_t^{(1)}\right)\right) = 1 \end{cases} \tag{29}$$

where $\pi\left(V_t, D_t^{(1)}\right)$ is the steady state probability distribution at the state $x_t = \left(V_t, D_t^{(1)}\right)$ and the transition probability $P_{\overline{x}_t|x_t}$ under the optimal action $a_t = (a_t^v, a_t^{d_1})$ under the optimal policy φ, is given by

$$P_{\overline{x}_t|x_t} = P\left\{\overline{x}_t = \left(\overline{V}_t, \overline{D}_t^{(1)}\right) \,\Big|\, x_t = \left(V_t, D_t^{(1)}\right), a_t = \left(a_t^v, a_t^{d_1}\right)\right\}$$

$$= \sum_{v_t=0}^{N_v-V_t} \sum_{d_t^{(1)}=0}^{N_{d_1}-D_t^{(1)}} P_\varphi\left\{\overline{x}_t = \left(\overline{V}_t, \overline{D}_t^{(1)}\right) \,\Big|\, x_t = \left(V_t, D_t^{(1)}\right), a_t = \left(a_t^v, a_t^{d_1}\right), v_t, d_t^{(1)}\right\}$$

$$\cdot P\left\{v_t, d_t^{(1)} \,\Big|\, x_t = \left(V_t, D_t^{(1)}\right), a_t = \left(a_t^v, a_t^{d_1}\right)\right\}$$

$$= \sum_{v_t=0}^{N_v-V_t} \sum_{d_t^{(1)}=0}^{N_{d_1}-D_t^{(1)}} P_\varphi\left\{\overline{x}_t = \left(\overline{V}_t, \overline{D}_t^{(1)}\right) \,\Big|\, x_t = \left(V_t, D_t^{(1)}\right), a_t = \left(a_t^v, a_t^{d_1}\right), v_t, d_t^{(1)}\right\}$$

$$\cdot P\left\{v_t, d_t^{(1)} \,\Big|\, x_t = \left(V_t, D_t^{(1)}\right)\right\}$$

$$= \sum_{v_t=0}^{N_v-V_t} \sum_{d_t^{(1)}=0}^{N_{d_1}-D_t^{(1)}} b\left(V_t, V_t + a_t^v - \overline{V}_t, p_{10}^v\right) \cdot b\left(D_t^{(1)}, D_t^{(1)} + a_t^{d_1} - \overline{D}_t^{(1)}, p_{10}^d\right)$$

$$\cdot b\left(N_v - V_t, v_t, p_{01}^v\right) \cdot b\left(N_{d_1} - D_t^{(1)}, d_t^{(1)}, p_{01}^d\right)$$

$$\tag{30}$$

Finally, after solving the aforementioned system of linear equations with $P_{\overline{x}_t|x_t}$ obtained from the above equation, we are able to calculate the steady-state probability distribution $\pi\left(V_t, D_t^{(1)}\right)$ of the state $x_t = \left(V_t, D_t^{(1)}\right)$. Since all voice and data users are statistically independent each other, the blocking rates of calls and high priority data messages under the optimal allocation policy φ are given by

$$P_B^v = \frac{\sum_{V_t} \sum_{D_t^{(1)}} \sum_{v_t} \sum_{d_t^{(1)}} \left(v_t - a_t^v\right) \pi\left(V_t, D_t^{(1)}, v_t, d_t^{(1)}\right)}{\sum_{V_t} \sum_{D_t^{(1)}} \sum_{v_t} \sum_{d_t^{(1)}} v_t \cdot \pi\left(V_t, D_t^{(1)}, v_t, d_t^{(1)}\right)} \tag{31}$$

$$P_B^{d_1} = \frac{\sum_{V_t} \sum_{D_t^{(1)}} \sum_{v_t} \sum_{d_t^{(1)}} \left(d_t^{(1)} - a_t^{d_1}\right) \pi\left(V_t, D_t^{(1)}, v_t, d_t^{(1)}\right)}{\sum_{V_t} \sum_{D_t^{(1)}} \sum_{v_t} \sum_{d_t^{(1)}} \pi\left(V_t, D_t^{(1)}, v_t, d_t^{(1)}\right)} \tag{32}$$

where the joint probability $\pi\left(V_t, D_t^{(1)}, v_t, d_t^{(1)}\right)$ is evaluated in a sequence of steps described below

$$\pi\left(V_t, D_t^{(1)}, v_t, d_t^{(1)}\right) = P_r\left(v_t, d_t^{(1)} \middle| V_t, D_t^{(1)}\right) \pi\left(V_t, D_t^{(1)}\right) = $$
$$P_r\left(v_t \middle| V_t\right) P_r\left(d_t^{(1)} \middle| D_t^{(1)}\right) \pi\left(V_t, D_t^{(1)}\right) \tag{33}$$

using the facts that

$$P_r\left(v_t \middle| V_t\right) = b\left(N_v - V_t, v_t, p_{01}^v\right) \tag{34}$$

$$P_r\left(d_t^{(1)} \middle| D_t^{(1)}\right) = b\left(N_{d_1} - D_t^{(1)}, d_t^{(1)}, p_{01}^d\right) \tag{35}$$

The throughputs of voice and data traffic are given by

$$\eta_v = \sum_{V_t} \sum_{D_t^{(1)}} V_t \cdot \pi\left(V_t, D_t^{(1)}\right) \tag{36}$$

$$\eta_{d_1} = \sum_{V_t} \sum_{D_t^{(1)}} D_t^{(1)} \cdot \pi\left(V_t, D_t^{(1)}\right) \tag{37}$$

The performance measure of voice only traffic can be derived similarly.

6 Data Traffic Analysis

According to the code allocation protocol considered in this paper the surplus of low priority data packets is queued, if (i) all $D2$ CDMA codes reserved for them are used and (ii) there are no codes left unsed by the voice and high priority data. This queueing will be necessary for both the return link [from mobile to satellite and gateway station (for satellite networks) or from mobile to base station (for cellular networks)] as well as for the forward link [from gateway to mobile or from base station to mobile] of CDMA networks. For the forward link our analysis and derivation of an efficient code allocation policy is complicated by the fact that the low priority data traffic and the desirable decision (admission) policy must be distributed at the mobile sites. Such a distributed multiple-access protocol will be very complex and its analysis is beyond the scope of this paper. However, for the return link all traffic is concentrated at the gateway or base station and there is a single decision maker: the network controller at that station.

For the scenario described in the previous paragraph there are two approaches that we can follow for modelling and performance evaluation. According to the first approach, the data traffic evolution is modeled by an M/D/c/B queueing model, where B is the buffer size and c the number of servers (e.g., CDMA codes) available for the low priority data traffic. We can then obtain the performance measures (throughput, delay, and packet loss probability) for low priority data traffic from an M/D/c/B queueing model with fixed c and then average these performance measures over the probability distribution $q(c)$ of c. Alternatively, we can obtain the transition probability matrix for the states of the system from scratch, without any reference to the M/D/c/B model, and work with the associated Markov chain. In the remaining of this section we follow this second approach.

For the single data type system, the probability distribution $q(c)$ of the number of available servers can be obtained as follows:

$$q(c) = \begin{cases} \pi^v_{K_{d-c}}, & \text{if } 1 \leq c \leq K_d \\ \sum_{s=K_d}^{K_v} \pi^v_s, & \text{if } c = 0 \\ 0, & \text{elsewhere} \end{cases} \tag{38}$$

If the low priority data is considered, then

$$q(c) = \sum_{\left\{ D_t^{(1)} \neq 0, M_1\left(D_t^{(1)}+c+1\right) < V_t \leq M_1\left(D_t^{(1)}+c\right) \right\}} \pi\left(V_t, D_t^{(1)}\right)$$
$$+ \sum_{\left\{ D_t^{(1)}=0, M_2(c+1) < V_t \leq M_2(c) \right\}} \pi\left(V_t, D_t^{(1)}\right) \tag{39}$$

where $M_2\left(D_t^{(1)}\right)$ is defined similar to $M_1\left(D_t^{(1)}\right)$ in eq.(7) with K_{d_1} replaced by K_{d_2}.

Assume that the data buffer at the gateway (for satellite networks) or the base-station (for cellular networks) has size B. Let the number of data packets in the buffer at the tth time-slot be n_t, where $0 \leq n_t \leq B$. The number of arriving data packets at each time-slot is Poisson with mean rate λ_{d_2}. Let $\bar{l}_t^d$ be the number of arrival packets at tth time-slot, which is independent of n_t; then we have

$$\bar{n}_t = n_t + \bar{l}_t^d - \bar{d}_t \tag{40}$$

where $\bar{d}_t$ is the number of departing data packets that are transmitted at the beginning of the next time-slot; these depend upon n_t and the number of available channels for data users as $\bar{d}_t = \min\{n_t, c + D2\}$. The states $n_t = 0, 1, ..., B$ follow a discrete Markov chain and their steady-state probability distribution π^{d_2} is obtained from the solution of the following set of $B + 1$ linear equations

$$\begin{cases} \pi^{d_2} = \pi^{d_2} \cdot P_{\bar{n}_t|n_t} \\ \sum_{k=0}^{B} \pi^{d_2}(k) = 1 \end{cases} \tag{41}$$

where $P_{\bar{n}_t|n_t}$ is the time-slot transition probability matrix for data under the threshold model. The entries of the transition matrix $P_{\bar{n}_t|n_t}$ are obtained from the relationships

$$P_{\bar{n}_t|n_t} = \sum_{\bar{l}_t^d=0}^{\infty} \sum_{c=0}^{K_{d_2}-D2} P\left(\bar{n}_t|n_t, \bar{l}_t^d, c\right) \cdot P\left(\bar{l}_t^d\right) \cdot q(c) \tag{42}$$

where $P\left(\bar{l}_t^d\right)$ is independent of n_t is given by

$$P\left(\bar{l}_t^d = k\right) = \frac{e^{-\lambda_{d_2}} \lambda_{d_2}^{k}}{k!}, k = 0, 1, 2... \tag{43}$$

for the Poisson model of arriving data traffic (at gateways or base stations). The random variable c only depends on the number of channels used for voice calls and high priority data and is independent of n_t and $\bar{l}_t^d$. Thus we have

$$P\left(\bar{n}_t|n_t, \bar{l}_t^d, c\right) = \begin{cases} 1, & \text{if } \bar{n}_t = \min\left\{B, n_t + \bar{l}_t^d - \min(n_t, c + D2)\right\} \\ 0, & \text{elsewhere} \end{cases} \tag{44}$$

After obtaining the time-slot transition matrix $P_{\bar{n}_t|n_t}$ we can compute the steady-state probability distribution $\pi^{d_2}(n_t)$, for all $0 \leq n_t \leq B$. ¿From the latter we can easily obtain the average data delay and the packet loss probability, which is due to the finite buffer size at the gateways or the base stations. First we obtain the average data throughput as

318

$$\eta_{d_2} = \sum_{c=0}^{K_{d_2}-D2} \sum_{n_t=0}^{B} \min(c+D2,n_t)\pi^{d_2}(n_t)q(c) \tag{45}$$

and the average total data delay

$$D_{d_2} = 1 + \frac{1}{2} + \frac{1}{\eta_{d_2}} \sum_{n_t=1}^{B} n_t \pi^{d_2}(n_t) \tag{46}$$

where the value of D_{d_2} is in packets (time-slots), 1 is the service time of one packet, $\frac{1}{2}$ is the average waiting time from the arrival time of one packet to the beginning of the next time-slot, and the last term in the right-hand side represents the average data queueing delay. For gateways connecting satellites and mobiles the round trip propagation time between earth stations and satellites may have to be added; the latter delay can be easily computed from the satellite altitude and translated to time-slots by dividing by the packet transmission time. Finally, the packet loss probability is given by

$$P_L = \left(\lambda_{d_2} - \eta_{d_2}\right)/\lambda_{d_2} \tag{47}$$

Under the graceful degradation model, the time-slot transition probability matrix is now determined by the following equations:

$$P_{\bar{n}_t|n_t} = \sum_{\bar{l}_t^d=0}^{\infty} \sum_{s=0}^{K_v} P\left(\bar{n}_t|n_t,\bar{l}_t^d,s\right) \cdot P\left(\bar{l}_t^d\right) \cdot \pi_s^v \tag{48}$$

where π_s^v is the steady-state distribution of voice traffic and

$$P\left(\bar{n}_t|n_t,\bar{l}_t^d,s\right) = \begin{cases} P_{gdm}\left(n_t,n_t+\bar{l}_t^d-\bar{n}_t,P_E(s+n_t)\right), & \text{if } \bar{l}_t^d \le \bar{n}_t < K, \\ \sum_{l=0}^{n_t+\bar{l}_t^d-K} P_{gdm}\left(n_t,l,P_E(s+n_t)\right), & \text{if } \bar{n}_t = K, \\ 0, & \text{elsewhere} \end{cases} \tag{49}$$

The throughput becomes

$$\eta_d = \sum_{s=0}^{K_v} \sum_{n_t=1}^{K} \sum_{l=1}^{n_t} l \cdot P_{gdm}\left(n_t,l,P_E(n_t+s)\right) \cdot \pi^d(n_t) \cdot \pi_s^v \tag{50}$$

7   Numerical Results

For our performance results we assume that the DS/CDMA system employs BPSK data modulation with coherent demodulation, binary convolutional codes with rate 1/2 and constraint length 7, and Viterbi decoding with hard-decisions. The parameter values E_b/N_0 = 10dB (signal-to-AWGN ratio) and N = 255 (processing gain, number of chips per bit) are used in the numerical computations. For the MAC indices K_v = 20 and K_d = 15, which we use for the single-data type system, the corresponding packet error probabilities are $P_E^v = 1.5 \times 10^{-3}$ and $P_E^d = 6 \times 10^{-5}$, respectively.

Figure 5 shows a 3-D plot of the optimal admission policy for voice calls. In this figure the number of accepted voice calls (vertical or z-axis) is plotted versus the number of newly arrived voice calls (x-axis) and the number of current (ongoing) voice calls (y-axis). Notice that the optimal admission policy lacks structure (i.e., a simple description in terms of inequalities or convexity), in particular for large numbers of current and new voice calls. This lack of structure necessitates the derivation of near-optimal admission policies which are simpler to construct (they obey certain structural rules) and maintain to a large degree the enhanced performance of the optimal admission policy.

The blocking probabilities of voice calls under the direct admission policy and the optimal admission policy are plotted vs. the offered voice load in Figure 6. The offered load of voice traffic is defined as the average of the active voice calls.

$$G_v = \sum_{s=0}^{K_v} \left[(M_v - s) p_{active}^v + s \right] \pi_s^v = M_v p_{active}^v + \left(1 - p_{active}^v \right) \sum_{s=0}^{K_v} s \pi_s^v \tag{51}$$

The blocking probability for the optimal policy is smaller than that in the direct admission policy, in particular, under heavy voice loads. In logarithmic scale this advantage is somewhat suppressed; however, in the following figure it is established with any doubt.

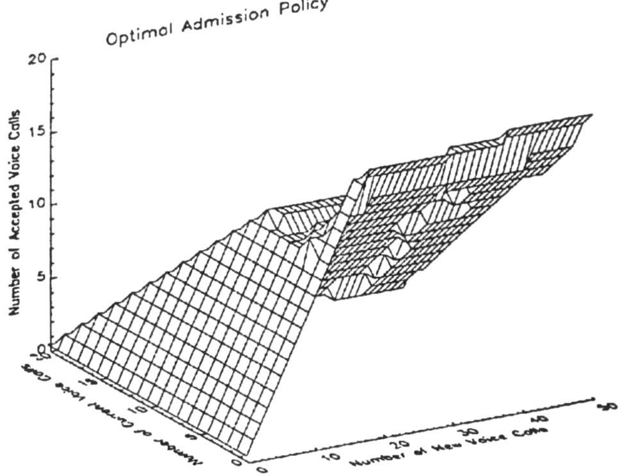

Figure 5. Optimal Admission Policy

320

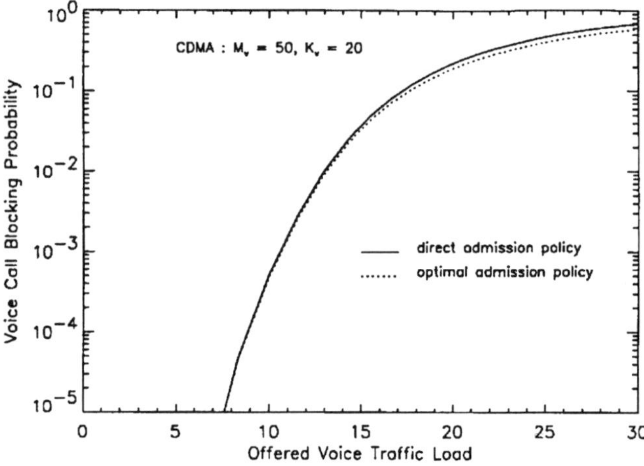

Figure 6. Blocking Probability of Voice Calls vs G_v

The advantage of the optimal admission policy over the direct admission policy is clearly shown in Figure 7 where the percentage reduction in the voice blocking probability (VBP) [= 100 x (VBP under direct policy - VBP optimal policy)/(VBP direct policy)] is shown as a function of the offered voice traffic load. The reduction in voice blocking probability provided by the optimal admission policy can be a high as 15 %.

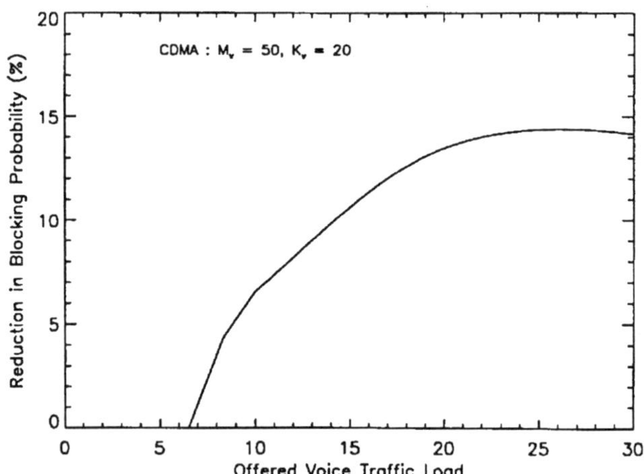

Figure 7. Percentage Reduction in Blocking Probability for Optimal
Admission Policy over Direct Admission Policy vs G_v

The performance measures of data traffic in the numerical results are evaluated for an offered voice load of $G_v = 11.54$. The average delay of data traffic vs. the offered data rate is displayed in Figure 8. There is no perceptible difference in the data delay under the direct and the optimal admission policies for voice traffic. However, the average data delay under the graceful degradation model is substantially smaller than that under the threshold model.

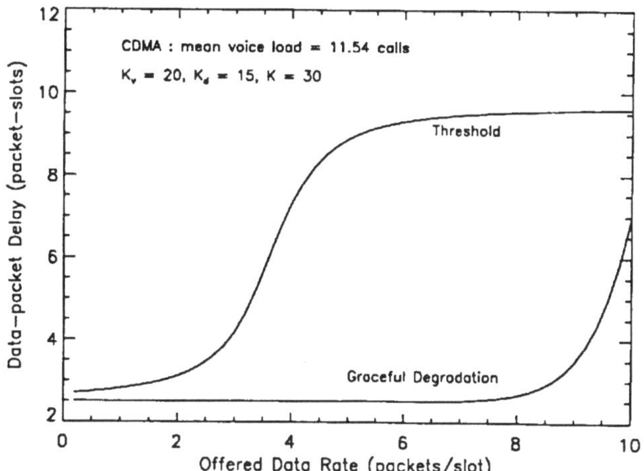

Figure 8. Average Data-Packet Loss Probability vs λ_d

Figure 9 shows the packet loss probability of data traffic vs. the offered data rate. From the log-scale figure we see that the packet loss probability under the optimal admission policy for voice is a little worse than that under the direct admission policy. Moreover, the packet loss probability under the threshold model is slightly smaller than that under graceful degradation model when the data traffic load is light. On the other hand, there is a substantially higher packet loss probability under the threshold model when the data traffic load is heavy.

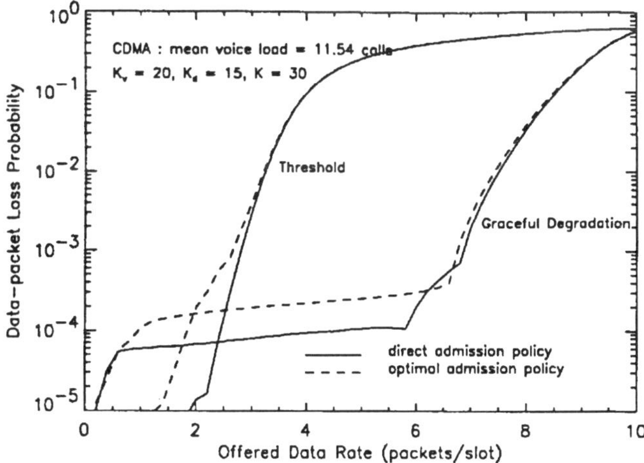

Figure 9. Average Data-Packet Loss Probability vs λ_d

Figure 10 shows the packet error probability (for voice and data packets) vs. the offered data load. The results under the optimal and direct admission policies are almost identical. However, the difference between the threshold model (that guarantees a packet error probability P_E of 8 x 10^{-5}) and the graceful degradation model (under which P_E increases with increasing data load) is very clear. Refer to [14] for additional numerical results for this formulation of the CDMA multi-media integration problem.

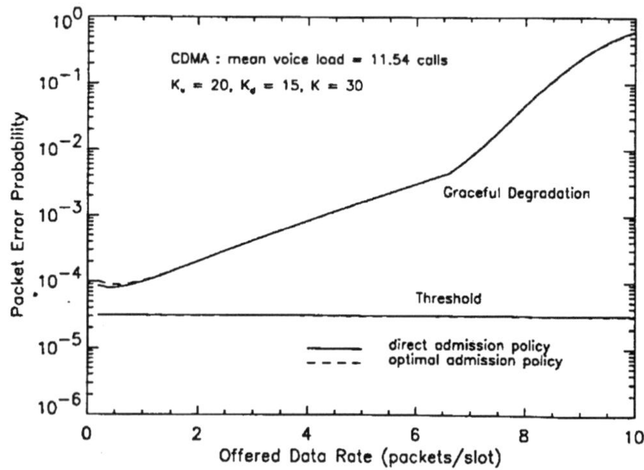

Figure 10. Average Packet Error Probability vs λ_d

In the code allocation problem for voice and two types of data traffic, we use $E_b/N_0 = 9$ dB and MAC indices $K_v = 15$ and $K_{d_1} = K_{d_2} = 10$ in the numerical computations. The corresponding packet error probabilities are $P_E^v = 3 \times 10^{-3}$ and $P_E^{d_1} = P_E^{d_2} = 1.5 \times 10^{-4}$, respectively. We present the numerical results for $D2 = 0$, that is, there are no CDMA codes reserved for the low priority data, and $D2 = 1$, that is, a single CDMA code is reserved. In Equation ??, we set $\delta = 1\%$; resulting in a probability of 0.99 that the number of active calls in talkspurt is less than and equal to $K_{d_1} - D_t^{(1)}$. Then, if up to $M_1\left(D_t^{(1)}\right)$ voice calls and $D_t^{(1)}$ data users are in the system, the above voice and data BER (and packet error) requirements are met. The values of $M_1\left(D_t^{(1)}\right)$ as shown in Table 1a when $\delta = 1\%$.

Table 1a

$D_t^{(1)}$	0	1	2	3	4	5	6	7	8	9	10
$M_1\left(D_t^{(1)}\right)$	24	13	11	9	8	6	4	3	2	1	0

For comparison purposes, notice that, if the voice activity characteristic is not taken into account, then the maximum number of active voice users is 9 rather than 13 when there is 1 active data user in the system and 15 rather than 24 when there is no active data user in the system. Table 1b and 1c present similar results for the smaller MAC indices $K_v = 12, K_{d_1} = K_{d_2} = 8$ and $D2$ values of $D2 = 0$ (Table 1b) and $D2 = 1$ (Table 1c) which correspond to packet error probabilities of approximately $P_E^v = 1.3 \times 10^{-3}$ and $P_E^{d_1} = P_E^{d_2} = 3.5 \times 10^{-5}$,

Table 1b

$D_t^{(1)}$	0	1	2	3	4	5	6	7	8
$M_1\left(D_t^{(1)}\right)$	18	9	8	6	4	3	2	1	0

Table 1c

$D_t^{(1)}$	0	1	2	3	4	5	6	7
$M_1\left(D_t^{(1)}\right)$	13	8	6	4	3	2	1	0

The offered loads of voice and data traffic are defined as the average of the number of active voice calls and the number of active data users, respectively,

$$G_v = N_v \frac{p_{01}^v}{p_{01}^v + p_{10}^v} \quad \text{and} \quad G_{d_1} = N_{d_1} \frac{p_{01}^d}{p_{01}^d + p_{10}^d} \tag{52}$$

In Table 2, we list some of the actions taken under the optimal policy and three different data weighting factors $w_d = 1, 10,$ and 100 (for voice $w_v = 1$). with the parameters $G_v = G_{d_1} = 1.43$.

Table 2

Current Users		New Users		$w_d = 1$		$w_d = 10$		$w_d = 100$	
V_t	$D_t^{(1)}$	v_t	$d_t^{(1)}$	a_t^v	$a_t^{d_1}$	a_t^v	$a_t^{d_1}$	a_t^v	$a_t^{d_1}$
3	2	3	0	3	0	3	0	3	0
3	2	3	1	3	1	3	1	1	1
3	2	3	2	3	1	1	2	0	2
3	2	3	3	3	1	0	3	0	3
3	2	3	4	3	1	0	3	0	3
6	0	1	0	1	0	1	0	1	0
6	0	1	1	1	1	1	1	1	1
6	0	1	2	1	2	1	2	0	2
6	0	1	3	1	2	0	3	0	3
6	0	1	4	1	2	0	3	0	3

In Figures 11 and 12 we show the blocking rates of voice and high priority data traffic versus the offered loads of voice traffic (G_v) for an offered high priority data load of $G_{d_1} = 1.43$ and two different data weighting factors ($w_d = 1$, and 100). As expected the blocking rates of voice calls (Fig. 11) and of high priority data traffic (Fig. 12) increase as G_v increases. The blocking rate of voice calls increases and that of high priority data decreases as the weighting factor increases. When $D2$ increases, both the voice and high priority data blocking rates increase.

In Figures 13 and 14 we show the blocking rates of voice and high priority data traffic versus the offered load of data traffic (G_{d_1}) for an offered voice load of $G_v = 1.43$ and two different data weighting factors ($w_d = 1$, and 100). As expected the blocking rates of voice traffic (Fig. 13) and of high priority data traffic (Fig. 14) increase as G_{d_1} increases. The effect of increasing w_d and $D2$ is the same as for Figures 11 and 12.

In Figures 15 and 16 we show the average packet loss probability and average system delay of low priority data traffic versus the offered low priority data rate λ_{d_2}, for a voice load $G_v = 1.43$, a high priority data load $G_{d_1} = 2.09$, and two data weighting factors $w_d = 1$, and $w_d = 100$. When a CDMA code is reserved for the low priority data traffic ($D2 = 1$), the data packet loss probability (Figure 15) and system delay (Figure 16) decrease; this change is more substantial for larger values of the weighting factor w_d ($w_d = 100$ versus $w_d = 1$).

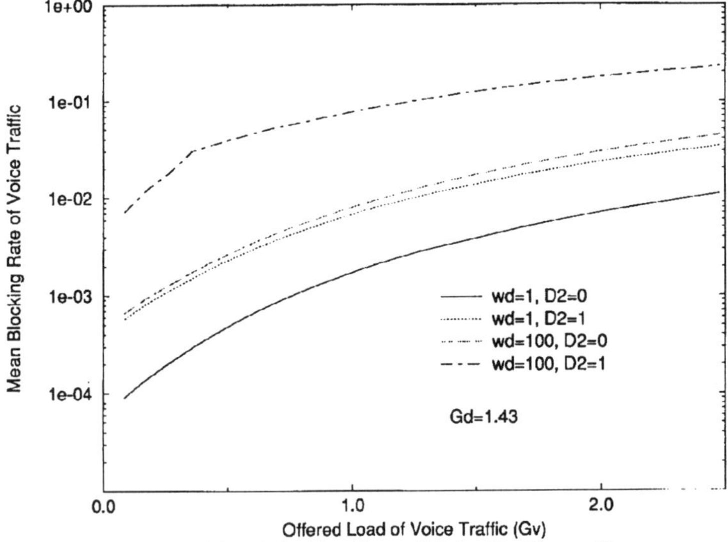

Figure 11. Mean Blocking Rate of Voice Traffic vs G_v

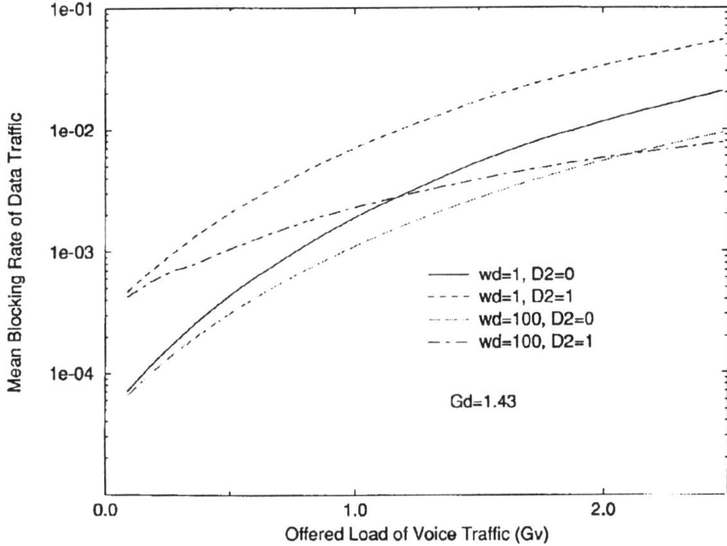

Figure 12. Mean Blocking Rate of Data Traffic vs G_v

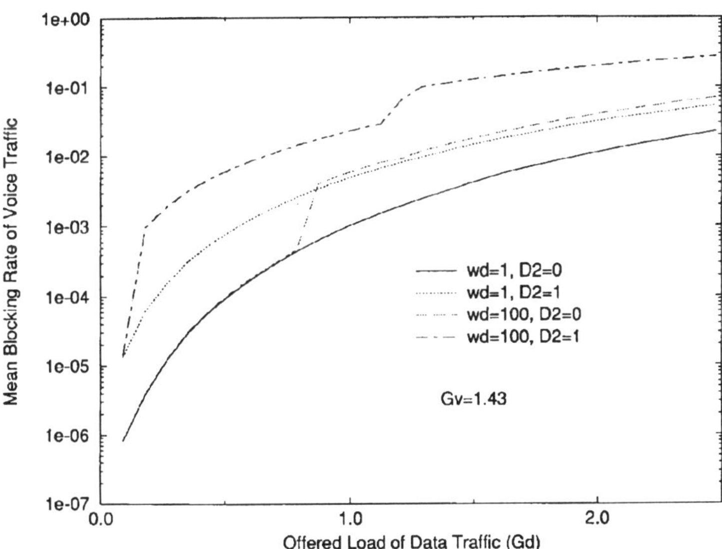

Figure 13. Mean Blocking Rate of Voice Traffic vs G_{d1}

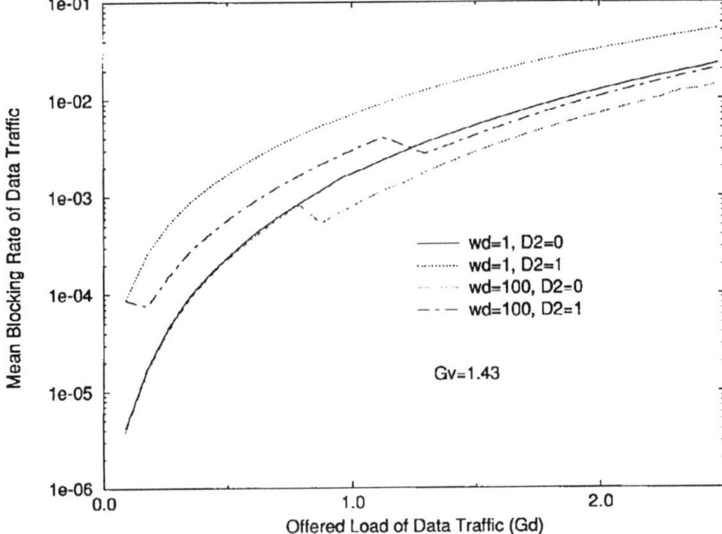

Figure 14. Mean Blocking Rate of Data Traffic vs G_{d1}

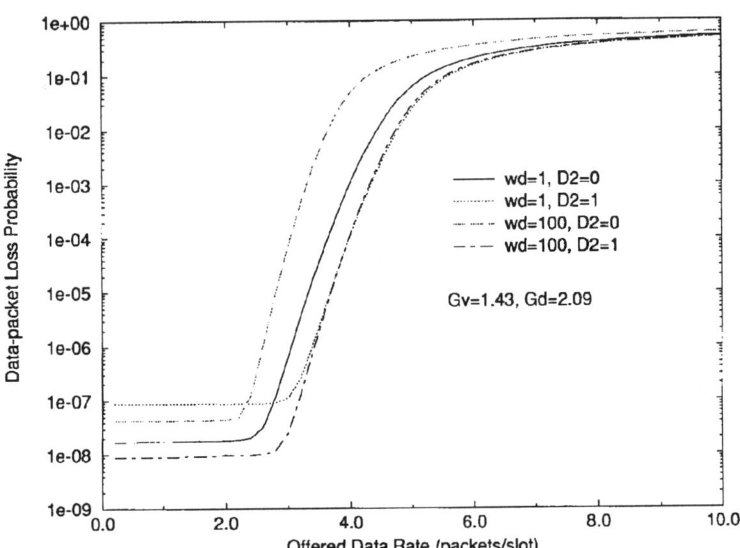

Figure 15. Mean Packet Loss Probability vs λ_{d2}

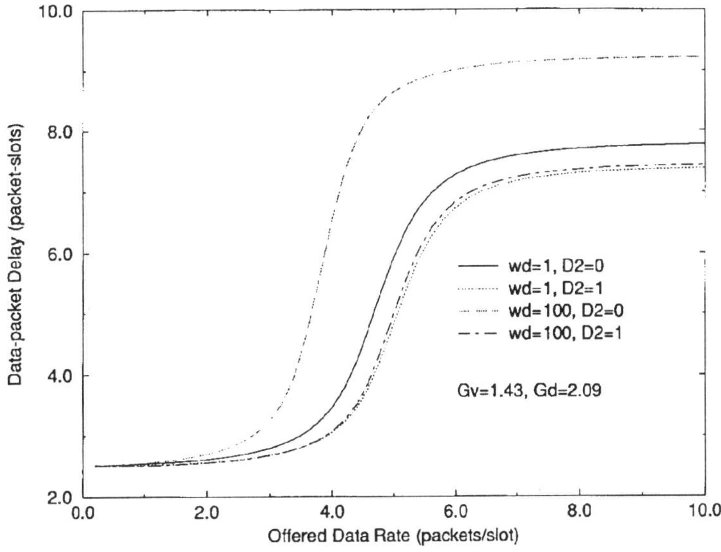

Figure 16. Mean Packet Delay vs λ_{d2}

8 Conclusions

In this paper, we have developed and analyzed the performance of admission policies for integrated voice and data traffic in wireless CDMA packet radio networks. In our first model, voice packets have priority over data packets. Therefore, the voice traffic analysis and derivation of admission policies was conducted first without any influence from data traffic considerations. A direct admission policy that accepts voice calls when there are sufficient unused resources (CDMA codes) and an optimal admission policy that minimizes the long-term rejection rate of arriving voice calls were derived and their performance analyzed in terms of the average blocking probability of voice calls. The optimal admission scheme was shown to outperform the direct one especially for heavy voice loads.

For data traffic two models for multiple-reception were analyzed the threshold and graceful degradation models for CDMA interference. An admission policy for data was derived which strongly depends on the voice admission scheme and the average throughput, delay, and packet loss probability of data were evaluated. The data performance was shown to change little with the different voice admission policies but it depends drastically on the interference model used; the selection of the threshold model versus the graceful degradation model depends critically on the data user specifications and priorities.

In the second model, voice calls and high priority data traffic are treated in the same manner and they share the CDMA capacity in a movable boundary fashion; real time delivery is guaranteed for both while their BER requirements are met, otherwise blocking occurs. The long-term blocking rates of voice and high priority data traffic are minimized.

Low priority data users use whatever capacity is left unused by the other two traffic types plus a small fraction of the capacity reserved for them. We showed that the optimal admission policy differs for varying weighting factors $w_d(w_v = 1)$ but the performance differs only when w_d becomes much larger than w_v (e.g., $w_d = 100$). The activity characteristic of voice traffic was taken into account and it helped to increase the number of users accommodated.

In the work described in this paper it was assumed that perfect power control was exercised by the network controller; neither fading nor shadowing of the CDMA signals was accounted for. Interference from adjacent cells (for cellular CDMA network applications) or satellite beams (for satellite CDMA network applications) was not accounted for. Moreover, all voice and data users were transmitting at the same data rate. We are currently incorporating these aspects and relaxing the unnecessary constraints; various extensions of the two basic schemes of this paper will be reported in forthcoming papers. Suboptimal CDMA code allocation schemes which are easier to implement are also under development.

References

[1]   David Jacobson, "Yankee Group Reports on Digital Cellular Technologies," *Wireless,* AT&T, vol. 1, Number 1, pp. 1-8, July 22, 1991.

[2]   K.S. Gilhousen, I.M. Jacobs, R. Padovani, and L.A. Weaver, Jr., "Increased Capacity Using CDMA for Mobile Satellite Communication", *IEEE JSAC,* May, 1990.

[3]   D.C. Cox, "Wireless Network Access for Personal Communications", *IEEE Commun. Magaz.,* Vol. 30, NO. 12, pp. 96-115, December, 1992.

[4]   L.C. Palmer, E. Laborde, A. Stern, and P.Y. Sohn, "A Personal Communications Network Using a Ka-Band Satellite", *IEEE JSAC,* February, 1992.

[5]   Loral/Qualcom Inc., " Globalstar—FCC Filing,, 1991.

[6]   TRW Inc., "Odyssey—FCC Filing, May, 1991.

[7]   M. Soroushnejad and E. Geraniotis. "Performance Evaluation of Multi-Access Strategies for an Integrated Voice/Data CDMA Packet Radio Network." To appear in the *IEEE Trans. on Commun.*

[8]   E. Geraniotis, M. Soroushnejad, and W.-B. Yang. " A Multi-Access Scheme for Voice/Data Integration in Hybrid Satellite/Terrestrial and Heterogeneous Mixed-Media Packet Radio Networks." To appear in the *IEEE Trans. on Commun.*

[9]   T. Ketseoglou and E. Geraniotis, "Multireception Probabilities for FH/SSMA Communications," *IEEE Trans. on Commun.*, Vol. COM-40, January 1992.

[10]   E. Geraniotis and T.-H. Wu. "The Probability of Multiple Correct Packet Receptions in Direct-Sequence Spread-Spectrum Networks." *IEEE JSAC,* pp. 871-884, June 1994.

[11]   M.B. Pursley and D.J. Taipale. "Error Probabilities for Spread-Spectrum Packet Radio with Convolutional Codes and Viterbi Decoding." *IEEE Trans. on Commun.*, Vol. COM-35, January 1987.

[12]   J. Conan, "The Weight Spectra of Some Short Low-Rate Convolutional Codes," *IEEE Trans. on Commun.*, Sept. 1984.

[13]   P.T. Brady, "A Statistical Analysis of on-off patterns in 16 conversations," *Bell System Technical Journal,* Vol. 47, No. 1, pp.73-91, 1968.

[14]   H. C. Tijms. *Stochastic Modeling and Analysis: A Computational Approach.* New York: Wiley, 1986.

[15]   W.-B. Yang and E. Geraniotis. "Admission Policies for Integrated Voice and Data Traffic in CDMA Packet Radio Networks." *IEEE JSAC,* pp. 654-664, May 1994.

CDMA for Mobile LEO Satellite Communications

Raymond L. Pickholtz and Branimir R. Vojcic

1. Introduction

Satellite orbiting the earth would appear to be the ideal way to obtain world-wide coverage to mobile users. During WARC'92[1] several frequency bands were established for that purpose [14]. Since then there have been numerous proposals for implementing such systems. There are proposals for using geosynchronous orbit satellites (GEOS) at an altitude 35,784 km, medium earth orbit satellites (MEOS) at 5,000 - 10,000 km and Low Earth Orbit Satellites (LEOS) at 150 - 1,500 km. Of these, the LEOS have attracted the most attention because of the novelty of having many satellites, handoffs and a cellular-like configuration. The advantages would be small propagation loss so that handsets could be used for direct communication from a mobile user and small propagation delay for better performance in voice, data and other interactive services. A disadvantage is that more satellites are required -- offset by cheaper launch costs -- and increased probability of shadowing. The design challenges include worldwide coverage, system user capacity, margins of fading, interaction delay, call handoffs, spectrum sharing, handheld battery life and user health hazards (transmit power). The frequency bands to be used for LEOS is 1616.5 - 1626.5 MHz for the uplink and 2483.5 - 2600 MHz for the downlink if frequency division duplex (FDD) is used and just the former band if time division duplex (TDD) is used. Of five major proposals for "big" LEOS, four have proposed CDMA in one form or another, and only one has proposed FDMA/TDMA. Because of this, and the general interest in CDMA in LEOS we have focused our effort in this paper on LEOS CDMA and, in particular, how it differs from terrestrial cellular CDMA which has already demonstrated its virtues and has become one of several standards in the United States. The network architecture of a typical CDMA based LEOS is shown in Figure 1 where the satellites act as "dumb transponders" or "bent pipes". Thus, while communications can take place between users within the footprint of the satellites, the signals are decoded at an earth-based gateway. Communications from users in different footprints access their respective satellites directly but communications between them involves both gateways to/from the satellites and terrestrial lines between gateways.

There are some issues that distinguish the application of satellite CDMA from its use in terrestrial cellular communications. This paper is an attempt

---

[1]World Administrative Radio Conference, Torremolinos, Spain, 1992.

331

S.G. Glisic and P.A. Leppänen (eds.), Code Division Multiple Access Communications, 331-350.
© 1995 *Kluwer Academic Publishers. Printed in the Netherlands.*

to summarize these differences and, to the extent possible in a short review, to quantify some key issues. Details are available in references cited.

We start with a description of the channel to include the effects of both shadowing and fading. Next, we describe the constellations of the five "big" LEOS. In Section 4 we elaborate on the differences between terrestrial and satellite CDMA. In Section 5 we show some quantitative results of applying the characteristics of the channel model and that of the CDMA use in LEOS. Section 6 outlines how we may be able, with novel techniques, to obtain greater capacity. Section 7 addresses a key issue relating to the possibility and desirability of spectrum sharing with multiple CDMA providers using interference sharing. Sections 8 and 9 wraps up with future trends and conclusions.

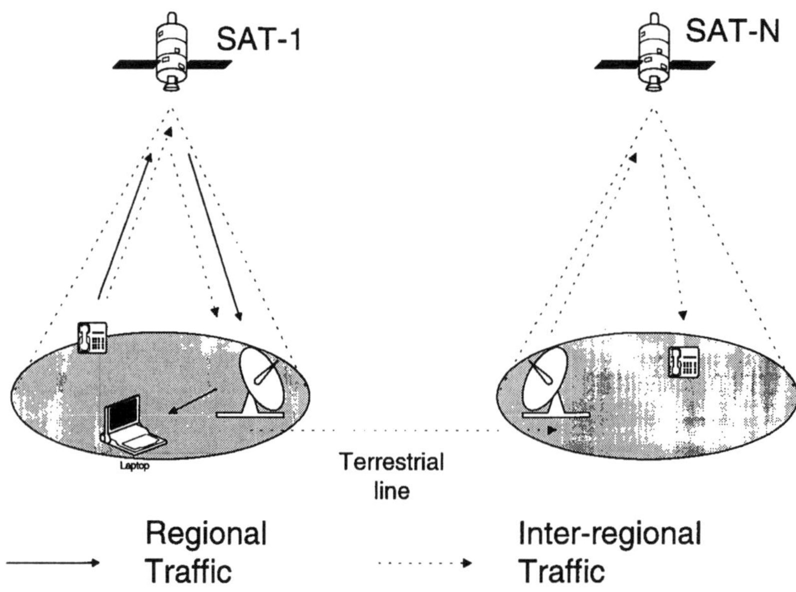

Fig.1 Mobile Satellite Network Architecture with Gateways.

2. Mobile Satellite Channel

Since the primary goal of this paper is to examine some characteristics of DS-CDMA for mobile satellite communications, we will concentrate on a discussion of channel properties of a mobile satellite (MS) channel. In general, it is possible to design and operate a MS system in different frequency bands, subject to the antennae size constraints and international/domestic radio frequency regulations. Since the L band has been of most interest recently for commercial MS communications, we will address mainly channel models based on measurements in this part of the radio spectrum.

In one of the first proposed MS DS-CDMA systems in L band [1], a frequency-flat fading model was assumed. That is, the fading process was modeled as

$$r(t) = s(t) + w(t), \tag{1}$$

where $s(t)$ corresponds to the direct path which experiences lognormal shadowing and $w(t)$ represents a zero-mean complex Gaussian random process, corresponding to the diffuse multipath (Rayleigh fading). The fading spectra corresponding to shadowing and scattering were modeled by 6th order Buterworth spectra with cutoff frequencies $F_c = 5$ cycles/m and $F_c = 10$ cycles/m, respectively. For such a combined model of shadowing and scattering, the probability density function (p.d.f.) of the envelope of $r(t)$, R, conditioned on $s(t) = s$, can be represented by

$$f_{Rice}(r|s) = \frac{r}{\sigma^2} \exp\left[-\frac{(r^2 + s^2)}{2\sigma^2}\right] I_0\left(\frac{rs}{\sigma^2}\right), \quad r \geq 0, \tag{2}$$

where $2\sigma^2$ is the average power in the process $w(t)$, and the resulting p.d.f. of R is given by

$$f_R(r) = \int_0^\infty f_R(r|s)\, f_{LN}(s)\, ds, \tag{3}$$

in which $f_{LN}(s)$ represent the lognormal p.d.f. of s

$$f_{LN}(s) = \frac{1}{\sqrt{2\pi}\, s\sigma_l} \exp\left[\frac{(\ln(s) - m_l)^2}{2\sigma_l^2}\right], \tag{4}$$

where m_l and s_l^2 are the mean and the variance of the signal amplitude of direct path (note that $s(t)^2$ has also lognormal distribution). In the absence of shadowing, that is when $s(t) = 1$, this model reduces to pure Rician fading.

The p.d.f. of the direct received signal phase can be approximated by a Gaussian p.d.f. [2], [5]

$$f_\phi(\phi) = \frac{1}{\sqrt{2\pi}\,\sigma_\phi} \exp\left[-\frac{(\phi - m_\phi)^2}{2\sigma_\phi^2}\right]. \tag{5}$$

According to a statistical fit of measurement reported in [2], [6], this model for the signal envelope statistic corresponds to rural and suburban areas where the main causes of shadowing are trees, houses and small buildings. The same references suggest that in open areas the received signal envelope experiences purely Rician fading (as in (2) without conditioning on s). In urban areas, on the other hand, where the line-of-sight between the mobile and the satellite is almost completely obstructed, the results in [2] indicate that the signal envelope behavior is appropriately described by a Rayleigh p.d.f. given by

$$f_{Rayl}(r) = \frac{r}{\sigma^2} \exp\left[-\frac{r^2}{2\sigma^2}\right], \quad r \geq 0, \tag{6}$$

whereas the signal phase is uniformly distributed in $[0, 2\pi]$. The Doppler spectrum was modeled by the usual mobile fading spectrum, the low pass equivalent of which has the form

$$S_r(f) = \left[A\pi f_m \sqrt{1 - \left(\frac{f}{f_m}\right)}\right]^2, \quad |f| \leq f_m, \tag{7}$$

where A is a constant which depends upon the type of receiving antenna and mobile's direction, and f_m is the maximum Doppler frequency defined as

$$f_m = \frac{v}{\lambda}, \tag{8}$$

in which v represents the vehicle speed and λ is the carrier wavelength.

Hence, different local propagation environments surrounding a mobile result in quite different statistical variations of the received signal envelope. To get yet another perspective on the delicacy of choosing a statistical model that describes appropriately the behavior of a MS channel, let us present the model used in [3]. In [3], based on a statistical analysis of measurements undertaken in Europe, the authors have postulated a model which can be characterized by the mixture density

$$f_m(r) = (1 - B)\, f_{Rice}(r) + B \int_0^\infty f_{Rayl}(r|\sigma^2)\, f_{LN}(\sigma^2)\, d\sigma^2, \tag{9}$$

where B represents the probability of shadowing. That is, in this model the channel can be characterized by a "good" (nonshadowed) state with Rician fading and a "bad" (shadowed) state with a combination of shadowing and scattering.

This brief description of reported models suggests the complexity of channel modeling based purely on the statistical analysis of measurements. Nevertheless, it appears that the best case propagation scenario corresponds to Rician fading and the worse case corresponds to Rayleigh fading, perhaps with lognormally distributed power. Since lognormal shadowing is relatively slow and can be effectively neutralized by either open-loop or close-loop power control, the authors in [7] considered a simplified model represented by a mixture of a Rician and a Rayleigh p.d.f.s. Because the performance is dominantly affected by Rayleigh statistics, a further simplification of the model was used in [8], where a Gaussian channel was assumed in the nonshadowed state and a Rayleigh fading channel in the shadowed state. This model is depicted in Figure 2, and will be used throughout the paper to represent some analytical results.

The reader may have already noticed that all models considered correspond to frequency nonselective fading. The multipath spread in MS communications is smaller, by at least one order of magnitude, than in land mobile communications. Indeed, the measurements conducted in the Chicago area, reported in [9], indicate that the multipath signal delay is typically less than 100 nsec, with power relative to the direct path of -15dB or less. For DS-CDMA systems spreading up to 10MHz, such a channel can be modeled as a frequency-flat channel.

To analyze the behavior of the physical and the data link layer over large areas (longer time intervals), a finite-state Markov model can be used [2]. That is, a large area can be divided into M subareas, each with a statistical model corresponding to an urban, suburban or open area. The channel evolution of the quasistationary model is defined by a transition probability matrix. Under the assumption that the Rayleigh fading effect is dominant, only a "good" and a "bad" state may suffice and the two-state Gilbert-Eliot model can be used [10].

For a more detailed description of models described, the reader is referred to the above mentioned references.

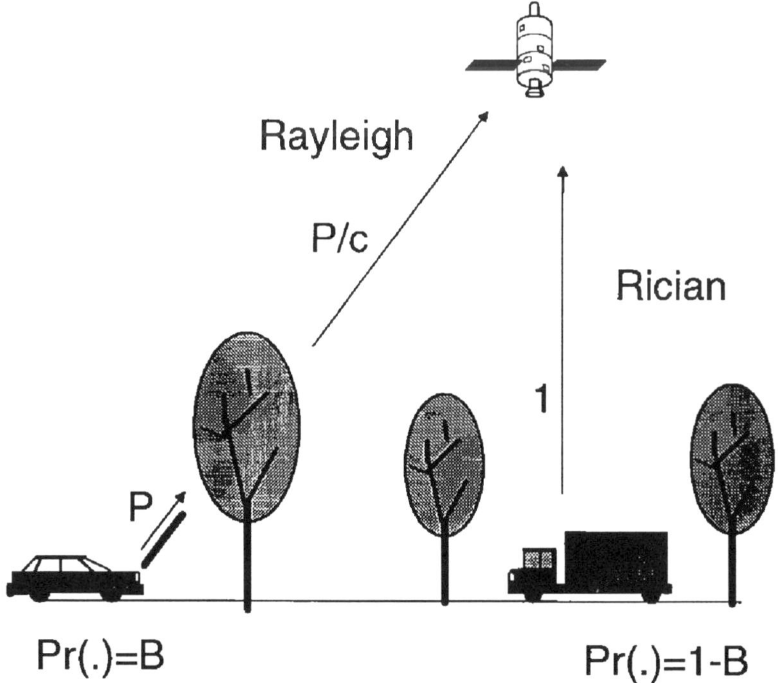

Fig. 2 Single system uplink shadowing scenario

3.   Satellite Constellations

Among the distinguishing features of mobile satellite communications is that coverage of the earth's surface depends on the orbital altitude and inclination, as well as on the number of satellites used. For example, only three satellites in geosynchronous earth orbit in the equatorial plane are sufficient to cover virtually the entire planet with the exception of two small polar caps. Furthermore, geosynchronous satellites exhibit very little Doppler shift and, for stationary satellite communications, tracking and handoffs are not necessary. However for mobile communications the advantages of low earth orbiting satellites (LEOS) are overwhelming:

- smaller range means lower mobile and satellite power
- propagation delay is much smaller
- launch costs are significantly lower

We will therefore concentrate on LEOS in this paper.

There have been 5 "big" LEOS proposed for world-wide mobile communications. The constellation parameters, taken from public information, are shown in Table 1.

Table 1

Company/System	Total No. of Satellites	Orbit Altitudes (Km)	Satellite Beams
Constellation/Aires	48	1020	7
Ellipsal/Ellipso	6-10	580/7800	8
LQSS/Globalstar	48	1414	16
Motorola/Iridium	66	780	48
TRW/Oddesey	12	10370	19

As one can see, the Oddesey systems is not really a LEOS but rather a medium earth orbiting satellite (MEOS) system. Nevertheless it has been included in the general class of LEOS because the objectives of the systems are similar, although not in detail. The designation "big" refers to the fact that many satellites will be in orbital use simultaneously to accomplish global coverage. Thus, for example, Iridium will have 66 satellites, all in polar orbit with 11 orbital planes each containing 6 equi-spaced satellites. The nominal altitude of 780 km means that launch costs can be reduced and round trip propagation delay is several milliseconds. Ellipso, as its name implies, has a very eccentric elliptical orbit so that the apogee/perogee ratio is more than 10:1. The orbits are inclined so that even with a small number of satellites, good coverage is achieved over the most likely initial markets. Globalstar has 48 total satellites with 8 differently inclined orbital planes of 6 satellites each to achieve global coverage. Aires has a similar constellation. Notice that all of these systems employ means of putting down several to many spot beams within the footprint of any single satellite. These spots (analogous to cells in a terrestrial system will move along the earth faster than any mobile, which means the satellites in orbit and their associated spot beams will be moving at great speed relative to the earth's surface.

4. Differences Between Terrestrial and Satellite Mobile Communications

Terrestrial mobile communications on a large scale has been achieved by cellularization. That is, base transceiver stations are deployed so as to geographically reuse radio resources. In conventional analog (and digital) systems such as AMPS, D-AMPS and GSM, adjacent "cells" avoid using the same frequency. However remote cells which are electromagnetically more isolated, can do so provided that their distance has sufficiently attenuated the signal to tolerable levels of co-channel interferences. In terrestrial CDMA it may be possible to reuse the same frequency in adjacent cells provided the signature sequences exhibit sufficiently low cross correlation and the signals in adjacent cells have been adequately attenuated with distance. This ability to reuse frequencies is crucial to making cellular radio work as such a grand scale. The major ally in accomplishing this is that, because of multipath and scattering, the signals from a transmitter at the surface attenuate close to an inverse 4th power with distance rather than the free-space 2nd power. This empirical fact is, of course, valid for reception when line-of-sight is not dominant and thus the statistical amplitude variation of a narrow band reception is Rayleigh distributed. In satellites, by contrast, the isolation of the spot beams (corresponding to the cells in terrestrial) is determined by the antenna beam flux isolation.. In addition, the statistical characterization of the fading is a mixed Rayleigh-Rice distribution as described in the previous section. For terrestrial signals which propagate at low elevation angles, this multipath spread is often measured in 10's of microseconds (up to 100 μs spread has been observed for macrocells in regions near mountains). For satellites, especially at moderate to high elevation angles (otherwise signals are blocked) the observed multipath spread is of the order of about 10's of nanoseconds. Thus in the use of CDMA, RAKE gain can be achieved with signals spreading of the order of 1 MHz in terrestrial but not in satellite systems. Tight power control is a critical technique for making a conventional CDMA system work for, otherwise, the "near-far" problem dissipates much of the capacity. In terrestrial systems, equalized power control within one dB can be achieved by means of closed loop over a large (80 + dB) dynamic range. In satellite systems, even in low earth orbit, the round-trip delays of 10 ms or more inhibit the use of closed loop systems. In addition, the use of open loop power measurements for power control implies that measurement of the downlink power infers the power required for the uplink. Unfortunately the up/downlink frequencies allocated are separated by 800 MHz so that, except for shadowing loss, this power measurement is likely to have a much larger error variance than that obtained by closed loop power control. Independent of power control, the additional delay in satellites inhibits the use of deep interleaving since interleaving itself introduces delays and the total allowable delay budget for interactive services is usually upper bounded. This problem could be especially acute in MEO systems.

An additional regulatory constraint in satellite communications is the so-called power flux density (PDF) limit imposed by international radio regulations. These regulations are imposed largely to protect existing terrestrial users that share the band with satellites. This makes the uplink and downlink asymmetrical.

In terrestrial systems, as capacity requirements increase, it is common practice to "sectorize" antennas or perform "cell splitting" by installing additional base stations for smaller cells thereby increasing potential capacity. Sectorization is particularly valuable in CDMA because each section has *almost* the same capacity as the entire original cell provided a reasonable degree of isolation can be achieved; and the same r.f. channel is used. It is difficult to envision an analogous concept in satellite systems. In terrestrial systems where the cells are defined by geographically separated base stations, the statistics of the fading of the interfering signals are independent because their scattering paths are totally different. In satellite systems when many spots are laid down from a single satellite, the paths to the user are all identical so that both the desired signal to a spot and the interference spillover destined for other spots are highly, if not completely, correlated. In LEO/MEO satellite systems, the spots are continuously moving so that handoffs will take place periodically even when the mobile is stationary. Finally, the Doppler shifts are much higher in satellite systems so that tracking loops must be designed differently.

5. Design Imperfections

A CDMA system operating in a mobile satellite environment may experience performance degradation relative to idealistic operation due to shadowing, fading, imperfect power control, finite interleaving, imperfect isoflux[2], etc. We emphasize only the potential degradors which are not present in terrestrial CDMA systems or have a more detrimental effect on the system performance than in its terrestrial counterpart.

One of the main differences between fixed and mobile satellite CDMA system is in that in fixed satellite systems there is no near-far problem, because all earth terminals are at approximately the same distance from the satellite. This is not true in a mobile system where a mobile can experience shadowing due to a low elevation angle to the satellite. One possible shadowing scenario is shown in Figure 2 for the uplink from a mobile to the satellite. We see that a mobile in the shadowed state needs to increase its

---

[2]Isoflux corresponds to antenna gain adjustment within a spot beam to compensate for different propagation path lengths within the beam.

transmit power to compensate for shadowing and Rayleigh (or nearly Rayleigh) fading. Since the change of shadowing conditions can be rather abrupt compared to a terrestrial shadowing scenario, and because of a relatively large propagation delay, a closed loop power control may not be feasible as discussed above. Thus an open loop power control mechanism, based on the measurement of average received power is assumed. Since such an approach can not provide tracking of fast fading due to uncorrelated fast fading on the uplink and the downlink, an additional margin must be provided to compensate for Rayleigh fading, in addition to the shadowing loss, c. Hence the transmit power is increased by a factor $P > c$ to provide the same performance as in the non-shadowed state. The increase in the MAI power at the satellite, with fraction B of users shadowed, will be (1 - B) + B • P/c. This would correspond to a degradation in the total capacity of the corresponding CDMA system which employs a conventional detector (say a single user matched filter receiver) [11]. As shown in [11], the capacity degradation on the downlink will be much more severe due to the PFD limit imposed on the downlink operation. This is because the PFD limit is measured at a non-shadowed location and an increase in the power by a factor of P to compensate a shadowed user will cause a capacity degradation by a factor of (1 - B) + BP. This may result in a quite significant capacity degradation when B is non-negligible.

The above estimates are based on the assumption of perfect average open loop power control. However, due to a relatively large propagation delay, abrupt changes in shadowing conditions and a finite time delay for an average power measuring device to detect a change in average received power, some imperfection even with open loop power control will result. This problem was thoroughly examined in [7]. The performance degradation of a shadowed user, experiencing Rayleigh fading, due to imperfect power control is shown in Figure 3 (reprinted from [7] with a permission from the IEEE). To obtain the results in Figure 3, a convolutional code of constraint length 8 and rate 1/3 was considered. Also, perfect interleaving and soft decision decoding with perfect fading state information were optimistically assumed. Nevertheless, these results indicate the very sensitivity of a CDMA system to an imperfect power control mechanism. The performance degradation with imperfect interleaving, due to a loss in coding diversity, can have a very dramatical consequences on the capacity of a CDMA system, as indicated in [7]. It was also shown in [7] how diversity significantly improves the capacity in a fading channel and enhances the tolerance to the power control error and imperfect interleaving. Moreover, if the satellite diversity is employed (dual coverage), in addition to the diversity gain in Rayleigh fading, a CDMA system will benefit from a reduced probability of shadowing, which will in turn contribute to an additional capacity enhancement. Thus, if the probability of shadowing with a single coverage system is B, with a dual coverage, and assuming independent shadowing to two satellites (which may be an optimistic assumption), the probability that a mobile is shadowed to both satellites simultaneously reduces to B^2. The main drawback of this approach is an increased system complexity (increased number of satellites in the orbit).

6.   Reduction of MAI of Improved Capacity in CDMA

Since CDMA is an interference-limited system, any method that reduces multiple access interference (MAI), or its effects, can potentially serve to increase capacity. Our focus here is on techniques suitable for satellites but some of them can be equally effective in terrestrial systems.

As mentioned in Section 4, the MAI from/to adjacent spots is determined by antenna beam overlap rather than by physical separation of the sources. Thus antenna design plays a crucial role. Currently, the spots are formed by using a large (planar) phased array with a large aperture. It is technically possible with such a configuration to use adaptive techniques to steer both beams and nulls dynamically so that desired users get individual beams with high antenna gain while undesired signals -- from different directions -- are nulled. The combination of these can increase the signal to MAI ratio by several orders of magnitude with corresponding significant gains to capacity and/or performance. Ideally, one need not even think in terms of beams and nulls but rather as a generalized array processing which maximizes the desired signal to MAI ratio. The number of degrees of freedom, determined by the number of array elements relative to the number of users in a footprint ultimately limits how large this ratio can be made. This idea will require significant on-board processing, but with technology improvements it may be feasible in the not-too-distant future. Another antenna technique which can more readily increase capacity by a factor close to two is the use of polarization isolation. The most promising is the use of circular clockwise and counterclockwise independent signals on the same frequency which can be isolated by a corresponding terrestrial receive antenna. Only two antenna elements are required, properly oriented and phased and it may be possible to do this even within a handset.

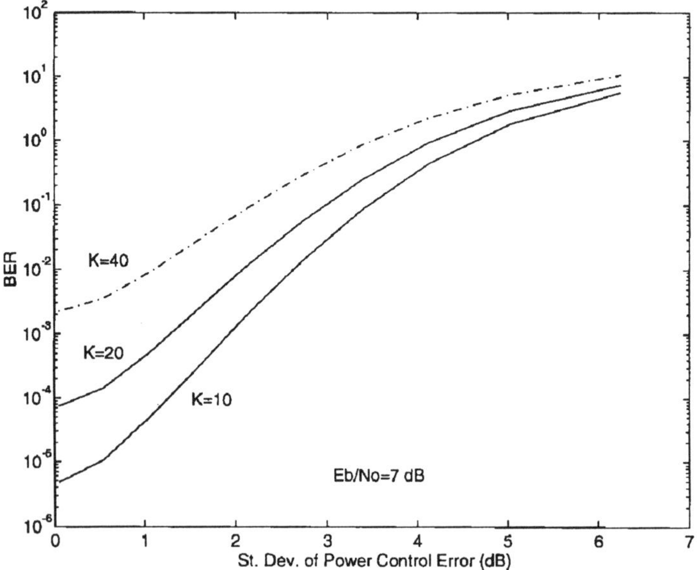

Fig. 3. Effect of imperfect power control; coded CDMA.

Diversity and forward error correction (FEC) are both powerful techniques of improving performance in unreliable channels. CDMA systems can and do use powerful FEC codes so that the required E_b/N_o per user can be held to a minimum and thereby reduce the net MAI. In CDMA, powerful, low rate FEC codes can be used without penalty in total throughput because the signals are spread anyway and the coding gain is traded off against the processing gain per coded symbol to achieve a maximum overall gain. The coding gain provides protection against random noise and fast fading if the interleaving depth is sufficient. For slow fading, and when accurate power control accuracy cannot be achieved, we have seen in Section 5 that diversity is necessary. An interesting use of such diversity is that of satellite diversity when at least two satellites cover an area. First, the likelihood of being totally shadowed is reduced; and second, when both satellites are in view, significant diversity gain is achieved. In the uplink from the mobile, diversity combining can take place on the ground at a gateway and full maximal ratio combining gain can be achieved. On the downlink to the mobile, since the paths are sufficiently separated, the mobile can exercise a RAKE receiver and achieve diversity combining gain without multiple antennas.

To the extent that the maximum number of orthogonal signals can be used in the downlink, it is possible in satellite systems to reduce the MAI from adjacent spot beams if there are sufficient orthogonal signals to assign to adjacent spots. In the satellite downlink, orthogonality is preserved because time shifts occur in synchronism in contrast with attempting to use orthogonal signals from adjacent terrestrial cell sites. If there are insufficient orthogonal signals then PN sequences can be chosen for adjacent spots which have minimal cross correlation.

Finally, one of the most promising techniques for improving capacity is that of multiuser detection. Most CDMA receivers use a correlator matched to the code of interest and ignore other user's signals which results in self-interference. However, it is well known that, if we know the other, undesired signals' signatures, that their use in a maximum likelihood detector achieves much better performance which can be traded up against capacity.

If the ith user knows all signature sequences, then he can employ a bank of matched filters where the output of each of the K matched filters can be written as

$$U_i = \sqrt{E_i}\, b_i + \sum_{l \neq i}^{K} \sqrt{E_l}\, b_l\, \rho_{il} + n_i$$

where E_i, $i = 1, ..., K$ is the energy per bit of the ith user, b_i is the data symbol, ρ_{il} is the cross correlation of the ith and lth signature sequences and n_i is noise. Normally, when only one matched filter (correlation) is used the second term is the MAI we must live with. However, if the ith user takes advantage of this a-priori knowledge of the cross-correlation of the signature sequence, the above equation can be written in matrix form as

$$\underline{U} = \underline{R} \, \mathrm{diag}\left(\sqrt{E_i}\right) \underline{b} + \underline{N}$$

where $\underline{R}$ is the K x K cross correlation matrix $\{\rho_{il}\}$, $\underline{b}$ is the data vector for all K users and $\underline{N}$ is a noise vector. Since R is symmetric, and assuming it is positive-definite, we can formally solve for an *estimate* of $\underline{b}$ by taking R^{-1} so that

$$\underline{y} = R^{-1} \, \underline{U} = \mathrm{diag}\left(\sqrt{E_i}\right) \underline{b} + \underline{Z}$$

and

$$\widehat{b} = \mathrm{sgn}\,(\underline{y})$$

This technique is known as the decorrelating detector (DD) [12] because it decorrelates and completely eliminates the MAI leaving only a noise vector (correlated and enhanced) and the desired signal. In an MAI-limited system such as CDMA this not only significantly improves performance -- and hence capacity -- but is near-far tolerant and little or no power control needs to be exercised. The DD is not a maximum likelihood detector (MLD) (it is when signal energies are not known) but its complexity is only polynomial in K rather than exponential as is the MLD and the improvements relative to a simple correlator receiver are significant and limited only by the signature cross correlations when the thermal noise is small. Here the DD was chosen only as an illustration and in practice a multistage feedback MAI cancellation may be the preferred method.

Notice that multiuser detection has significant advantages in uplink satellite communications compared to terrestrial. This is because all of the signals from adjacent spots are collected at one point (either in the satellite on-board processor or at the gateways). In addition, although the signal energy of the MAI from adjacent spots are different, their relative energies are reasonably constant and can be estimated. In contrast, signals arriving at a base station from adjacent terrestrial cells are weak and highly variable because they are distance-dependent. Errors in estimating these can lead to significant errors in the MAI cancellation.

7. CDMA Spectrum Sharing

One of the important issues in mobile satellite systems for PCS is spectrum sharing. As indicated in Section 3 of this paper, there are several proposed CDMA LEOS systems in addition to the IRIDIUM system which uses TDMA. The question is how to divide, or share, the allocated frequency spectrum for mobile satellite communications among these systems. Sharing between the TDMA and CDMA systems is not feasible unless the spreading bandwidth is sufficiently large. When the spreading bandwidth is not sufficient, a viable approach is to segment the total available bandwidth for exclusive TDMA and CDMA operation. On the other hand, it has been suggested that multiple, satellite-based CDMA systems for mobile communications can co-exist in the same frequency band [13]. This wouldn't be possible in a terrestrial scenario, due to the near-far effect, unless the base stations of different systems are collocated and all systems have the same cell size. On the other hand, in satellite systems without shadowing and fading and where orbits are near uniform, it is possible, because, in that case, there is no near-far effect. However, the performance of CDMA systems in a sharing scenario depends heavily on the channel conditions, the design of the systems involved in sharing and radio regulations which constrain the PFD's from particular systems.

International agreements on the operation of satellite communication downlinks have been made to protect terrestrial, fixed, line-of-sight microwave systems which share the same frequency band [14, 15] with the satellite system in question. These radio regulations specify a so-called "coordination trigger level" on the downlink power flux density (in $dBW/m^2/4KHz$) from each satellite. That is, when the PFD reaches a certain limit, the regulations require coordination between the satellite and terrestrial systems. Because the coordination between fixed terrestrial services and LEOS systems is not possible due to the mobile nature of the latter, these trigger levels limit the amount of power flux density (PFD) that a satellite transmitter may illuminate the earth in an unobstructed area. As such, they essentially impose a power limited operation on LEOS systems, with corresponding consequences on their system capacities. The regulations, originally adopted by CCIR (now called ITU-R) over a two decades ago for protection of terrestrial microwave systems against geosynchronous earth orbit satellites (GEOS), have been recently modified at WARC'92 [14]. Specifically, for the frequency band 2483.5-2500 MHz, which is allocated for downlinks of mobile satellite services and regulated by ITU regulation RR2566, WARC'92 set in Footnote 753F a coordination trigger of $-142 dBW/m^2/4KHz$ (previously $-144 dBW/m^2/4KHz$), and a lower PFD level for low elevation angles. However, the operational scenario of multiple LEOS systems that share the same frequency band is different from that of GEOS, in that several LEO satellites can be in the main beam of the antenna of a fixed terrestrial microwave link. In that case, the specified PFD limit on a per satellite basis may not be adequate protection for terrestrial links. Indeed, if the interference protection is the objective of PFD limits, it would appear that a PFD limit from the aggregate of *all* satellites in view ought to be the operational criterion.

There are no equivalent formal radio regulations for the uplink, i.e., limits on the total irradiance from, say, one square kilometer on the ground, designed to protect the satellite receiver, although several such proposals [13] have been made recently. Similar equations about per system versus aggregate uplink PFD limits need to be addressed.

This problem of the spectrum sharing by multiple CDMA system providers was thoroughly analyzed in [8]. It was shown, even under optimistic assumptions of perfect open loop power control, identical and perfectly coordinated systems, and perfect interleaving, that spectrum sharing, compared to band segmentation, results in the total capacity degradation for realistic values of fading channel parameters. To illustrate some of the ideas and results presented in [8] we will present here some results corresponding to the uplink.

A shadowing scenario for two systems sharing the spectrum is shown in Figure 4. It can be seen that a mobile belonging to the system A that is shadowed to its own satellite "OWN" increases the MAI to its own system by a factor of P/c, while if it has a clear line of sight with the other satellite "OTHER" increases the MAI to that system by a factor P. It is apparent that a significant near-far effect can incur in such a sharing scenario, due to the shadowing and corresponding power control compensation.

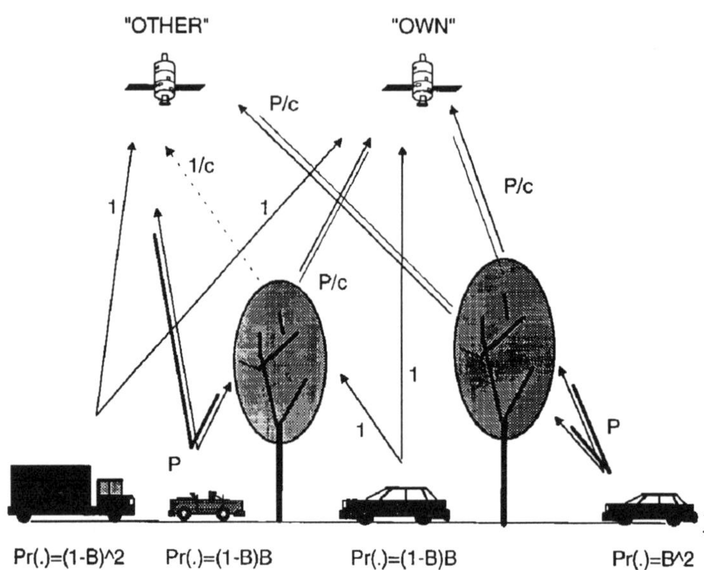

Fig. 4. Multiple systems uplink shadowing scenario.

Now, consider n satellite systems sharing the same total bandwidth, each with processing gain L. It is straightforward to show that SNR_n is given by

$$SNR_n = \frac{1}{\dfrac{N_o}{2E_n} + \dfrac{K_n N_o}{2L} A_1 + E(I^2)},$$ (10)

where the subscript n corresponds to the system parameter value when n systems are operational. Specifically, $\dfrac{E_n}{N_o}$ is the energy-per-bit-to-noise spectral density ratio of a single user, K_n is the number of users per spot beam per system, L is the processing gain, and I_o is the ratio of the total multiple access interference that a user in a given spot beam experiences due to all spot beams in its own satellite system to the interference the user experiences only from its own spot beam, I is identified to be the interference seen by the user-of-interest due to mobiles belonging to all of the other n - 1 satellite systems. To compute $E[I^2]$, note that there are four distinct situations which can arise (depicted in Figure 4, whereby all four vehicles shown are communicating with the satellite labeled "OWN", and, as a consequence, are interfering with the satellite labeled "OTHER"); the interfering user on the ground can be shadowed both to its own satellite as well as to the satellite-of-interest, the interfering user can be shadowed to neither satellite, or it can be shadowed to one, but not to both of the satellites. Realizing that a user shadowed to its own satellite has its power augmented by the factor P, that a user shadowed to the satellite-of-interest has its power attenuated by the factor c (note that these are not mutually exclusive events), and that the probability of the interfering user being shadowed to one satellite is taken to be independent of the probability of its being shadowed to any other satellite, we have for the second moment of I

$$E[I^2] = \frac{(n-1)\,K_n I_o}{2L}\left[\frac{P}{c}\,B^2 + PB(1-B) + \frac{1}{c}\,(1-B)B + (1-B)^2\right]$$

$$= \frac{(n-1)\,K_n I_o}{2L}\,A_2\,A_3,$$

(11)

where

$$A_2 \triangleq 1 - B + PB \tag{12}$$

and

$$A_3 \triangleq 1 - B + \frac{B}{c} \tag{13}$$

Note that $A_2 \geq 1$ and $A_3 \leq 1$; note further that A_2 corresponds to a user shadowed to its own satellite but not to the victim satellite, whereas A_3 corresponds to a user with a clear path to its own satellite but with a shadowed path to the victim satellite.

For the n satellite-system case, we have to consider separately the PFD limit per system and the aggregate PFD limit. In the former case, we have

$$I_o\,K_n\,P_n\,A_2\,A_3 \leq P_{max} \tag{14}$$

where we define the PFD as that imposed on any one system by any of the remaining n - 1 systems. If we assume that in (14) the equality sign holds (for all $n \geq 1$), we see that

$$P_n = \frac{K_1}{K_n} \frac{A_1}{A_2 A_3} P_1. \tag{15}$$

Therefore, upon setting

$$SNR_1 = SNR_n, \tag{16}$$

where SNR_1 is given by

$$SNR_1 = \frac{1}{\dfrac{N_o}{2E_1} + \dfrac{K_1 I_o}{2L} A_1}, \tag{17}$$

that is, upon demanding that the performance of a given user in a single service provider system be the same as when n service providers coexist, we have

$$\frac{nK_n}{K_1} = \frac{1}{A\left\{\dfrac{1}{n} + \dfrac{K_1}{K_{1,\infty}}\left[1 - \dfrac{1}{n}\left(2 - \dfrac{1}{A}\right)\right]\right\}}, \tag{18}$$

where

$$A \triangleq \frac{A_2 A_3}{A_1} \tag{19}$$

It should be noted that $A > 1$ when $B > 0$; it should also be noted that the net effect of fading caused by the shadowing is a degradation in performance, since the increase in MAI because of other users compensating for shadowing in their systems outweighs the power increase in the systems-of-interest due to its own shadowing. With no shadowing, $B = 0$ and $A = 1$.

It is of interest to know when the right-hand side of (19) exceeds unity. That is, it is of interest to know when the total capacity of n service providers exceeds that of a single service provider. For the denominator of (18) to be less than unity, we have[3]

$$n > \frac{1 - \left(2 - \dfrac{1}{A}\right)\dfrac{K_1}{K_{1,\infty}}}{\dfrac{1}{A} - \dfrac{K_1}{K_{1,\infty}}}, \tag{20}$$

---

[3] The normalized capacity, $K_1/K_{1,\infty}$ for a single system of monotonically related to the PFD limit by (14) and (15).

assuming that $\dfrac{1}{A} > \dfrac{K_1}{K_{1,\infty}}$. Note that if there was no shadowing in any system (and thus no fade compensation), $A = 1$, (20) reduces to $n > 1$, and the total capacity of n systems always exceeds that of a single system. When shadowing is present, the result depends upon the specific values of A, which is a function of B, P, and c. A characteristic result is shown in Figure 5. It can be seen that a capacity improvement is possible to achieve with multiple system only when the capacity of a single system is very small to begin with. This result corresponds to a scenario when a PFD limit on a per system basis is artificially imposed. On the other hand, when the aggregate PFD limit is imposed, which is a more meaningful constraint for the uplink, it was shown in [8] that spectrum sharing always results in the total capacity degradation, compared to band segmentation. For a more detailed discussion and analysis of the downlink sharing, with and without diversity, the reader is referred to [8].

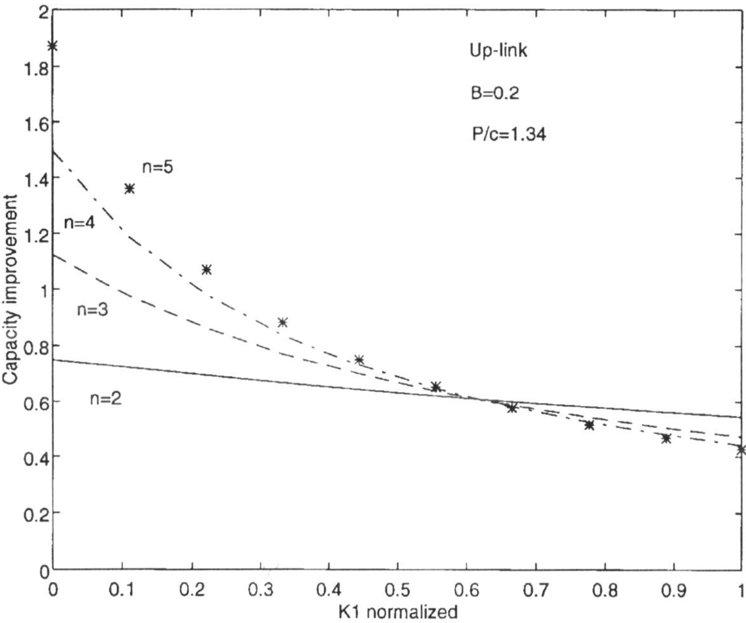

Fig. 5. Uplink capacity improvement due to spectrum sharing; PFD constraint on a per system basis.

In general, whether a single system or multiple systems achieves greater capacity is a function of a variety of parameters, among which are the degree of shadowing, the type of PFD limit, and the specific link under consideration (i.e., the uplink or the downlink). To succinctly summarize some of our key results in [8], we list the following conclusions:

1. For an AWGN channel, when the PFD limit is on a per system basis, multiple systems always outperform a single system, on either the uplink or the downlink.

2. For AWGN channel with an aggregate PFD limit, there is no difference in total capacity between multiple systems and a single system, on either the uplink or the downlink.

3. When there is shadowing present on the uplink, a single system always yields the highest capacity when an aggregate PFD is used; for a per system PFD, whether a single system or multiple systems results in greater capacity is a function of, among other things, the degree of shadowing.

4. When shadowing is present on the downlink, and when orthogonal sequences are used on a per spot beam basis, for either a per system or an aggregate PFD constraint, whether multiple systems achieve a higher net capacity than that of a single system is, again, a function of the degree of shadowing (among other things).

If the intention of introducing a PFD limit is to control the amount of interference to terrestrial systems for the downlink or to space systems for the uplink, it appears that an aggregate PFD limit is more appropriate. Under this condition, a single system provider is *always* more spectrally efficient on the uplink, and is more efficient for system parameters of most practical interest for the downlink. Indeed, even when the PFD limit is imposed on a per system basis, a single system still often outperforms n systems in terms of capacity.

8. Future Trends

The major problems confronting global LEOS for personal communications are reduced cost, greater spectral efficiency, higher data rates and inexpensive, low power-low battery drain handsets. The challenge is that should LEOS be successful, the solutions to these problems will take on even greater significance. Although the higher data rates necessary for multimedia communications can be achieved by expanding the available bandwidth, the use of more bandwidth will exacerbate the cost and power requirements. This leads naturally to solutions involving both source and channel coding, including trellis coded modulation, transmitter/receiver shaping, multiuser detection and adaptive arrays. There are some basic issues of deciding the optimum mix of coding gain and processing (spreading) gain in CDMA and this problem becomes more interesting when high bandwidth efficiency codes such as trellis coded modulation are introduced. If multiuser detection techniques can be made robust, not only will this increase capacity of CDMA, but it will also mitigate the need of very tight power output. We can also expect to see the introduction of "smart antennas" to improve performance while reducing transmitter power requirements but the jury is still out as to whether comparable results can be acheived by other means.

It will be a major challenge to coordinate both satellite and terrestrial coverage, especially with the proliferation of many disparate standards and multi-band operation. A creative solution is the development of an ultra-broadband, high dynamic range front-end and high speed digital signal processing for signal separation, processing and conversion. Such "software radios" will provide opportunities for multi-standard compatibility as well as the implementation of novel techniques discussed here.

9. Conclusions

We have presented a summary of the key factors that characterize mobile satellite communications with an emphasis on CDMA in LEOS systems. The distinguishing characteristics of satellite versus terrestrial applications were discussed and illustrated by using the results of several studies and data sources for channel characterization. The prospects of spectral sharing of CDMA multiple service providers were described and a skeptical conclusion is reached. Nevertheless, with the introduction of new techniques, it is believed that CDMA has bright prospects and opportunities for the introduction of new services delivered by satellite.

References

[1] K. Gilhousen et al., "Increased capacity using CDMA for mobile satellite communications", *IEEE J. on Sel. Areas in Commun.*, Vol. 8, pp. 503-514, May 1990.

[2] B. Vucetic and Jun Du, "Channel modeling and simulation in satellite mobile communication systems", *IEEE J. on Sel. Areas in Commun.*, Vol. 10, pp. 1209-1218, October 1992.

[3] E. Lutz et al., "The land mobile satellite communication channel-recording, statistics, and channel model", *IEEE Trans. on Veh. Tech.*, Vol. 40, pp. 375-386, May 1991.

[4] C. Loo, "A statistical model for a land mobile satellite link", *IEEE Trans. on Veh. Tech.*, Vol. 34, pp. 122-127, August 1985.

[5] C. Loo, "Measurements and models of a land mobile satellite channel and their application to MSK signals", *IEEE Trans. Commun.*, Vol. 36, pp. 114-121, August 1987.

[6] M. Miller, B. Vucetic and L. Berry, *Satellite Communications-Mobile and Fixed Services*, Kluwer Academic Publishers, Norwell, MA, 1993.

[7] B. Vojcic, R. Pickholtz and L. Milstein, "Performance of DS CDMA with imperfect power control operating over a low earth orbiting satellite link", *IEEE J. on Sel. Areas in Commun.*, SAC-12, pp. 560-567, May 1994.

[8] B. Vojcic, L. Milstein and R. Pickholtz, "Total capacity in a shared CDMA LEOS environment", to be published in the *IEEE J. on Sel. Areas in Commun.*, in January 1995.

[9] N. Kleiner, "The affect of multipath propagation in the Iridium system", Iridium Technical Note ITN-98, Motorola Satellite Communications, Chandler, AZ, 1991.

[10] E. Eliot, "Estimates of error rates for codes in burst-noise channels", *Bell Syst. Tech. J.*, Vol. 42, pp. 1977-1997, September 1963.

[11] B.R. Vojcic, L.B. Milstein and R.L. Pickholtz, "Power control versus capacity of a CDMA system operating over a low earth orbiting satellite link", *Proceedings of the CTMC'93* (held in conjunction with GLOBECOM'93), Houston, TX, pp. 40-44, December 1993.

[12] R. Lupas and S. Verdu, "Linear multiuser detectors for synnchronous code-division multiple-access channels", *IEEE Trans Inform. Th.*, COM-38, pp. 123-136, 1989.

[13] Documents of FCC's MSS Above 1 GHz Negotiated Rulemaking, Washington, DC, 1993.

[14] ITU, Final Acts of the World Administrative Radio Conference (WARC), Torremolinos, Spain, February, 1992.

[15] ITU Radio Regulations, ITU, Geneva.

STANDARDIZATION IN A WIRELESS ENVIRONMENT

Donald L. Schilling, Jack Taylor
and Joseph Garodnick

ABSTRACT

The mobile communications industry was built upon a fundamental technological stability in the radio transmission link: analog FM. For nearly forty years this stability supported a comfortable environment in which uniform standards were feasible. Single standards supported the manufacturer's goal of lowering production costs. Consumers benefitted as competition forced these lowered costs to be passed through to the public.

This comfortable equilibrium was shattered in the 1980's with the introduction of digital radio into the commercial marketplace. Digital radio will end, forever, the era of a single standard and a complacent marketplace.

This paper presents the authors' opinion of the role of standardization in a wireless environment.

THE DECADE OF DIGITAL WIRELESS COMMUNICATION

Looking back over seven short years, one is struck by the enormity of change that digital technology has brought to wireless communications. The first commercial digital radio placed in use in the terrestrial telephony network was an InterDigital (then IMM) time division multiple access (TDMA) system, placed in service in rural Wyoming by U.S. West (then Mountain Bell). This system provided high quality voice grade wireless telephone service to 4 separate subscribers on a single 25 kHz voice channel.

Since then, digital radio has made inroads into all manner of new wireless communications systems. Cellular radio, the most widespread and well known wireless radio system in the world, which is currently undergoing a technology change from analog FM to digital Cellular, is just 10 years old and already it is beginning to exchange the existing technology for advanced digital systems.

Likewise, specialized mobile radio systems (SMRs) are replacing existing analog systems with digital. SMR is a trunked radio system which provides spectrum efficiency through the trunking efficiency achieved by linking

351

S.G. Glisic and P.A. Leppänen (eds.), Code Division Multiple Access Communications, 351-358.
© 1995 *Kluwer Academic Publishers. Printed in the Netherlands.*

multiple channels in a computer controlled access scheme. With the introduction of digital technology, SMRs, like Nextel (formerly Fleet Call), are able to take advantage of the spectrum efficiency of digital technology to compete directly with cellular radio systems. In fact, many attribute cellular's hasty move to digital to be a response to the introduction of digital technology in the SMR industry.

Digital technology is also found in primarily fixed wireless applications. The new local multipoint distribution service (LMDS) uses digital radio technology operating in the 28 GHz band to provide multichannel video and voice channels, in competition to cable TV systems and wireline telephone systems. Another fixed application of digital radio is the Interactive Video and Data Service (IVDS) which will operate in conjunction with all video delivery technologies.

The largest and by far the most publicized use of digital radio is the recently authorized Personal Communication Service (PCS). PCS was the recipient of the second largest spectrum allocation ever given by the Federal Communication Commissions (FCC). Second only to the allocation to the broadcast industry in the 1950's. The FCC allocated 120 MHz of spectrum to PCS for licensed service.

THE STANDARDS QUESTION

The question of standards has been brought to the forefront of the telecommunications debate primarily because of the introduction of digital radio. As noted earlier, standards development in an analog world was much more orderly. It was driven mainly by the manufacturer's desire to lower manufacturing cost and the operators (and consumers) desire for interconnections and interoperability.

Now, standards are more difficult. More difficult to bring about and more difficult to rationalize in the world of fastpaced digital technology change. However, there remains, at least in the view of some, many benefits of existing and future standards:

> Enhanced Availability of Service. Many believe, and rightfully so, that the widespread availability of cellular service is directly attributable to a single standard codified in the FCC rules. However, the rigid analog standard also locked cellular to a technology that many felt was technologically obsolete on the day it was implemented. It may be remembered that at that time Cooper (Purdue, ktd) and Nettleton (US West) proposed a FH CDMA approach, which was promptly discarded by the two major FDMA manufacturers, in favor of FDMA, a technology in which they had already made substantial investment. The changeover from analog to digital will be expensive and lengthy. Hybrid analog/digital cellular systems will probably be in existence well into the 21st century.

<u>Enhanced</u> <u>Competition.</u> The theory behind this benefit is that uniform standards level the competitive playing field and don't allow a single manufacturer to monopolize the marketplace, due to the fact that it is the first or only manufacturer creating a <u>de facto</u> standard. In fact, even with the single analog standard, one manufacturer was able to build and sell a narrowband, spectrally efficient analog cellular product, that then became a <u>de facto</u> standard for the industry. Technology is moving too fast to try to harness competition within a single uniform standard!

<u>Lower</u> <u>Costs.</u> If all manufacturers build to a single standard the obvious result would be lower costs, but there are other factors at work here. Part of the price of lower costs is the curtailment of technological progress. Why invest in R&D if the product is rigidly standardized. Standards by definition are antiinnovative. The drive to lower costs, abetted by a rigid standard, will force the manufacturing of the product to the lower cost region and stifle further innovation. The U. S. television industry is a perfect example of a highly standardized technology sector that was driven off-shore in search of lower cost production. Digital technology has brought about a reinvigoration of this industry in the U.S. and with the help of the FCC (and Congress'), that industry will be allowed to continue to bring technological innovation to the next generation of digital TV.

<u>Interoperability.</u> The fear is that without uniform standards for equipment and software, wireless networks would proliferate and balkanize into separate networks, each unable to work with the other and all to the detriment of the consumer. This brings into focus the issue of how much standardization is needed. The cellular industry started with a monolithic standard that covered all aspects of the system. In fact, for future systems there are real advantages to developing interface standards to allow various (yes, even balkanized) systems to interface into the switching networks of the wireline carriers.

The existence of multiple wireless networks, wing various <u>de facto</u> and <u>de jure</u> standards will become a fact of life soon. PCS is about to become a reality in the U.S. and the one predictable fact about PCS is that there will be multiple radio technologies, operated by hundreds of independent operators, and they will all be on line in less than 12 months.

PCS: WILL THERE (SHOULD THERE) BE STANDARDS?

The answers to both questions are yes. The critical question is how many and when. Recently, the two pre-eminent standards bodies in the U.S., the Telecommunications Industry Association (TIA) and the Alliance for Telecommunications Industry Solutions (formally Exchange Carriers

Standards Association) came together to work on developing a common air interface standard for PCS. The committee (Joint Technical Committee -- JTC) placed a call for proposals from interested industry representatives and received 17 proposals. The JTC has set a schedule to place PCS standard(s) on an industry ballot by mid-1994. With 17 proposals and 7 months to work through the selection process no one is predicting success.

The standards process within the JTC is based on consensus and with almost all of the companies participating in the process being either a proponent of a specific proposal or the licensee of one of the proponents, consensus will be difficult to achieve. The likely output of this standards process will be several different PCS standards. The hope is that they will try to develop some degree of compatibility or interoperability through the use of interface standards. However, time is the enemy in this case.

The schedule for equipment availability is being set, not by the manufacturers or the standards process but rather, by the regulatory process. The FCC is under a Congressional mandate to conduct auctions for PCS licensees and to have the first auction by May 7, 1994. That means that PCS licensees will require equipment by the end of 1994 to begin initial installation. A further inducement to move quickly to construct PCS systems comes from the FCC rules that require PCS licensees to offer service to one-third of the population of their serving territory by the end of the 5th year or risk losing the license.

Some prospective PCS licensees are already chafing at the bit and intend to circumvent the standards process. It was recently announced that one of the large consortiums of prospective PCS operators have decided to conduct their own technology evaluations and select the preferred vendor to build the equipment. They intend to invite vendors to make presentations and demonstrations of their technologies at a central location. After evaluation and review, the consortium will attempt to select a single approach, which all consortium members would commit to, for the purchase of their systems. This could result in at least one de facto standard for PCS.

However, there are dozens of combinations of spectrum which will be licensed and until the auctions are completed no manufacturer will confidently know exactly what the final licensed spectrum block will look like. One commentor added up the various ways in which the seven frequency blocks could be combined within the overall constraint of the aggregation limit of 40 MHz. The final total was a theoretical 166 possible ways of allocating the PCS licenses throughout the country.

That means that the "safe" approach for manufacturers will be to concentrate on the large 30 MHz block allocations and wait for the market to work out the actual allocation strategy for the 20 MHz and 10 MHz allocations. This will result in a delay of equipment for the smaller block licensees.

MULTIPLE STANDARDS A REALITY

The FCC;s decision to allocate spectrum for PCS in 3 discrete spectrum blocks ensures that a multitude of standards will result. Moreover, because cellular operators will be allowed to bid on the 10 MHz blocks and preference categories for the 20 and one 10 MHz block are being proposed, there is little chance that the 10 MHz licensees will be able to aggregate these smaller blocks in any rationale manner. The 40 MHz aggregation limit will restrict cellular 10 MHz PCS licensees from further aggregating their spectrum. And the 20 and 10 MHz licenses that will attract the preference holders are in different bands. As a result, multiple equipment formats will be available and multiple <u>de facto</u> standards will develop.

However, multiple standards are already the norm in the U.S. as well as in the rest of the world. In the U.S. we already live in a multiple standard world. As am example, the cellular industry has four existing standards, two analog and two digital. Moreover, a third digital proposal is currently being reviewed by the industry standards setting body. The proliferation of standards in the cellular environment may cause some short term dislocations in the marketplace. We note that the two digital standards currently in place are incompatible with each other. The third digital proposed standard: Broadband CDMA, has so many advantages over narrowband CDMA and TDMA (the other two digital standards) that many operators may opt for the broadband approach. This situation, incompatibility among multiple technologies within the same market, would have been unheard of just a few short years ago.

DIFFERENCES IN TECHNOLOGY REQUIRE DIFFERENT STANDARDS

To illustrate the benefits of multiple standards for multipath technologies, we can briefly compare the various technologies:

First, let us compare the analog standards, AMPS and NAMPS. NAMPS has 2.5 to 3 times the capacity of AMPS without any noticeable degradation in voice quality. However, neither system provides privacy or high speed data. Further, due to multipath fading and the relatively high signal-to-voice ratio required for FM, both systems must transmit using extremely high power.

Secondly, let us compare the two digital standards: TDMA and NCDMA, with the third proposed digital standard, BCDMA. TDMA offers about 3 times the capacity of AMPS, while NCDMA appears to offer a capacity improvement of 6 to 10 times AMPS. BCDMA has a capacity improvement of 20 to 30 times AMPS because of its immunity to multipath fading. While both TDMA and NCDMA generate synthetic voice (with its accompanying objectionable delay) and only low rate data, BCDMA uses ADPCM to generate wired line quality voice and data to 144kb/s. Further, BCDMA, using a much wider bandwidth than the other technologies, is significantly more resistant to fading and can operate at significantly less power than either NCDMA or TDMA. BCDMA can also share the spectrum with AMPS or TDMA systems, thereby providing for more efficient use of the spectrum.

There are several companies that have filed CDMA patents. These include InterDigital Communications Corporation in Great Neck, Long Island, Omnipoint Corporation in Colorado Springs, Colorado, Qualcomm Corporation in San Diego, California and Cylink Corporation in Sunnyvale California. The unique patents for broadband CDMA held by InterDigital and the narrowband CDMA patents held by Qualcomm seem to hold the most promise for future developments in the wireless industry.

Thus, we see that limiting ourselves to a single standard would stifle new technological development which is required to meet the ever changing needs of the consumer. This trend toward multiple standard is reflected in the existence of multiple technology approaches now being pursued by various standards bodies in the U.S. The Telecommunications Industry Association (TIA), an ANSI accredited standards organization has established a technical standards group (TR-46) to develop U.S. standards m the PCS arena. Similarly the Alliance for Telecommumications Industry Solutions (ATlS), also an ANSI accredited standards group is working on PCS standards. For practical reasons, the two groups have joint Technical Committee (JTC) to pursue common air interfaces for PCS. At the moment, the JTC meets monthly for 1-2 weeks at a time attempting to rationalize the eight different technical approaches brought forward by the various companies involved in the standards work.

Several of the technical approaches are based on existing and planned techniques currently being advocated for U.S. and European cellular networks. The TDMA standard IS-54, the CDMA standard IS-95, the GSM-based DCS-1800 and the BCDMA are the most prominent of the technical approaches under review by the JTC. The multiplicity of technologies that is reflected in the U.S. cellular market, and will soon appear in the U.S. PCS market, is mirrored in the rest of the world.

In fact, spread spectrum technology is quickly being adopted throughout the world. Most Latin American countries have adopted U.S. for example, Argentina adopted the Federal Commumications Commission's (FCC) approach for unlicensed spread spectrum systems in September 1992, Brazil in January 1993 and in September 1993 Mexico adopted a modified version of the FCC rules in September 1993.

Several Pacific Rim countries and Russia have adopted similar rules for spread spectrum systems. Australia, New Zealand, Taiwan, the Philippines and Russia all allow spread spectrum systems under rules similar to those in place in the U.S. However, these rules should lead to more specific applications standards.

China, with its huge population and land mass should have standards to meet its unique needs. PCS in Europe may come in several flavors, and certainly the Japanese approach is different from that taken in Europe. Similarly, within Europe there are a multitude of different cellular approaches. The UK's TACS system differs from the Ericsson Group's NMT system which, in turn, comes in three different configurations.

The chaos caused by multiple standards is balanced, by the flexibility afforded operators, in not being linked to a single monolithic *de jure* standard. The rate of generational technological change in digital technology is about 3-5 years, which mirrors the amount of time it took the U.S. standards industry to agree and publish the first digital TDMA standard. The value, therefore, of *de jure* standards are questionable.

THE FUTURE: FLEXIBLE INTERFACES, NOT RIGID STANDARDS

The wireless industry is facing a future of multiple standards, a future that can be secured with attention to the establishment of interface standards, especially in the interface between radio cells and the switching network. The rate of exchange in technology, especial in the voice coding and modulation areas, is so rapid, that fixed radio link standards could not survive in any event. However, flexible interface standards which would allow different switches to operate with different trasmission systems supplied by many different vendors would bring a dazzling array of wireless services, at affordable cost to the customer.

The future of wireless communications, dominated by digital technology, will see a full break from the past. The rigid monolithic standard that dominated the first generation of U.S. cellular will not ever be repeated. This vision of multiple standards reflecting the demands of the marketplace was articulated nearly six years ago by Hiroshi Kojima, a spokesperson for the Japanese MPT. He said: Standardization was initially a concept of the 19th Century age of mass production. However, we are now entering an era of customer-orientated standards. To put it in a nutshell, as long as networks can be linked up, they can be utilized according to how individuals wish to use them. This may sound strange, but I believe that standards may have to become "destandardized standards" that is to say, standardization will be based on users needs and shifted from a single standard to sophisticated standards

CONCLUSIONS

The age of multiple and sophisticated standards has arrived. The future of the wireless industry lies in satisfying customer demands. The manufacturers (and to an extent the operators) goal of lowering production costs and end user costs through a single standard is no longer operative.

Technological progress has made obsolete the single standard in much the same manner as the simple black telephone desk set passed into the telephone industry history books.

The future of the standards process will be more a reflection of the marketplace needs of the customer and less of the specific desires of the operator/manufacturer. The myriad of choice that the customer has in the wireline customer premise equipment sector will extend out into the wireless networks of the future.

Subject Index